Advances in Underwater Communication and Networking

Other related titles:

You may also like

- SBRA537 | Huang and Gill | Ocean Remote Sensing Technologies: High frequency, marine and GNSS-based radar | 2022
- SBRA525 | Ehlers | Autonomous Underwater Vehicles: Design and practice | 2020
- PBTR011 | Sanjay Sharma and Bidyadhar Subudhi | Navigation and Control of Autonomous Marine Vehicles | 2019

We also publish a wide range of books on the following topics:
Computing and Networks
Control, Robotics and Sensors
Electrical Regulations
Electromagnetics and Radar
Energy Engineering
Healthcare Technologies
History and Management of Technology
IET Codes and Guidance
Materials, Circuits and Devices
Model Forms
Nanomaterials and Nanotechnologies
Optics, Photonics and Lasers
Production, Design and Manufacturing
Security
Telecommunications
Transportation

All books are available in print via https://shop.theiet.org or as eBooks via our Digital Library https://digital-library.theiet.org.

IET RADAR, SONAR & NAVIGATION SERIES 563

Advances in Underwater Communication and Networking

Edited by
Syed Ali Hassan and Huma Ghafoor

The Institution of Engineering and Technology

About the IET

This book is published by the Institution of Engineering and Technology (The IET).

We inspire, inform and influence the global engineering community to engineer a better world. As a diverse home across engineering and technology, we share knowledge that helps make better sense of the world, to accelerate innovation and solve the global challenges that matter.

The IET is a not-for-profit organisation. The surplus we make from our books is used to support activities and products for the engineering community and promote the positive role of science, engineering and technology in the world. This includes education resources and outreach, scholarships and awards, events and courses, publications, professional development and mentoring, and advocacy to governments.

To discover more about the IET, please visit https://www.theiet.org/.

About IET books

The IET publishes books across many engineering and technology disciplines. Our authors and editors offer fresh perspectives from universities and industry. Within our subject areas, we have several book series steered by editorial boards made up of leading subject experts.

We peer review each book at the proposal stage to ensure the quality and relevance of our publications.

Get involved

If you are interested in becoming an author, editor, series advisor, or peer reviewer please visit https://www.theiet.org/publishing/publishing-with-iet-books/ or contact author_support@theiet.org.

Discovering our electronic content

All of our books are available online via the IET's Digital Library. Our Digital Library is the home of technical documents, eBooks, conference publications, real-life case studies and journal articles. To find out more, please visit https://digital-library.theiet.org.

In collaboration with the United Nations and the International Publishers Association, the IET is a Signatory member of the SDG Publishers Compact. The Compact aims to accelerate progress to achieve the Sustainable Development Goals (SDGs) by 2030. Signatories aspire to develop sustainable practices and act as champions of the SDGs during the Decade of Action (2020–2030), publishing books and journals that will help inform, develop, and inspire action in that direction.

In line with our sustainable goals, our UK printing partner has FSC accreditation, which is reducing our environmental impact to the planet. We use a print-on-demand model to further reduce our carbon footprint.

Published by SciTech Publishing an imprint of The Institution of Engineering and Technology, London, United Kingdom

The Institution of Engineering and Technology (the "**Publisher**") is registered as a Charity in England & Wales (no. 211014) and Scotland (no. SC038698).

First published 2025

The Institution of Engineering and Technology
Futures Place,
Kings Way, Stevenage,
Herts, SG1 2UA,
United Kingdom

www.theiet.org

British Library Cataloguing in Publication Data

A catalogue record for this product is available from the British Library

ISBN 978-1-83953-968-8 (hardback)
ISBN 978-1-83953-969-5 (PDF)
ISBN 978-1-83724-978-7 (EPUB3)

Typeset in India by MPS Limited

Cover image: *Abstract big data inside of a digital sea; Yuichiro Chino/Moment via Getty Images*

Contents

Preface

Underwater communication systems have been gaining significant interest in the past two decades to address different applications, ranging from the depths of the ocean to its surface. To implement these applications such as oil monitoring, earthquake and tsunami detection, marine animal study, military and defence surveillance, and more, underwater communications are equally crucial as terrestrial communications in establishing reliable connections between sensor nodes. The underwater environment is complex and challenging because both natural and artificial nodes are involved in the network which uses the same acoustic band for communication. In the era of intelligent technology, similar to other environments such as terrestrial, aerial, and space, it is necessary to have 6G underwater communications and networking. This book covers 6G-based underwater acoustic communications that integrate modern technologies and concepts such as software-defined networking (SDN), network function virtualization (NFV), non-orthogonal multiple access (NOMA), reconfigurable intelligent surfaces (RIS), unmanned aerial vehicles (UAVs), autonomous underwater vehicles (AUVs), and artificial intelligence (AI). In the opening chapter, 6G internet of underwater things (IoUT) is introduced to permit successful connections of underwater vehicles and sensors, which allows for end-to-end monitoring of the environment. The discussion covers recent 6G technologies that involve the use of UAVs and AUVs. With 6G technologies, an SDN-based underwater architecture is provided, which is the foundation for adaptive and mission-based underwater networks that have the capacity to enable sustained exploration and coordinated underwater operations. Furthermore, SDN and AI is combined with a comprehensive TCP/IP protocol stack to supply cutting-edge solutions for future underwater networks. The integration of SDN and NFV is followed by a marine distributed underwater network (MDUN) that provides a solid understanding to develop energy-efficient, scalable, and resilient underwater networks.

As previously stated, the acoustic bandwidth is limited and is shared between both natural and artificial users. After examining recent technologies, the book investigates NOMA as a solution to enhance spectral efficiency and user fairness in order to tackle acoustic band issues. By using RIS-based systems, the next chapter will further resolve the spectrum problem by highlighting a technological path to durable high-speed acoustic networks. Further, the optimal route is analysed by examining different routing protocols that incorporate 6G technologies to guarantee data delivery reliability from source to destination nodes. As UAVs and AUVs are involved in the network, energy-efficient paths and optimization of battery usage for these types of autonomous vehicles are then discussed. Additionally, security is guaranteed by ensuring secure acoustic communication between underwater nodes through 6G technology. To guarantee data integrity, AI is utilized to establish intelligent, secure, and stable communication links. The book then goes into detail about various machine learning and deep learning techniques for exploring underwater acoustic environments, optimizing acoustic signal transmission, and using hybrid intelligent models to power acoustic communications. The next chapter explore few underwater applications that cover AI-based seismic monitoring, detection of tsunamis and earthquakes using AI-based models, and AI-based sonar image classification and analysis. In the final two chapters, the focus is on testing sound propagation in water and using various sonar systems to acquire underwater images and investigate real-life applications such as mine-detection, seabed mapping, and AUV navigation. Finally, a modelling scenario is developed and verified using the UPPAAL model checker for sonar-based subsystems. The book discusses a variety of recent concepts that open new avenues for researchers in this field to explore underwater environments using 6G concepts and technologies.

About the editors

Syed Ali Hassan serves as a professor at the National University of Sciences and Technology (NUST), Pakistan, where he directs the Information Processing and Transmission Laboratory. His research primarily addresses theoretical aspects of wireless communications, with over 300 published articles in this filed. He obtained a PhD in electrical engineering from Georgia Institute of Technology, United States.

Huma Ghafoor holds the position of assistant professor at the School of Electrical Engineering and Computer Science (SEECS), National University of Sciences and Technology (NUST), Pakistan. Until January 2019, she worked as a postdoctoral researcher at Inha University, South Korea. Her research interests include vehicular ad hoc networks, maritime networks, and underwater sensor networks. She completed her PhD in electrical engineering from the University of Ulsan, South Korea, in 2018.

Chapter 1

The 6G internet of underwater things (6G-IoUT)

Mian Tahir Nadeem[1], Muhammad Danish Khattak[1] and Mahrukh Liaqat[2]

Sensor networks underwater are faced with extremely high signal attenuation, bandwidth, energy, and complexity due to multipath propagation, as well as high latency, making timely and guaranteed data collection from underwater sensors difficult. To overcome these challenges, autonomous vehicles have emerged as an inevitable means of effective data collection, negating the design constraints typical in conventional acoustic systems. The chapter discusses current trends in underwater networks and their communication with different autonomous vehicles. The focus is on the most recent 6G technologies, which involve the utilization of unmanned aerial vehicles (UAVs) and autonomous underwater vehicles (AUVs), their obstacles and potential solutions. With the advent of 6G technology, ultra-high data rates, ultra-reliable low latency, and artificial intelligence-based intelligent management significantly enhance underwater operational efficiency. Technological advances, such as hybrid acoustic-optical communication networks and 6G-supported underwater charging stations continue to support sustained, stable, and long-endurance missions even further. Using the idea of the internet of underwater things (IoUT), the chapter focuses on the successful connectivity of underwater vehicles and sensors, thus allowing end-to-end monitoring of the environment. Additionally, novel solutions to battery challenges in autonomous vehicles are discussed. Finally, the discussion proves that the incorporation of 6G into underwater networks is a groundbreaking opportunity to establish underwater communication and exploration systems that are intelligent, green, and resilient.

1.1 Introduction

Underwater communication is critical in a number of applications, such as environmental monitoring, the safety of marine life, and underwater exploration. However,

[1]School of Electrical Engineering and Computer Science (SEECS), National University of Sciences and Technology (NUST), Pakistan
[2]College of Electrical and Mechanical Engineering (CEME), National University of Sciences and Technology (NUST), Pakistan

underwater communication faces very serious challenges, such as limited transmission power, extensive attenuation, high energy levels, low speed, and multipath propagation complexities. This environment poses additional challenges, as underwater sensor node communication is primarily centered around an acoustic band of 1–40 kHz. These issues prevent the underwater sensor nodes from collecting data for the various underwater applications mentioned above. AUVs are certain robot systems with the ability to explore underwater areas on their own, playing a major role in underwater communication to recover real-time and energy-efficient data from underwater sensors. AUVs offer solutions to conventional acoustic communication problems, such as massive propagation delay, bandwidth limitations, and memory limitations of sensor nodes, which result in data loss or overwrites [1].

Over the past almost 10 years, IoUT has been introduced to connect underwater devices and address the challenges of the underwater environment, taking inspiration from the advancements of terrestrial networks through the concept of the Internet of Things (IoT). The IoUT is driven by the fact that underwater equipment, sensors, and AUVs must be connected in a manner that becomes helpful in data collection and multi-aggregated observation in remote underwater areas. The technology focuses mainly on solving underwater communication challenges, including limited transmission power and bandwidth, energy constraints in sensor nodes, dynamic conditions at water depth, and the pressing need for real-time data acquisition. The motivation for the use of this technology arises from the inefficiency of conventional acoustic communication that causes problems in multi-hop routing which increases network overhead and delay due to the movement of sensor nodes with the ocean waves, in making stable node-to-node links in single-hop communication, etc.

The focus of these single- or multi-hop communications was primarily to increase data throughput and identification of the shortest path, particularly in delay-sensitive underwater applications. However, due to the challenges of the acoustic band, data were lost or overwritten due to communication delays and sensor memory constraints. In this chapter, we focus on the integration of AUVs in IoUT networks to provide possible ways to enhance the performance of the underwater network. The chapter concentrates on ways to improve underwater networks and solve existing issues, using 6G technology. The remainder of the chapter is arranged in the following manner. Section 1.2 discusses related work, Section 1.3 provides 6G-based solutions to challenges with existing techniques in underwater networks, and Section 1.4 concludes the chapter.

1.2 Related work

The literature outlines the idea of combining multiple AUVs with IoUT to collect information and guarantee stable real-time monitoring. Zhang *et al.* [2] presented a solution to resolve the issue of how to make multiple AUVs fetch data from ocean sensors in an environmentally and energy-aware way, balancing dynamic upload priorities of data. The existing literature employed pre-calculated AUV routes or static optimization techniques that ignored environmental turbulence and data urgency. These approaches were optimized mainly for the shortest path or least energy consumption but not for real-time requirements and device-specific preferences. To account for these loopholes, a novel data acquisition method based on a multi-agent learning-based path-searching algorithm called multi-agent independent soft actor-critic (MAISAC) was developed in the literature [2] on decentralized training and deployment. This reinforcement learning model allowed every AUV to autonomously make decisions on the basis of real oceanic conditions, device priority levels, and communication quality.

The model was defined as a Markov decision process, and the objectives included balancing data rate, throughput, energy consumption, and timeliness. Their simulations showed that the introduced model outperforms baselines as it can perform well in turbulent environments, enhancing data efficiency and system scalability. In our opinion, their system is highly innovative in integrating environmental modeling in reinforcement learning and has sufficient adaptability. However, there is still room for improvement that can open doors for future researchers. This improvement can include dynamic environment-based testing in the real world, implementation of energy harvesting modules to extend the duration of the AUV flight, and further investigation of hybrid communication technologies (e.g., optical transmission in shallow water areas). The system can also be improved by including predictive models to predict ocean turbulence and adjust AUV trajectories ahead of time.

AUVs have previously been utilized by other researchers for passive or active sonar tracking. But these kinds of systems were normally constrained by interoperability, were high-energy consumers, or were not capable of real-time data processing. Ghafoor and Noh [3] discussed the application of unmanned underwater vehicles (UUV) and acoustic networking technologies to provide target tracking. This technology is significant because submarine surveillance and defense operations are increasingly faced with stealth targets and a sophisticated oceanic communications environment. Target detection and tracking must be real-time, accurate, and energy-efficient, which cannot be done with traditional passive sonar or fixed sensor networks. Therefore, their focus was on proposing an integrated network using advanced technologies together in an underwater environment. Hence, the main research question explored in their study was how the forthcoming technologies of underwater cognitive acoustic networks (UCANs), software-defined networking (SDN), network function virtualization (NFV), and fog or cloud computing can be utilized to enhance target detection, tracking, and decision making in underwater systems.

They proposed a multilayer architecture with IoUT devices, AUVs, fog or cloud computing, and SDN or NFV to handle control and data exchange [3]. Data were collected from underwater sensors in their setup and fed into the local AUVs, surface buoys, and then to fog nodes for real-time computation and kept on cloud servers for longer-duration analysis. They also cited a myriad of ray tracing models that have been used in simulating acoustic signals such as BELLHOP, HARPO, and TRACEO3D, with the conclusion that none of them individually adequately accommodates speed, accuracy, and flexibility in realistic applications. This architecture accommodates a wide variety of applications ranging from surveillance to environment monitoring and disaster relief. We believe that this combination architecture is an important step toward the creation of fault-tolerant, real-time, and smart underwater detection systems. However, for further improved performance, future systems can accommodate adaptive learning at fog node levels, modularity-integrated acoustic model capacity, and power-saving routing strategies to retain minimized latency and decreased energy consumption.

Additionally, the stackable protocol that facilitates underwater SDN and NFV capability could be applied toward scalable missions and environmental assistance. In total, both research efforts discussed above help significantly toward the creation of next-generation underwater systems, with each serving a complementary purpose: one optimizing data collection and the other crafting detection and network control. Additionally, a robust analytics methodology is necessary for effective handling and analysis of the huge volume of data generated by sensor nodes in underwater environments. The study by Albekairi [4] presented a comprehensive mutable sensor data analytics (MSDA) technique for successfully separating valuable sensor data from undesired or noisy data. This method ensured fast data validation and transmission from underwater sensors to land facilities by processing and optimizing data exchange securely using blockchain technology. Precisely, pattern recognition improved with greater overall reliability and fewer processing errors through the application of gradient descent learning methods.

This technique directly solved the issues identified in earlier studies, including sensor malfunctions, ambient turbulence, and communication limitations experienced by IoUT systems. The accuracy, timeliness, and scalability of underwater detection and communication systems were ultimately improved by the successful isolation of meaningful data using the MSDA methodology, which greatly enhanced current approaches such as integrated underwater infrastructure and AUV-based environmental awareness strategies. All the schemes discussed in this section provide ways to solve underwater problems, but none of them focus on 6G technologies, which could benefit from this new technology and make underwater networks more stable with fewer challenges. In the next section, we first introduce the role of 6G in underwater networks and then provide possible ways to solve existing problems in underwater communication systems.

1.3 Role of 6G in underwater networks

6G wireless networks form a quantum leap in communication technology with the aim of building a smart, globalized society. In contrast to other previous generations, 6G envisions multi-terabit-per-second data rates, ultra-low latencies, and complete coverage of land, space, air, and underwater networks. The 6G vision is not just to achieve higher speeds, but a vision of the future where an autonomous highly adaptive network platform is built based on the emerging modern artificial intelligence (AI) and machine learning (ML) technologies. This would be an edge for all but instantaneous and full-dimensional wireless coverage feasible communication, extending networks into areas previously difficult to reach, such as deep underwater locations. With the help of 6G, autonomous vehicles, underwater sensor nodes, and monitoring systems will operate more effectively and independently in larger networks. The application of 6G technologies have developed underwater wireless networks extensively. With very high data rates, it is now possible to transmit high-definition pictures, live videos, and a significant amount of sensor data even in very remote areas of the sea [5]. Higher latency enables real-time command and control, which is important in applications such as remote-operated vehicles and underwater sensors.

Furthermore, AI-based network control will enable underwater communications systems to adapt dynamically according to the environment, energy optimization, and increase overall reliability. Such advances can improve deep-sea mining, oceanography, and disaster relief. Terrestrial and aerial infrastructures have been widely explored in 6G architectures, but underwater communication has not been explored much despite its potential significance. The state-of-the-art in underwater communication technologies categorizes them as acoustic, optical, and radio-frequency (RF) communications, which presents numerous challenges in terms of balancing transmission distance, reliability, and data rate. Several architectures have been proposed to embed underwater communication in 6G, including onshore interconnects through cable and satellite interfaces, as well as offshore underwater grids, along with a detailed discussion of the improvements required in localization, energy sustainability and security [6]. Some critical application scenarios for military, scientific, and industrial applications argue that underwater networks have the potential to expand the reach and scope of 6G. It is of the need of the hour that many technological and infrastructure-related challenges, such as the absence of stable water–air interfaces and power limitations, should be discussed before this vision comes into existence. We will explain each one in the following sections and provide possible ways to resolve these issues using 6G technologies.

1.3.1 Infrastructure-related underwater challenges and the possible solutions in 6G

Underwater networks present a unique set of infrastructure challenges because it is very difficult to communicate underwater. Water greatly attenuates most electromagnetic waves, and most traditional wireless communication technologies, such as RF or light signals, are extremely useless. Acoustic communication is also bandwidth and latency-limited, particularly in deep water or highly unstable environments, as stated previously. Limited reliable communication between AUVs and surface stations limits continuous monitoring or operation over long distances. Navigation is also compromised in underwater vehicles, as they are unable to utilize GPS signals. They use sonar-based systems for navigation, which are prone to errors due to environmental factors, such as turbulence or temperature changes. Therefore, underwater communication and navigation systems must be properly planned and designed, and underwater operations are complex and expensive.

To address such limitations, hybrid communication systems that involve the combination of acoustic and optical communication technologies are being investigated [7]. Acoustic communications remain active as the chosen modality for extended-range missions, with optical and high-frequency RF-based systems likely to be used for high-data rate short-range data transfer under ideal conditions. These hybrid systems allow AUVs to take advantage of the latest communications technology available in their environment and environmental state to achieve maximum data throughput and integrity. In addition to communication, underwater docking stations are being developed to allow AUVs to be charged without surfacing. Such stations should be carefully located so that the coverage area is optimized, and 6G-based smart energy grids would have AUVs charged at all times, reducing downtime and maximizing mission time.

With the advancements in 6G technology, the underwater infrastructure can be changed through transparent communication between AUVs, surface nodes, UAVs, and global satellite systems (see Figure 1.1). The ultra-reliable low-latency communication (URLLC) of 6G allows AUVs to transmit and receive data and commands virtually in real time, even from deep areas [8]. This increases coordination between underwater vehicles, surface stations, UAVs, and remote operators with maximum energy use, route utilization, and mission success. In addition, intelligent underwater networks consisting of acoustic relays, AUVs, UAVs, and surface nodes can be dynamically reconfigured as a function of real-time monitoring of the environment. Such an infrastructure improves the network wisdom of 6G to provide end-to-end data routing optimization and energy minimization for AUVs and UAVs, improving both underwater operations' efficiency and reliability. This network infrastructure, powered by the worldwide connectivity of 6G, must be more scalable, more capable, and more robust underwater networks to carry out demanding long-duration missions in challenging environments.

The applications of 6G to underwater infrastructure and communication are an exceptional opportunity to automate underwater operations and make them scalable [9]. 6G-enabled infrastructure greatly increases the bandwidth and energy efficiency of communications, solving many of the challenges that underwater vehicles are facing today and allowing operations to be autonomous and sustainable on previously untapped underwater territory. With technological advancements, wireless power transfer and network recharge stations also improve the ability of the AUV to operate uninterrupted without surface intervention. Future underwater networks depend on

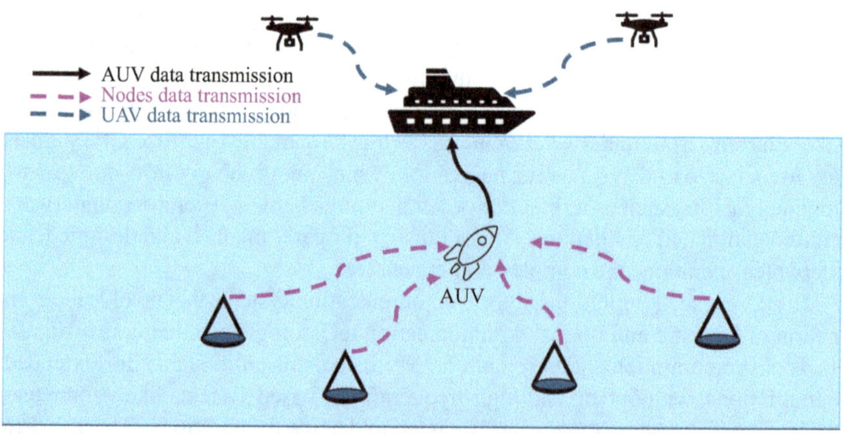

Figure 1.1 The 6G internet of underwater things network

the convergence of the latest communications, energy, and 6G network intelligence into a unified format, creating a new exploration paradigm and ocean monitoring.

1.3.2 Battery-related issues of AUVs in underwater networks and their possible solutions in 6G

AUVs are faced with severe limitations in battery life, as previously stated, which have a direct impact on their ability to operate in underwater networks. AUVs use batteries onboard to power propulsion devices, sensors, and communication systems, which limits their range and mission time. Unreliable power sources, particularly for long mission durations, are some of the inherent limitations of underwater surveillance and exploration. Although there is improvement in battery technology, constraints, including onboard power capability to determine exploration duration, continue to be a limiting factor for long-duration operations [10]. For mitigation, AUVs tend to visit the surface to recharge, disrupting the continuous data collection or remote monitoring mission process. The energy limitation is further exacerbated by the requirement, in deep-sea operations, to insulate batteries against high-pressure and extreme-temperature conditions by special, and therefore expensive designs.

To resolve energy issues, some methodologies have been proposed, such as hybrid power systems through the hybridization of batteries with other means of alternative power. New advances in solar-powered AUVs have the capability to increase operating time by using renewable energy. The disadvantage of utilizing such systems is that they have to surface the AUVs at regular intervals for recharging based on sunlight, thus not being able to utilize them for deep-sea operations. More accurate methods involve environmental energy harvesting techniques such as wave or thermal power to power AUVs during mission operations because they have the ability to provide endless power without surfacing [10]. High-tech batteries such as lithium-polymer and semi-fuel cell systems are also being integrated into AUVs to increase endurance [11]. Although more efficient, these also require charging devices that are capable of handling long-duration operations [12].

For example, AUVs may employ underwater docking facilities for replenishment so that they are charged when submerged. Such docking facilities do have concerns about docking precision, electrical safety, and power transfer efficiency, which all need investigation and enhancement to prepare them for remote and deep-water deployments. For 6G networks, all these power problems can be addressed by placing AUVs in a global communications infrastructure. With 6G's URLLC, the AUVs can be monitored remotely, and battery levels can be monitored in real time. This allows mission planners to dynamically adjust operations and properly control that energy. For example, scheduling efficient energy usage can maximize the utilization of AUVs by planning the direction of the route according to the battery levels of the AUVs and their proximity to recharging stations. Cooperative operations can also be part of the system in which a collection of AUVs works together in data exchange and task sharing, energy-efficient routes are shared, and recharged [12].

Moreover, wireless power transmission can also be a key feature in a phase-out of physical charging points, with AUVs recharging without the requirement to dock.

Through 6G-enabled networks, AUVs are not only more energy efficient but also enjoy a degree of autonomy which reduces dependency on surface infrastructure. 6G connectivity also enables autonomous power management since AUVs can be autonomously recharged by automatically identifying charging points near each other according to their power capacity. Underwater charging stations with high-speed wireless connectivity due to an ongoing development of 6G technology can enable effective data transmission and power recharge, enhancing the efficiency and performance of AUV. Sensor technology integrated with AUVs can deliver more precise energy consumption analysis, increasing operational efficiency and mission success rates in extreme underwater environments. Such advances in energy management, along with 6G connectivity, can greatly enhance the battery life of AUV, reducing the impacts of energy constraints and facilitating longer, unbroken operations in underwater networks.

1.3.3 Battery-related issues of UAVs in underwater networks and their possible solutions in 6G

UAVs also suffer from the battery life issue, which restricts their range and endurance. Most UAVs, particularly quadcopters, employ lithium-ion batteries to avoid them flying around perpetually, thus being most suitable for short-term use. The limited battery life of UAVs, also known as drones, restricts their payload capacity and range, which is a tremendous disadvantage in uses such as delivery, reconnaissance, and data collection [13]. Battery replacement, as a viable solution, still involves clumsy logistical challenges. UAVs generally have to come back to base after each mission or a pre-defined charging point, which translates into too much downtime. Even more advanced systems, such as fuel cells, have long run times and come with complexity and expense. Therefore, more conservative energy management methods and equipment continue to be necessary for the continuation of UAV operation in the long term, especially for long-range or commercial applications.

Battery swap is one of the proposed options that allows UAVs to replace spent batteries with charged ones in a matter of minutes, thus going on without long recharging durations. The joint routing and charging strategy has also been suggested to utilize batteries effectively by having charging stations on delivery routes or UAV mission routes. Through this, downtime is minimized and the efficiency of UAV fleets in operations is enhanced, especially in such large delivery operations. Cooperative dispatch of UAVs with offloading/sharing batteries can also work as a remedy. This technique allows a convoy of UAVs to be cooperative, consuming power and flight cycles to prevent stressing individual UAV batteries. These systems are very useful where UAVs have to fly continuously throughout the day and for a long period of time, that is, in the case of UAV networks in the middle of the sea to monitor underwater applications and maintenance of wind farms for terrestrial application provision [14].

With 6G networks, UAV battery limitations can be removed with wireless links with high bandwidth and low latency for real-time monitoring of data and battery sharing. Real-time route planning and energy consumption by 6G-enabled UAVs can be regulated with predictive analytics and energy demand forecasting. Intelligent charging stations on 6G networks can automatically plan recharging or replacement of UAVs, and UAVs can be prepared for subsequent flight without human interference [15]. UAVs are able to inform such stations about battery status to enable preemptive maintenance and reduce downtime. In addition, wireless power transfer technology can also provide power to UAVs in flight through 6G ultra-high-speed networks to provide best-in-class power transmission in UAV location and power usage. Such efficiency and connectivity can greatly increase the flying distance and time of UAVs, further increasing their value for critical missions such as disaster rescue, logistics, and environmental monitoring.

The integration of 6G's extremely high communication reliability and new energy management methods can change UAV operation entirely [16]. Thus, UAVs can operate independently, with the network itself ensuring battery and task management. Additionally, edge computing in 6G networks can also be potentially used to decrease payload on onboard processing systems of UAVs, thereby saving power and maximizing flight duration. For example, by offloading computationally intensive data processing to the edge nodes or cloud, UAVs can gain the ability to concentrate more on mission-critical communication and flight protocols at reduced battery drain. This increases UAV efficiency, autonomy, and dependability and becomes simpler to execute in city and rural settings where energy management and battery life otherwise limit them.

1.4 Conclusion

This chapter discussed the 6G technologies that can enhance underwater communications through the integration of various technologies. Innovation in AUV and UAV technology can be used to transform exploration of the seafloor with unmatched operational efficiency, cost-effectiveness, and safety over conventional operations. Despite development, the challenges of energy autonomy and communication networks remain a major impediment. Battery technology, as much as it continues to improve, continues to constrain the operation time of AUV and UAV, specifically long-duration operations under deep-sea, sea surface, or polar operations. Hybrid power systems, solar charging and energy harvesting technologies, along with 6G developments, provide promising methods for increasing AUV and UAV capability. The high-speed and low-latency communication capability of 6G can enhance AUV and UAV energy management and support adaptive control in real time according to environmental conditions. The future of AUVs, UAVs, and underwater communications with their integration is in a fusion of next-generation technologies that not only solve energy and infrastructure problems but also foresee new possibilities for sustainable and efficient underwater operations.

References

[1] M.Y.I. Zia, J. Poncela, and P. Otero, "State-of-the-art underwater acoustic communication modems: Classifications, analyses and design challenges," *Wireless Personal Communications*, vol. 116, pp. 1325–1360, 2021.

[2] Z. Zhang, J. Xu, G. Xie, J. Wang, Z. Han, and Y. Ren, "Environment- and energy-aware AUV-assisted data collection for the internet of underwater things," *IEEE Internet of Things Journal*, vol. 11, no. 15, pp. 26 406–26 418, 2024.

[3] H. Ghafoor and Y. Noh, "An overview of next-generation underwater target detection and tracking: An integrated underwater architecture," *IEEE Access*, vol. 7, pp. 98 841–98 853, 2019.

[4] M. Albekairi, "A comprehensive mutable analytics approach to distinguish sensor data on the internet of underwater things," *IEEE Access*, vol. 12, pp. 95 007–95 019, 2024.

[5] N.-N. Dao, N. H. Tu, T. T. Thanh, V. N. Q. Bao, W. Na, and S. Cho, "Neglected infrastructures for 6G—underwater communications: How mature are they?" *Journal of Network and Computer Applications*, vol. 213, p. 103595, 2023. https://www.sciencedirect.com/science/article/pii/S1084804523000140.

[6] Z. Zhang, Y. Xiao, Z. Ma, *et al.*, "6G wireless networks: Vision, requirements, architecture, and key technologies," *IEEE Vehicular Technology Magazine*, vol. 14, no. 3, pp. 28–41, 2019.

[7] Z. Wang, J. Lian, H. Liu, *et al.*, "Review on repair technologies for underwater concrete structure damage of infrastructures," *Water*, vol. 17, no. 1, pp. 1–28, 2025. https://www.mdpi.com/2073-4441/17/1/35.

[8] E. Santos, "Innovative solutions for coastal and offshore infrastructure in seawater mining: Enhancing efficiency and environmental performance," *Desalination*, vol. 575, p. 117282, 2024. https://www.sciencedirect.com/science/article/pii/S0011916423009141.

[9] J. Franken, T. Reinhold, T. Dörnfeld, and C. Reuter, "Hidden structures of a global infrastructure: Expansion factors of the subsea data cable network," *Technological Forecasting and Social Change*, vol. 215, p. 124068, 2025. https://www.sciencedirect.com/science/article/pii/S004016252500099X.

[10] A. Bounyasith, B. Mueller, D. Jin, *et al.*, "Exploring the feasibility of solar-powered autonomous underwater survey vehicles for extended underwater exploration," in *SoutheastCon 2024*, IEEE; 2024, pp. 1490–1495.

[11] K. Hasan, S. Ahmad, A.F. Liaf, *et al.*, "Oceanic challenges to technological solutions: A review of autonomous underwater vehicle path technologies in biomimicry, control, navigation, and sensing," *IEEE Access*, vol. 12, pp. 46 202–46 231, 2024.

[12] S. Fan, N. Bose, and Z. Liang, "Polar auv challenges and applications: A review," *Drones*, vol. 8, no. 8, 2024, pp. 1–51. https://www.mdpi.com/2504-446X/8/8/413.

[13] H. Jahani, Y. Khosravi, B. Kargar, K.-L. Ong, and S. Arisian, "Exploring the role of drones and UAVs in logistics and supply chain management: A novel text-based literature review," *International Journal of Production Research*, vol. 63, no. 5, pp. 1873–1897, 2025. https://doi.org/10.1080/00207543.2024.2373425.

[14] T. Fan, L. Fu, C. Guo, Y. Zhang, and L. Sun, "Multi UAV inspection optimization for offshore wind farms considering battery exchange process," *IEEE Transactions on Intelligent Vehicles*, vol. 10, no. 2, pp. 1–11, 2024.

[15] F.A. Silva, V. Barbosa, L.N. Lima, *et al.*, "Efficient strategies for unmanned aerial vehicle flights: Analyzing battery life and operational performance in delivery services using stochastic models," *IEEE Access*, vol. 12, pp. 144544–144564, 2024.

[16] Z. Pei, W. Pan, K. Weng, and T. Fang, "A branch-and-price-and-cut algorithm for the unmanned aerial vehicle delivery with battery swapping," *International Journal of Production Research*, vol. 62, no. 19, pp. 7030–7055, 2024. https://doi.org/10.1080/00207543.2024.2318017.

Chapter 2

Software-defined underwater acoustic communication

Ahmad Nasir[1], *Muhammad Attique*[1] *and Salman Ghafoor*[1]

Underwater acoustic communication is still among the technically most constraining domains in modern networking, with high propagation delays, low bandwidth, and dynamically varying environmental conditions. To meet the challenges of such environments, this chapter presents a modular, software-defined architecture well adapted for underwater environments based on the principles of software-defined networking (SDN) and programmable acoustic modem platforms. Tiered architecture is adopted, including application, remote control, and data tiers, where adaptive routing, load balancing, and security management are divorced from hardware constraints and are centrally controlled. The most notable innovations include the application of autonomous underwater vehicles in the control plane, dynamic clustering based on fuzzy logic for efficient data aggregation, and leveraging distributed multi-controller topologies for scalability, resiliency, and power optimization. These aspects, when integrated into a cross-layer optimization framework, ensure synchronization, low latency, and improved data throughput. In this chapter, we also discuss how several researchers have performed simulation-based analysis and verified the performance of the system, thus exhibiting an enhanced delivery ratio, lifetime, and stability in dense and heterogeneous underwater deployments. The SDN-based underwater architecture is the foundation for clever, adaptive, and mission-based networks according to the vision of the 6G underwater networks and has the capacity to enable sustained exploration, marine monitoring, and coordinated underwater operation.

2.1 Introduction

Underwater communications are confronted with enormous technical challenges because their environment is harsh and fluctuating. Unlike terrestrial radio frequency (RF) and optical communication, the underwater channel limits the use of such technologies. RF signals are absorbed in a matter of meters and optical signals suffer from high scattering and absorption. As such, acoustic communication, as

[1]School of Electrical Engineering and Computer Science (SEECS), National University of Sciences and Technology (NUST), Pakistan

bandwidth-limited and noisy as it is, is the sole realistic method for medium-to-long-range communications. Underwater acoustic communication (UAC) is subject to special and perpetual constraints that strictly limit system performance, mainly as a result of the intrinsic difficulty of the underwater channel. The underwater acoustic channel experiences a high propagation delay due to the low speed of sound in the water, along with time-varying multipath propagation caused by seabed and surface reflections. Other issues like Doppler spreading, artificial and marine animal-generated sea ambient noise, and changing environmental conditions (pressure, salinity, and temperature) also affect signal integrity. They cause sporadic packet loss, reduced data rates, and synchronization problems, rendering real-time high-rate data transmission difficult. Underwater communications play an important role in most mission-critical scenarios in applications, such as underwater surveillance, monitoring of the marine environment, and inspection of offshore platforms with the coordination of an autonomous underwater vehicle (AUV) [1].

Software-defined networking (SDN) has emerged as a transformative paradigm in the case of underwater communication networks, yielding centralized control, architectural flexibility, and dynamic software-programmable configurability. Compared to conventional networking models, where control logic is mapped onto forwarding functions, SDN decouples the control plane from the data plane, allowing network intelligence to be concentrated within a centralized controller. This decoupling delivers global network visibility and dynamic real-time control over communication parameters, which are absolutely required in very dynamic and resource-constrained underwater environments. SDN abstracts itself over a three-layered structure of the application, control, and infrastructure layers and enables frictionless resource orchestration and network management with high granularity. It also facilitates the quick introduction of new services, reduces the dependence on vendor-specific hardware, and facilitates interoperability through standardized interfaces. For submarine networks such as submarine platforms and AUV fleets, SDN provides the flexibility to dynamically tune communication parameters to improve their ability to cope with adverse and time-varying channel conditions. In addition, SDN provides an adaptive testbed on which optimization strategies and experimental protocols can be tested without the need for upgrading equipment, a vital capability in systems where physical access is prohibitively expensive or unavailable. Figure 2.1 shows the UAC based on SDN technology that is divided into three layers.

Although it has immense advantages, the applications of SDN in underwater acoustic networks are associated with several technical and operational challenges. Underwater communication networks have conventionally used fixed hardware configurations that cannot handle the uncertainties of the acoustic channel. Networks typically cannot accommodate multipath propagation-induced variations, temperature and salinity gradients, and physical perturbations, leading to high signal degradation, latency, and unreliable data transfer. Although SDN introduces flexibility, oceanic networks are plagued with built-in constraints such as limited bandwidth, massive propagation delay, Doppler spread, and interference due to ambient noise. Traditional acoustic modems also include vendor-specific hardware and vendor-locked stacks of software as a fault; thus, it is not easy to integrate them with SDN

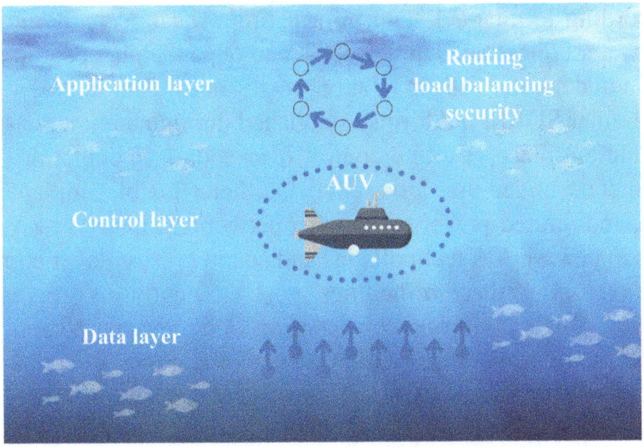

Figure 2.1 The SDN-based underwater acoustic communications

controllers and scale the network. Static routing policies and static management schemes lead to resource inefficiency. Complementing the above issues, the deployment of underwater networks requires computationally intensive embedded platform signal processing, traditionally with size, power, and thermal budget issues. Real-time adaptive modulation, error control, or multi-input multi-output (MIMO) techniques necessitate co-design in hardware and software to consume performance efficiently in the acceptable range. Thus, although SDN offers an infrastructure of the future for underwater communication networks, its potential can only be realized in the form of customized solutions prepared with the understanding of certain underwater physical and structural constraints. We talked about a traditional acoustic modem in the discussion above. In the next section, we will first go over the need for a software-defined acoustic modem (SDAM) and its limitations in underwater environments.

2.2 Why SDAM? Challenges and possible solutions

SDAM is a device for underwater communication that uses software to control the modem's functions, such as modulation and coding schemes, rather than relying on hardware fixes. When combined with SDAMs, SDN-based systems enable unprecedented flexibility in underwater communications. But both SDN and SDAMs are faced with serious technical and operational issues, primarily due to the very hostile nature of the underwater acoustic channel that strongly constrains reliability and data rate, particularly in real-time or high-data-rate applications such as video transmission and swarm AUV coordination. However, SDN enables dynamic reconfiguration of communication parameters, modulation scheme, coding scheme, and transmit power, in a way to render the system more adaptable to varying channel conditions. One-way travel-time (OWTT) ranging and high-accuracy oscillators enhance localization and synchronization, even in noisy or mobile deployments. In addition, MIMO methods, with scalable video coding and unequal error protection, enable optimal video quality and throughput with dynamic spatial diversity and data rate balancing according to channel state information. These technologies, apart from making underwater connections more robust and dependable, open up the operating envelope of the underwater system to support advanced, data-intensive, coordinated operations. Modular SDAM platforms established through commercial-off-the-shelf hardware equipment place absolute control over coding, modulation, and network protocol parameters under the designer to facilitate flexible implementation over dense, mobile, and limited-power underwater networks [2].

Advances in new SDAM technologies and adaptive cross-layer communications schemes are overcoming these difficulties more efficiently on progressively larger scales. Modular SDAM platforms consisting of common component pieces allow for fine-granularity control of adaptation of network protocol, coding, and modulation parameters for efficient, dynamic deployment in dense, low-power, mobile underwater networks. The application of unequal error protection, MIMO-based transmission, and scalable video coding within adaptive communication systems also allows real-time high-fidelity video transmission even in acoustically degraded channel conditions. These technologies not only make underwater connections more solid and stable, but also enlarge the operating envelope of underwater systems so that more advanced, data-intensive, and coordinated missions can be facilitated and made more reliable. Last but not least, SDN deployment in underwater networks should be accomplished through hardware-software co-design, carefully with high-resilience channel-aware algorithms to overcome the trade-off among flexibility, complexity, and performance. Future work on this topic should be targeted toward scalable and low-energy models that are able to adapt autonomously in order to deal with demanding and dynamic underwater conditions. In exchange, adaptive cross-layer communication systems utilize scalable video coding, unequal error protection, and dynamic MIMO to enable high-quality video and throughput despite underwater acoustic channel limitations. They overcome serious barriers such as low bandwidth, high propagation delay, and channel uncertainty to enable reliable real-time underwater operations in a range of deployment environments.

Notwithstanding that SDAMs and adaptive communication models facilitate substantial enhancements for underwater networks, numerous challenges hinder their scaling to large-scale. The current SDAM model is hampered by low power and limited communication distance, owing mainly to low-power transmission as well as narrowband, which disqualify their possibility of being scaled to large-scale or deep-sea networks. Accurate oscillators to be employed for high accuracy, like an oven-controlled crystal oscillator (OCXO) or atomic clocks, are costly, make systems more complex, and present scalability issues. Furthermore, the adaptive video transmission method, while robust in low-to-moderate channel conditions, is susceptible in low-SNR and highly dynamic environments as it has a long packet duration and low coherence time for channels, which degrades decoding capability. To replace these limitations, future systems can require adaptive power management, low-cost amplifier topologies, and wideband transducer arrays over a period of time to provide maximum range and economy. Software-based or network-based timing may use hardware-expensive method substitutes for clock synchronization at reduced cost. In addition, optimized video transmission with increased packetization, low-latency compression algorithms, and real-time feedback control can render it more robust. Lastly, integration of predictive channel estimation and machine learning decision-making can render system agility in time-varying acoustics feasible to enable robust performance for a wide range of challenging underwater tasks.

2.3 How SDN can overcome underwater challenges?

In the previous section, we talked about SDAM in underwater networks. In this section, we focus on how SDN handles underwater challenges. Below the waves is a dynamic and colorful world, in which the internet of underwater things (IoUT) is beginning to transform our understanding and interaction with ocean systems. Along the ocean margin, huge numbers of small, energy-limited devices collaborate to gather data on ocean state, patterns of biodiversity, and underwater motion, while facilitating navigation and monitoring. Compared to land networks with high-speed, high-bandwidth wireless links, underwater networks enjoy a much harsher environment. Acoustic signaling is the dominant means of communication, but is necessarily bound to low data rates, high latency, and sporadic signal loss, as previously stated. These are consequences of complex interactions between the signal and the underwater environment, for example, surface scattering, seabed reflection, and water composition variation. These events halt packet delivery, interfere with synchronization, and halt the regular operation of collaborative networks. To counteract this untrustworthy network environment, SDN has been suggested as a possible architectural revolution, allowing decoupled control logic, centralized decision making, and real-time flexibility. However, most conventional SDN frameworks optimized for terrestrial infrastructures are poorly suited to react to underwater acoustic links' intermittent, high-latency conditions. They rely on high bidirectional traffic and high-throughput assumptions, which comes at the cost of thin acoustic channels and bleeding energy budgets.

The design in Giarrè and Casari [3] avoid these limitations by providing a communication model that is more appropriate for acoustic propagation. They focused on a southbound interface, designed for broadcasting, that batched routing commands into dense, multi-node bursts. By converting from unicast control flows to tactical broadcasting, they reduced redundancy signaling, saved energy, and used the shared medium more efficiently. Their model supported three modes of operation: reactive, which supports on-demand control with higher overhead; proactive, which pre-disseminates flows based on expected patterns; and hybrid, which dynamically switches between the two approaches to balance responsiveness of control against efficiency. They deployed and tested the system on the UnetStack platform in different underwater environments. They showed that proactive and hybrid modes were more throughput stable and energy efficient than pure reactive modes, particularly in lossy environments. Their broadcast-aware protocol reduced control overhead with reserved channel capacity for valuable data and increased operating life for resource-constrained devices. By reinterpretation of the SDN principles in terms of underwater physics, they opened the way to scalable, long-term IoUT deployments that worked incredibly well in the Earth's most challenging communication environment.

Although static nodes are the foundation of most underwater networks, but cannot stay static due to ocean currents, the future of IoUT also relies on the ability of mobile platforms to be wireless and adaptively controlled in real time. Real-time discovery and dynamic data collection, along with coordinated underwater operations, require untethered, software-controllable underwater vehicles (UVs) to execute autonomously and responsively. Nevertheless, most UVs today are inevitably bound by physical limitations, such as tethering or rigid communication topologies, that introduce range, flexibility, and scalability limits. To avoid such limitations, an SDN infrastructure was designed based on software-defined underwater acoustic network (SEANet) modems implemented within the blue robotics casing (BlueROV2) interior [4]. Their solution integrated acoustical hardware that was modular and convenient, real-time signal processing, and SDN-optimized flexibility. A skid-mounted pressure vessel was considered that offloaded the processing core to programmable logic functions to provide real-time zero-padded orthogonal frequency division multiplexing (ZP-OFDM) transmission and adaptive demodulation. The acoustic chain comprised high-speed analog-to-digital and digital-to-analog converters that digitally controlled gain circuits, and a directional transducer, enabling the vehicle to form robust links in poor conditions.

Critically, the platform offloaded a generic ethernet interface, and the high-level programs used the acoustic medium transparently through the transmission control protocol (TCP) or the user datagram protocol (UDP). There were various modulation formats ranging from binary phase-shift keying (BPSK) to 16-quadrature amplitude modulation and special raw waveform modes so that UV dynamically adjusted its communications stack to accommodate high-rate and low-power missions. They validated system performance in real-world sea trials and demonstrated the effect of Doppler shifts caused by motion, channel variation, and environmental noise on acoustic link performance. Static tests employed quadrature PSK over ZP-OFDM with consistent bit error rates greater than signal-to-noise thresholds, and mobile testing revealed Doppler scaling shifts at center frequency, thus showing critical results toward dependable UV coordination. The UV maintained adaptive connectivity along scripted paths, and the platform captured detailed waveform datasets for post-processing and protocol optimization. Real-time modulations of schemes, transmit power, and sampling parameters also illustrated the responsiveness of the software-defined stack. Hence, the coupling of acoustic hardware and reconfigurable logic with user-level control offers new avenues to coordinated UV swarms, in which vehicles can negotiate, forward, and reconfigure their behavior as part of an extended underwater acoustic SDN.

By applying SDN principles from rooted sensors to mobile, smart platforms, the vision of adaptive underwater systems for learning, responding, and cooperating in remote and foreign areas is brought closer to reality. This mobility-driven expansion validates the general purpose of IoUT, to provide adaptive, autonomous sea networks as adaptive and responsive as the seas that it travels. The dynamic and random underwater environment poses insurmountable obstacles to conventional underwater communication systems, especially in terms of addressing the limitations of current hardware to accommodate changing acoustic channel conditions.

Conventional fixed-hardware-based underwater networks cannot manage the randomness of the underwater channel since it varies with temperature, salinity, and pressure. These anomalies, and others such as multipath propagation and Doppler effects, are amenable to high delay, low rate, and non-reliable communications links. As the need for an adaptive and dynamic environment is growing, the application of SDN, which can be controlled by central control, real-time network administration, and dynamic tuning of communication parameters based on different conditions, has become a reality. Hence, as discussed above, underwater scenarios pose challenges to network scalability, dependability, and energy saving in the form of adverse channel conditions, scarce bandwidth, and node mobility.

Hence, the discussion highlights that classic centralized control-based underwater acoustic sensor network structures suffer from low scalability, susceptibility to single-point failures, and communication bottlenecks. To mitigate these shortcomings, another hierarchical structure named multi-controller-based software-defined underwater acoustic sensor networks (SM-UASN) was introduced in Shi *et al.* [5]. This architecture was divided into three planes: a data plane of switch and sensor nodes in clusters for in situ data accumulation, a distributed control plane with redundant surface-level controllers, and an application plane for central policy installation. Controllers sent messages to underwater switch nodes on out-of-band channels for traffic control and observation of network statistics like node residual energy. Using underwater acoustic network energy models, such as environmental conditions of temperature and distance attenuation, more cognitive, energy-aware routing was guaranteed. Data and control signals were allocated reserved frequencies to minimize interference, and a partitioning technique in spaces allocated each controller a cylindrical space, reducing communication distance and enhancing manageability. Their simulation results for large-scale networks confirmed the effectiveness of SM-UASN over conventional architectures like traditional single-sink (TS)-UASN, traditional multi-sink (TM)-UASN, and single-controller software defined (SS)-UASN.

SM-UASN improved the network lifetime and data throughput over TS-UASN, resulting in a substantial performance gain. Nine controllers were optimal with respect to energy consumption and traffic overhead. Over controllers generated excessive control signal traffic, leading to premature energy consumption in switch nodes, while under controllers generated inefficiencies in data aggregation and management. By facilitating SDN-enhanced traffic engineering and load balancing, their architecture allowed for responsiveness to traffic bursts and environmental factors. This dynamic, distributed, and domain-aware architecture was particularly robust in high-density, high-traffic environments, meeting the need for scalable, responsive underwater network infrastructure. The authors of SM-UASN stated that the integration of a distributed SDN controller with energy-aware acoustic communication design was a viable avenue for efficient and resilient underwater sensor networks. In addition, as discussed above, UASNs have long been critical for marine exploration, but traditional centralized architectures face significant scalability and reliability challenges. These networks, which often suffer from bottlenecks and single-point failures, struggle to handle the increasing complexity of underwater operations. The introduction of SD-UASN revolutionized this landscape [6]. By integrating SDN with energy-sensitive design principles, SD-UASN leveraged the benefits of SDN that separated the control plane from the data plane and provided a more dynamic, agile, and scalable platform.

Their architecture also consisted of three planes: the data plane, the redundant surface-level control plane, and the application plane for policy deployment. This architecture enhanced network resilience with fewer single point failures and more efficient, energy-conscious communication. The SD-UASN architecture also considered energy consumption, a critical consideration in underwater environments, by dynamically adapting communication paths based on real-time environmental conditions such as temperature and distance attenuation. Moreover, the use of out-of-band channels for control messages had less interference and partitioning techniques divided the network into regions that were easy to manage, thus scaling the system itself and its power efficiency. Their simulation results proved that SD-UASN outperformed traditional architectures in terms of data throughput, network lifetime, and robustness. The optimal number of controllers was nine that balanced energy consumption against overhead in traffic. In addition to these, their architecture did not cover some of the principal concerns currently still to be worked through when it comes to actual deployment on the ground. One such issue, for example, is the problem of fault tolerance in respect of a controller failure, particularly under a distributed SDN configuration, which has yet to be properly addressed. SM-UASNs, and it makes a compelling case for transitioning from centralized, monolithic topologies to a distributed control structure.

Hence, dividing into data, control, and application planes enabled scalability, policy abstraction, and energy-efficient routing. The use of environment-sensitive metrics, spatial controller domains, and dedicated acoustic channels for control signaling in their architecture reflected thoughtful architectural planning. Their simulations also affirmed that a nine-controller deployment struck a balance between traffic overhead and energy consumption, outperforming legacy setups such as TS-UAN and TM-UAN. Yet, while their architecture was sound, several real-world complexities were absent from the discussion. For instance, underwater environments are highly dynamic, with varying node mobility, salinity, and pressure affecting signal quality, none of which were considered in the routing strategy. The coordination between multiple distributed controllers assumes reliable inter-controller communication, yet the impact of synchronization drift, message loss, or partitioning of network segments was not explored. The design also lacked a fallback mechanism for controller failure, which could compromise resilience. Furthermore, while acoustic models were leveraged for energy-aware routing, the overhead of real-time environmental sensing and its integration into routing decisions remained unquantified. Practical challenges, such as the constraints of embedded underwater hardware, the cost of real-world deployment, and how energy harvesting could support long-term autonomy, were also not touched upon. There was also limited insight into how SM-UASN can adapt during rapid topology changes or node failures.

Despite its strengths in concept and simulation, translating SM-UASN into a field-deployable solution requires addressing these physical-layer variabilities, reliability trade-offs, and the overhead of control-plane synchronization under realistic underwater conditions. Another scheme that considered fuzzy logic was proposed in Wang *et al.* [7] to overcome underwater network challenges by providing an optimized routing path. Fuzzy logic is a rule-based formalism that is programmed to deal with imprecise, uncertain, or vague information characteristics of the real world. It differs from binary logic systems in that it accommodates reasoning based on degrees of truth and not merely truth values, which makes it very suitable for underwater networks that are subject to constant data changes and noisy inputs. Its application in the SDN control plane enables more context-aware and adaptive routing decisions in instances where rigid thresholds are insensitive to react. With the increasing size and complexity of underwater networks, classical routing methods cannot provide consistent performance in heterogeneous environments with a high delay. So far, research in the literature has demonstrated how programmable frameworks such as SDN and modular SDAMs provide flexibility and real-time control, but routing in these networks is still vulnerable to node condition variability, link variability, and topology evolution. To meet this, recent studies have integrated fuzzy logic into the control plane of SDN-based underwater acoustic networks. Unlike fixed-threshold models, fuzzy logic can handle uncertainty by reasoning on imprecise and vague parameters.

Hence, this enables more subtle decision-making in dynamic scenarios where nodes move, residual energy is changing, or the acoustic paths change. A layered architecture with separate sensing, control, and application planes was employed in Wang *et al.* [7], with controllers making energy-conscious environmental inputs based on fuzzy inference and quality-conscious routing decisions. The outcome was an intelligent routing layer that reacted to both short-term disturbances and long-term degradation. Fuzzy logic was executed using membership functions and rule-based reasoning, where appropriateness amounts were obtained in place of binary values. Fuzzy logic was applied to classify candidate nodes as cluster head nodes based on residual energy, hardware utilization, and communication reliability. The selected nodes then constructed the roots of the network topology so that controllers could build minimum spanning trees among clusters. This framework routing responded graciously to node failure, load bursts, or connection loss by re-planning communications without flooding the network with control traffic. At the same time, the separation of control planes from data planes made re-planning real-time feasible without compromising throughput or adding substantial latency. Hybrid routing behavior was also allowed, dynamically switching between proactive and reactive modes depending on the stress in the environment. This type of dynamic flexibility represented an expression of a global trend toward decision-making systems that both processed numerical data and comprehended context and patterns of behavior in an underwater environment.

Their simulation results showed that fuzzy logic-based routing in SDN greatly improved communication performance in multi-layered networks. Experiments with different node densities and packet loads showed improvements in delivery reliability, delay reduction, and energy conservation. Compared to reactive-only approaches, the fuzzy logic controller was already selecting candidate paths based on network health, reducing the likelihood of retransmissions and route failures. Meanwhile, control commands were strategically batched and multicasted to prevent redundant signaling on the acoustic channel. Their design also enabled multi-hop clustering and adaptive relay mechanisms to ensure coverage and stability under sparse deployment or link failure. With the provision of computation distribution without sacrificing centralized policy enforcement, their system attained the ideal balance between global consistency and local autonomy. This type of design is suited to larger objectives of scalable underwater networking under the conditions of performance being guaranteed with resource constraints, dynamic topology, and time-varying acoustic environments. Its fault tolerance also qualifies it for integration into hybrid static-mobile networks with time-varying link quality. The response with minimal intervention makes this routing framework a prospect for long-term oceanic deployments.

Their approach enabled acoustic networks to self-optimize and self-correct, particularly when deploying over vast oceanic regions with non-uniform node distributions. By incorporating fuzzy reasoning into the SDN control plane, their system maintained synchronization among nodes, reduced protocol overhead, and saved energy in mission critical applications such as environmental monitoring, disaster surveillance, and autonomous AUV coordination. In doing so, this architecture sees the end-to-end vision of the IoUT as networked intelligence not physically networked but logically coordinated, context-aware and capable of adapting to environment. Smart architectures such as these are the basis for reliable marine infrastructure capable of handling long-duration deployments. As underwater operations become larger and more complex, these systems become front-and-center in ensuring data availability, coordination, and autonomous decision-making during periods of uncertainty. Hence, the use of fuzzy logic in SDN-based underwater networks gives an extensible and flexible platform for smart routing. By virtue of its ability to process nuanced information and reason its way out of uncertainty in the environment, the controller is less a rule enforcer and more of a decision engine one that can fine-tune approaches to maintain mission continuity.

2.4 Conclusion

The developments showcased throughout this chapter present a growing trend in underwater networking toward programmable, intelligent, and highly dynamic frameworks that can run in unstable and resource-constrained environments. Through the integration of SDN with modular and reconfigurable acoustic hardware, underwater communication systems are progressing toward more adaptive, coordinated, and energy-efficient systems. These advances address key obstacles of underwater settings, including propagation delay high enough to cause major difficulties, small acoustic bandwidth, dynamic topology, and impact from environmental factors such as salinity and pressure. Scalable architectures, hierarchical control frameworks, and adaptive routing protocols have been discussed to effectively reduce delay, promote robustness, and optimize system-wide efficiency. These advances provide a concrete path to maintain reliable underwater communication links in varying mission environments. Integration of fuzzy logic-based intelligent clustering algorithms, adaptive south-bound control, and multi-controller SDN deployment is a robust, fault-tolerant architecture. These systems offer congestion mitigation, balanced energy usage by nodes, and scalable synchronization across the dispersed domains. The evaluation through simulation carried out by researchers in the literature proves that the existing solutions can offer better throughput, less latency, and increased network lifespan with a few modifications in comparison to the traditional models if they are tested in a real-world underwater environment and hence can also validate the suitability of deployment at scale. As the underwater network continues to evolve, these adaptable and intelligent systems can become the lower-level layer of underwater exploration, monitoring, and coordinated autonomy. These programmable structures can offer a path to connected, but context-aware networks that can adapt to environmental shifts and support fault-free operations. Real-world deployment, minimizing synchronization overhead, and designing robust recovery mechanisms for controller-level failures can be included as future work. With continued evolution, such software-defined systems have the potential to become the infrastructural backbone for long-duration missions, reliable sensor operation, and intelligent underwater decision-making in ever more complex oceanic environments.

References

[1] M. Rahmati, Z. Qi, and D. Pompili, "Underwater adaptive video transmissions using MIMO-based software-defined acoustic modems," *IEEE Transactions on Multimedia*, vol. 25, pp. 473–485, 2023.

[2] F. Campagnaro, R. Francescon, E. Coccolo, A. Montanari, and M. Zorzi, "A software-defined underwater acoustic modem for everyone: Design and evaluation," *IEEE Internet of Things Magazine*, vol. 6, no. 1, pp. 102–107, 2023.

[3] F. Giarrè and P. Casari, "Enhanced south-bound communications for underwater acoustic software defined networks," in *OCEANS 2024 – Halifax*, IEEE; 2024, pp. 1–8.

[4] D. Unal, S. Falleni, E. Demirors, K. Enhos, S. Basagni, and T. Melodia, "A software-defined underwater acoustic networking platform for underwater vehicles," in *ICC 2022 – IEEE International Conference on Communications*, 2022, pp. 2531–2536.

[5] Y. Shi, X. Huang, Q. Jiang, and Q. Yang, "A multi-controllers architecture for software-defined underwater acoustic sensor networks," in *2022 IEEE International Performance, Computing, and Communications Conference (IPCCC)*, Austin, TX, USA. 2022, pp. 240–245.

[6] C. Lin, G. Han, T. Wang, Y. Bi, J. Du, and B. Zhang, "Fast node clustering based on an improved birch algorithm for data collection towards software-defined underwater acoustic sensor networks," *IEEE Sensors Journal*, vol. 21, no. 22, pp. 25 480–25 488, 2021.

[7] J. Wang, Q. Feng, J. Ma, and Y. Feng, "FL-SDUAN: A fuzzy logic-based routing scheme for software-defined underwater acoustic networks," *Applied Sciences*, vol. 13, no. 2, pp. 1–18, 2023. https://www.mdpi.com/2076-3417/13/2/944.

Chapter 3

Exploring the TCP/IP protocol stack in underwater networks: adapting to SDN-based 6G underwater networks

Ayesha Khanam[1] and Chih-Hsien Hsia[2]

Underwater communication networks introduce unavoidable challenges because of harsh media, constrained bandwidth, limited energy of the communicating nodes, and high latency. The applications of underwater communication networks range from adventurous voyages to mineral exploration and military missions. All of these applications require low-latency, reliable, and secure data transmission. As technology continues to evolve, frameworks, such as 6G, software-defined networking (SDN), and artificial intelligence (AI) bring cutting-edge solutions for next-generation networks. There is a wide range of innovation and customization across each layer of the standard transmission control protocol/internet protocol (TCP/IP) stack. In light of evolving technological capabilities, smarter and more adaptable underwater communication networks can be built. This chapter details the vulnerabilities faced by underwater sensor networks and presents appropriate solutions tailored to each layer of the TCP/IP model.

3.1 Introduction

Ocean surveillance, research, and resource management are moving in new directions with the emergence of new communication technologies. However, the physical constraints of the underwater medium pose considerable threats to the security and reliability of communication channels. Limited battery power, the use of low-frequency acoustic signals, and continuous link failures require us to look for alternative networks for traditional networking protocols, employed in significantly more reliable terrestrial networks.

[1]School of Electrical Engineering and Computer Science (SEECS), National University of Sciences and Technology (NUST), Pakistan
[2]Department of Computer Science and Information Engineering, National Ilan University, Taiwan

3.1.1 Importance of underwater communication

Underwater communication networks are coming up as vital contributors for multiple applications such as marine environment monitoring, oil exploration, strategic surveillance, and coordination of autonomous underwater vehicles (AUVs). As the demand for such applications grows, the need for state-of-the-art underwater communication networks assumes greater importance. Traditional wireless networks designed for terrestrial setups are not suitable for underwater environments. This is primarily because such designs are characterized by high propagation delays, limited bandwidth, severe multipath effects, hostile medium, and impact of water currents on node mobility underwater. Underwater networks depend on optical, acoustic, or magnetic induction-based communication, but these also have their own trade-offs in terms of communication range, energy consumption, and data rate [1]. Optical communications offer high data rates, but are prone to turbidity in water, and magnetic or radio signals also suffer from attenuation in underwater media. Acoustic communications, however, offer long-range propagation and are also suitable for non-line-of-sight links, which is a preferred choice for underwater communications. As modern maritime operations continue to grow, underwater communication has to develop beyond point-to-point acoustic links. The emerging applications of underwater communication networks require smart, secure, and easily scalable systems. These networks must be able to handle data transmission seamlessly and maintain real-time control. In order to fulfill these requirements, traditional network designs must be questioned and altered in line with underwater conditions.

3.1.2 Limitations of traditional TCP/IP

The TCP/IP protocol stack was originally designed for stable and high-bandwidth land-based networks. Conditions such as high latency, frequent link failures, and constrained energy resources, which are characteristic of underwater environments, do not align well with the protocols designed for the TCP/IP stack. A key limitation arises from the attenuation of the signal in underwater media, which follows the exponential model shown in (3.1).

$$S(d) = S_0 \cdot e^{-\alpha d}, \tag{3.1}$$

where $S(d)$ is the signal strength at distance d, S_0 is the initial signal strength, and α is the attenuation coefficient, which depends on both the type of medium, that is, water, and the signal frequency. This significant loss in signal strength is highly dependent on distance and reduces link quality, leading to high packet loss and disturbed connectivity. For example, TCP works on acknowledgment-based feedback and perceives the loss of a packet as a sign of congestion.

In underwater acoustic networks, long propagation delays and high bit error rates can cause TCP to produce reduced throughput and a packet delivery ratio. Also, traditional IP-based routing protocols are well-suited for terrestrial network topologies; however, this is not the case in underwater environments. Likewise, traditional signal authentication techniques in the physical layer are highly susceptible to interference, making them unsuitable for acoustic or optical media. Additionally, high-latency underwater channels cannot work with typical MAC protocols proposed for the terrestrial environment, which depend on frequent handshakes and acknowledgments. Standard transport layer protocols require rapid retransmissions and reliable end-to-end links, which are impractical to achieve in underwater communications. Encryption algorithms and centralized access control are heavily resource-intensive and not optimal for the constrained budget of underwater networks. Hence, the rigid layered architecture of the TCP/IP model does not align with the requirements of general underwater networks, particularly, dynamic AUV networks. This forms a basis for the need for a flexible and programmable protocol stack well-suited for underwater networks.

3.1.3 The role of SDN, 6G, and security in underwater networks

To overcome the limitations of traditional protocol stacks, novel communication architectures for underwater environments can be designed by exploring technologies such as SDN, the upcoming 6G paradigms, and intelligent control planes that AI can be driven. These technologies bring flexibility, scalability, and robustness. The provision of a centralized control plane in SDN isolates the network correspondence from data forwarding. This provision can be exploited for enabling dynamic routing updates, coordinated vehicular (or underwater) mobility, and energy-efficient protocol selection. A ship on the surface or a strategically positioned buoy with 6G connectivity can serve as an SDN controller, maintaining coordination with underwater nodes, and information can be relayed to an edge data center or to the cloud. Similarly, 6G communication envisions ultra-low latency, extreme reliability, AI-driven networking, and the use of intelligent reflecting surfaces. All of these features exhibit possible exploits that align with the constraints of underwater media. As shown in Table 3.1, the traditional TCP/IP protocol stack fails in the context of underwater networks, making it necessary to look for adaptations at every layer. Another pressing concern is the vulnerability of underwater networks to spoofing, eavesdropping, and jamming, particularly in open-access oceans. Each layer of the protocol stack must be fortified with appropriate techniques to ensure secure end-to-end communication [2].

3.2 Layer-by-layer discussion

In this section, we discuss the challenges of layer-by-layer underwater networks, the protocols proposed for each layer, and how intelligent 6G technologies can be utilized to make these protocols well-suited for underwater environments.

3.2.1 Application layer

The application layer is the interface between the network and the end node. In terrestrial topologies, this layer is based on protocols such as the hypertext transfer protocol (HTTP), the file transfer protocol (FTP), and the teletype network (Telnet). However, the uniqueness posed by ocean environments makes these protocols less efficient, or in certain cases, even useless [3]. This difference is fundamentally due to the

Table 3.1 Comparison of the traditional TCP/IP model and the current underwater communication requirements

Features	Traditional terrestrial TCP/IP networks	Underwater communication networks
Propagation medium	Electromagnetic (RF)	Mostly acoustic
Bandwidth availability	High (Mbps–Gbps)	Low (bps–kbps)
Latency	Low and stable (predictable)	High and variable (unpredictable)
Topology	Relatively stable and structured	Dynamic and mobile (drifting sensors)
Energy constraints	Ample energy (grid-powered or charging stations nearby)	Severe energy constraints (charging stations located on-shore)
Routing protocols	fixed or mobility-aware predictable routes	delay-tolerant, mobility-aware unpredictable routing
Protocol flexibility	Static, layered stack	Needs adaptive, cross-layer design
Error rates	Low due to reliable links	High due to noise and fading
Security focus	Encryption, firewalls, and IPsec	Lightweight and adaptive security essential due to limited resources

reliance of underwater communication on acoustic waves rather than electromagnetic or radio wave signals, leading to severe propagation delays, constrained bandwidth, and frequent link failures. These factors create barriers to reliable application layer correspondence that are not negligible.

3.2.1.1 Challenges in underwater application layer communication

One of the primary challenges in underwater communication across the application layer is the propagation delay. Radio waves travel at the speed of light, which is approximately 300,000,000 m/s, while acoustic waves travel at around 1500 m/s, marking a difference of approximately five orders of magnitude. This latency affects applications that require quick feedback. Also, the bandwidth available in underwater channels is highly limited and variable. Heavy application layer protocols, such as HTTP or the simple object access protocol (SOAP), are ruled out for underwater communication due to these constraints. Another challenge is the intermittent connectivity between the underwater sensor nodes and the AUVs. This produces a high packet loss, leading to incomplete session states. Underwater nodes are usually battery powered and deployed in inaccessible locations, making it essential to minimize protocol overhead and optimize data transmission. This reflects the energy inefficiency of such topologies, hence proving the conventional protocols to be ill-suited for underwater environments.

3.2.1.2 Protocols suitable in underwater networks for the application layer

To address the issues mentioned in the previous section, various application layer protocols have been proposed that are lightweight and tolerant to disruptions. One of them is constraint application protocol (CoAP) [4], which follows a RESTful architecture that is similar to HTTP but offers several benefits for low-power devices and limited-bandwidth. CoAP allows confirmable and non-confirmable message types, which makes it reliable for dynamic underwater links. Another widely accepted protocol is message queuing telemetry transport (MQTT) [5]. It is based on a publish–subscribe model that reduces the overhead of point-to-point communication. MQTT is essentially employed in scenarios where a central node such as a lighthouse, as shown in Figure 3.1, globally views subscriptions and is responsible for relaying data between end nodes. Likewise, the delay-tolerant networking (DTN) protocol presents another suitable architecture for underwater networks. It operates in a way that messages are kept at the relay nodes until a reliable communication link is available. This approach is useful in dynamic underwater environments with frequent link failures.

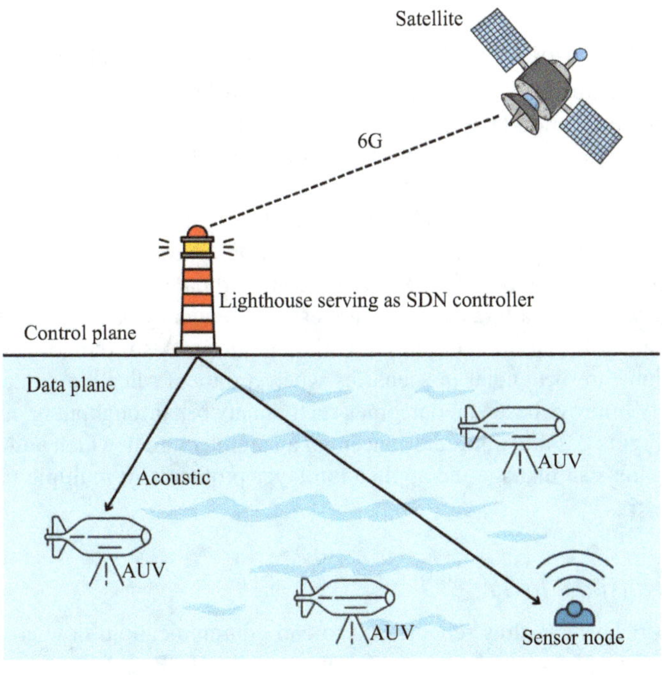

Figure 3.1 Illustration of an SDN-based acoustic underwater communication architecture incorporating 6G, and AUVs

Table 3.2 Comparison of application layer protocols for underwater communication

Protocol	Architecture	Delay tolerance	Bandwidth efficiency
CoAP	Client–server	Medium	High
MQTT	Publish–subscribe	Low	High
DTN	Store-carry-forward	High	Medium

A mathematical representation of the delay of the message in DTN-based networks is shown in (3.2).

$$T_{delay} = \sum_{i=1}^{n} (T^{(i)}_{store} + T^{(i)}_{prop} + T^{(i)}_{proc}), \quad (3.2)$$

where $T^{(i)}_{store}$ is the time the message is stored at node i, $T^{(i)}_{prop}$ is the propagation delay between nodes i and $i+1$, and $T^{(i)}_{proc}$ is the local processing time at node i. These mechanisms are adaptive to underwater scenarios and significantly improve the reliability of underwater communications. A comparative summary of these protocols is presented in Table 3.2.

3.2.1.3 Integration of SDN and AI for protocol adaptation

The integration of SDN and AI techniques can revolutionize the selection of protocols at the application layer in AUV networks. The SDN-enabled central control plane can monitor the network status in real-time and dynamically adjust the application layer protocols according to the network scenario. For example, in low-latency scenarios where links are relatively stable, MQTT can be employed, and DTN can be used in high-delay, disrupted environments. AI also provides predictive analysis. A deep learning (DL) model trained on pre-collected data from a conventional network can predict link failures and adjust message frequency or the type of protocol to be used according to the current network conditions. Also, reinforcement learning algorithms are beneficial in scenarios where protocol selection strategies must continuously improve based on outcomes such as network throughput or energy consumption. Figure 3.2 illustrates the concept of an architecture in which an AI-enabled SDN controller can manage the application layer protocols in multiple underwater nodes.

3.2.2 Transport layer

The transport layer ensures reliable end-to-end communication, flow control, congestion control, and error handling. Conventionally, TCP and the user datagram protocol (UDP) are widely used protocols at the transport layer for terrestrial networks. However, several constraints posed by underwater environments make TCP

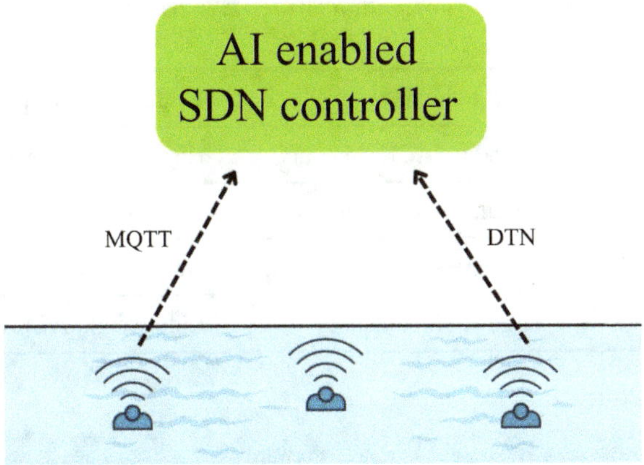

Figure 3.2 AI and SDN-enabled adaptive application layer protocol management in underwater networks

or UDP inadequate. Underwater media offer extremely low bandwidth, high propagation delay, and high bit error rates that cannot be neglected. These issues make it unavoidable to look for specialized protocols for underwater networks at the transport layer.

3.2.2.1 Limitations of traditional transport protocols in underwater networks

The TCP protocol assumes that packets that have congestion are lost, leading to a reduced packet delivery ratio. This assumption, however, fails in underwater networks because of channel modeling errors and temporary disconnections. Also, TCP relies on acknowledgment (ACK) and retransmission schemes, which result in considerable delays and energy consumption. A complete message transmission is characterized by the round-trip time, expressed in (3.3).

$$RTT = 2 \times \frac{d}{v}, \tag{3.3}$$

where *RTT* is the round-trip time and v is the speed of acoustic wave propagation (almost 1500 m/s). For long distances, *RTT* can reach several seconds, making TCP timeouts inefficient. This again is not preferable in underwater networks. Table 3.3 presents a comprehensive analysis of TCP in the context of underwater environments.

3.2.2.2 Alternative transport protocols

The underwater transport protocol (UWTP) [6] is an interesting protocol that minimizes control overhead in a way that passes only selective ACKs and does proactive error estimation. The reduced overhead makes it lightweight and perfect for acoustic links. Another suitable protocol is DTN. DTN can be configured at both the application and the network layer. It makes use of the store-carry-forward scheme and is ideal for thinly populated, connected AUV networks. This has also been discussed earlier

Table 3.3 Comparison of TCP with underwater requirements

Parameter	TCP behavior	Underwater requirement
Loss detection	Based on ACK timeouts	Error-aware detection
Congestion control	Assumes loss = congestion	Needs error-aware strategy
Bandwidth utilization	Inefficient in high latency	Delay-tolerant mechanisms needed
ACK overhead	High ACK frequency	Lightweight ACK or no ACK

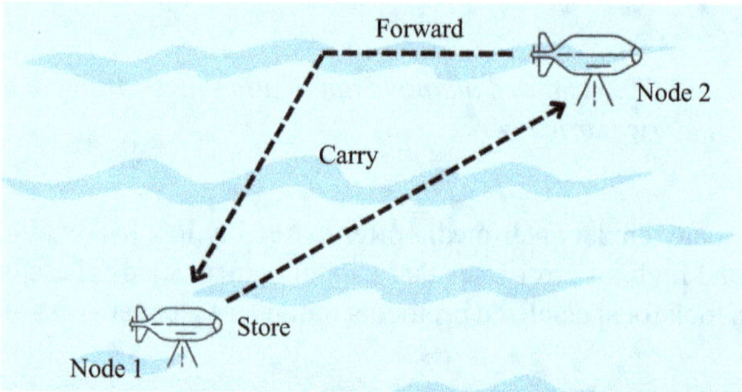

Figure 3.3 DTN store-carry-forward communication model for sparse underwater networks

in the application layer. Figure 3.3 shows a schematic detailing of the DTN communication model for sparse underwater networks. Unlike end-to-end TCP, maintaining reliability at each node pair hop-by-hop can also reduce retransmission delays.

3.2.2.3 Dynamic transport protocols via SDN

SDN enables dynamic control of packet flow strategies. This is significant for the transport layer correspondence. In underwater networks, SDN offers a wide range of benefits, some of which are:

- Choose an appropriate protocol at the transport layer, based on the network conditions, such as link quality and latency.
- Automatically adjust the waiting time for retransmission, the size of the sliding window, and the ACK frequency, as illustrated in Figure 3.4.
- Enable cross-layer optimization by maintaining a feedback system with lower layers.

3.2.2.4 Role of 6G and AI in transport layer optimization

6G networks can help underwater transport layers by allowing low-latency surface-to-underwater (data to control and vice versa) coordination. We can employ suitable

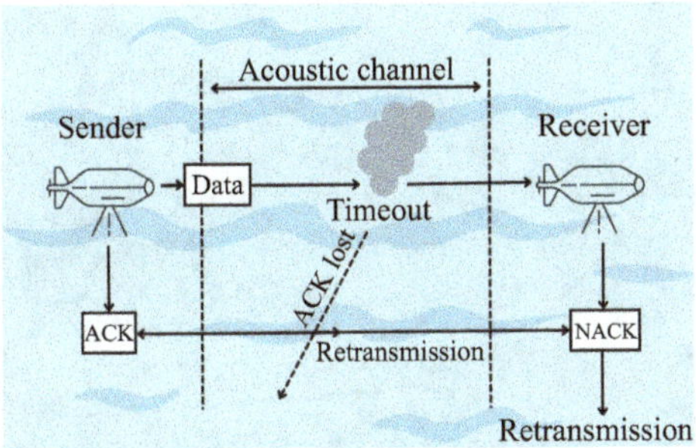

Figure 3.4 Transmission, rejection, and retransmission of ACKs among underwater AUVs

machine learning (ML) models to detect congestion at a certain point and predict random loss using simple classification algorithms. For example, a neural network-based classifier can classify whether packets being lost are due to congestion or to channel errors. This can be achieved by utilizing features, such as packet delivery ratio, packet loss rate, signal-to-noise ratio, and delay, depending on the outcome, the SDN controller can decide whether to allow retransmission or not.

3.2.3 Network layer

The network layer is responsible for delivering data from the source to the destination. Unlike land-based networks, where topologies are considerably stable and have dense connectivity, underwater networks have to deal with sparse deployments, highly mobile nodes (e.g., AUVs or movement of sensor nodes due to ocean currents), and unavoidable link failures. These characteristics make routing a complex task in underwater environments. Hence, the need for developing adaptive and intelligent routing protocols arises.

3.2.3.1 Routing challenges in underwater networks

Underwater communication networks encounter various challenges at the network layer. The constant movement of AUVs and the drifting of sensor nodes make the topology highly dynamic and continuously changes the structure of the network (see Figure 3.5). Usually, the deployment of nodes is sparse and scattered over large areas. This limits connectivity, and often communication loopholes are created; these are areas with almost zero availability of relay nodes. Another important concern is the slow speed of acoustic waves that causes high latency and leads to severe propagation delays. This significantly deteriorates real-time communication. Equation (3.4) shows the relationship between propagation delay and distance between nodes.

$$T_{\text{propagation}} = \frac{d}{v}. \quad (3.4)$$

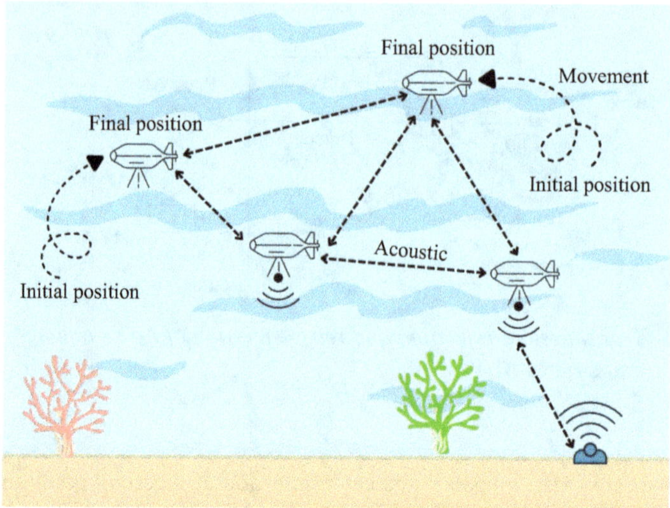

Figure 3.5 Node mobility and dynamic topology in underwater environments

We cannot neglect that most underwater nodes, either AUVs or sensor nodes, operate on limited battery power. This poses strict constraints on energy that necessitate the use of energy efficient routing protocols to extend the network lifespan.

3.2.3.2 Dynamic routing protocols for underwater networks

Various routing protocols are suitable for underwater environments. One such protocol is the vector-based forwarding (VBF) protocol [7], an extended version of which is the energy-balanced VBF, designed specifically for underwater networks. Figure 3.6 shows an illustration of the VBF protocol. VBF uses a virtual pipeline that forwards packets to only those nodes that are near a virtual vector between the source and destination. This selective forwarding of packets significantly reduces network overhead, thus avoiding redundant data transmissions. This contributes significantly to energy conservation. However, there is a downside to VBF in sparse network topologies, as virtual pipelines are difficult to maintain in such scenarios. On the contrary, energy-efficient localization-based (EEL) [8] utilizes node localization and link quality to identify reliable and energy efficient routes. EEL best suits network mobility and sparsity; however, it requires exact position information, which is again a challenge in underwater networks.

3.2.3.3 The role of SDN, AI, and 6G in the network layer adaptation

The underwater vehicular nodes are not fully capable of making autonomous routing decisions due to their constraints discussed earlier. The provision of separate planes for data and control traffic by SDN plays a crucial role in this context. Surface stations or surface buoys that serve as SDN controllers can maintain a global view of the whole topology, which helps to push routing policies to underwater nodes. This is how distributed and sparse networks can benefit from the centralized control plane

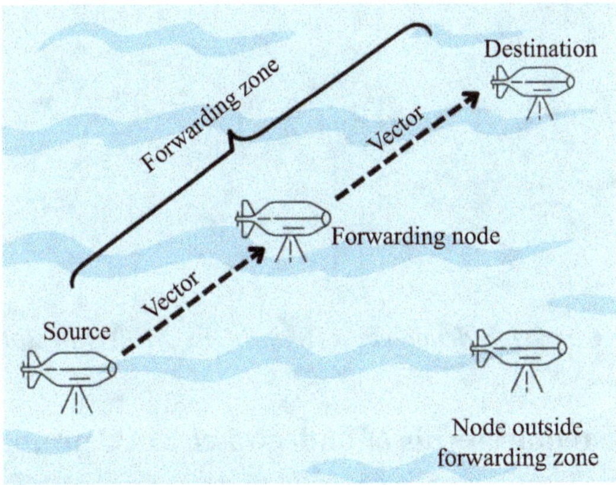

Figure 3.6 VBF zone in an underwater network

of SDN. Ultra-reliable low-latency communication (URLLC) is a primary concern in underwater sensor networks. The integration of 6G is beneficial for surface-to-underwater signaling at the network layer. Moreover, AI/ML techniques, especially reinforcement learning and graph neural networks (GNN), can assist in optimizing the path by predicting stable links and reliable relay nodes.

3.2.4 Data link layer

The data link layer is important for error detection, framing, and medium access control (MAC), hence it is also known as the MAC layer. In underwater environments, traditional MAC protocols are hard to implement due to high latency, variable propagation delays, and limited bandwidth. The MAC layer has to ensure reliable and sequenced communication over an unreliable underwater acoustic medium.

3.2.4.1 Link layer protocols for underwater communication

The Tone-Lohi protocol (T-Lohi) [9] is designed to minimize collisions and conserve energy. It works in a way such that a node, which can be a sensor node or an AUV, transmits a short tone before sending the actual data. If multiple nodes send tones together, the node that detects the least number of collisions precedes. The remaining nodes undergo a backoff algorithm and retransmit after a predetermined time interval. The characteristic asymmetric propagation delays in underwater environments can be exploited to achieve seamless communication. Another key protocol employed at the MAC layer is the flow acquisition multiple access (FAMA), known as slotted-FAMA [10]. In this conservative protocol, reservations are made based on a two-way correspondence between nodes. The nodes can acquire a channel by sending a request-to-send (RTS) packet in a predefined slot. Once a clear-to-send (CTS) packet is received, the data are transmitted, as shown in Figure 3.7. Slotted FAMA reduces hidden terminal problems in this way, but the entire process can still produce considerable delay.

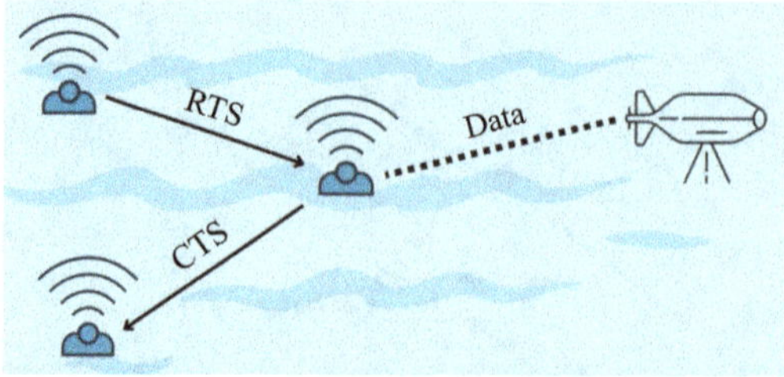

Figure 3.7 Correspondence via RTS and CTS packets in slotted-FAMA

3.2.4.2 Unique requirements of underwater MAC protocols

Due to the slow propagation speed of sound in water (around 1500 m/s), the magnitude of propagation delays is variable. Also, AUVs are typically dependent on batteries, which are difficult to recharge or change rapidly. Therefore, MAC layer protocols to be deployed in underwater networks must be able to operate efficiently despite extremely variable delays [11]. Another important factor is to minimize control signaling and retransmissions due to high energy consumption and delay.

3.2.4.3 SDN-based link layer configuration and control

The centralized intelligence offered by SDN allows for a responsive control of MAC layer operations, critical in the context of AUV networks. An SDN controller can monitor the ongoing quality of the link, traffic load, and network topology, unlike static and distributed MAC configurations in terrestrial topologies. Based on real-time observations of node placement and overall density, the SDN controller can switch between contention-based MAC protocols, such as T-Lohi and reservation-based (the slotted-FAMA). Time slots and contention window parameters can also be resized dynamically to minimize collision probability and delay, at depths with high sensor node density or AUV node density. To save energy, nodes can be programmed to sleep or wake up based on predicted traffic patterns and scenarios, reducing idle listening. Using a centralized controller, time-synchronous wake-up patterns can be implemented across different node clusters. This can enable less conflict access to the medium and improve energy efficiency. Flow monitoring and link diagnosis are also characteristic tasks of the MAC layer. The SDN controller can detect faulty, energy-restrained, or malicious nodes and reconfigure the MAC layer so that failed nodes can be isolated from the active topology. This can help maintain the overall reliability of the system.

3.2.4.4 6G and AI/ML driven enhancements at the link layer

Future 6G provisions can allow for satellite and aerial backhaul to AUVs. This will enable decentralized delegation of tasks and orchestrated control at the global level. Federated learning can be employed at the MAC layer, allowing for distributed training of models at different nodes without the need for centralized data collection. This can also preserve time, energy, and network resources. Also, link quality can be predicted using supervised learning and the MAC layer parameters can be set based on reward functions dependent on throughput, energy, and delay. This can be done using reinforcement learning. (3.5) can be worked on in this regard.

$$\eta = \frac{T_{data}}{T_{data} + T_{control} + T_{propagation}}, \tag{3.5}$$

where η is the channel efficiency, T_{data} is the data transmission time, $T_{control}$ is the MAC control overhead, and $T_{propagation}$ is the propagation delay.

3.2.5 *Physical layer*

The foundation of underwater communication lies in the physical layer. This ensures the successful transmission of binary bits over complex aquatic media. Terrestrial networks can leverage high-frequency radio waves; however, the unique physical constraints of underwater environments, such as attenuation, multipaths, and low propagation speeds, obligate us to look for an alternate medium.

3.2.5.1 Underwater communication modalities

1. Acoustic communication: Acoustic waves are suitable for underwater communication because of their ability to traverse long distances with significantly low attenuation. However, acoustic waves do suffer from limited bandwidth, typically in the range of a few kilohertz, because they encounter absorption at higher frequencies. The mobility of the nodes can also cause rapid frequency changes that can give rise to Doppler shifts. Reflections from the ocean surface, bottom, and AUVs or other objects in the depths of the ocean also cause a severe multipath effect.
2. Optical communication: Communication over optical lines offers high data rates, up to Gbps. Still, the downside to this is that optical communication is only possible over short distances, around tens of meters, and this requires a line-of-sight path as well. A preferred approach is to employ blue–green light, as it offers low attenuation in water. Table 3.4 shows a comparison of different physical layer technologies adaptable to underwater communications.

3.2.5.2 Channel modeling in underwater environments

The underwater acoustic channel is mathematically represented in (3.6).

$$r(t) = \sum_{i=1}^{L} a_i s(t - \tau_i) + n(t), \tag{3.6}$$

where a_i is the attenuation of the ith path, τ_i is the delay of the ith path, L equals the number of multipath components and $n(t)$ is the additive noise (modeled as colored Gaussian or non-Gaussian). Advanced channel models can be designed considering Doppler spread, time-dependent path delays, and noise underwater.

3.2.5.3 Modulation and coding techniques

Due to severe channel impairments, robust modulation and coding schemes are essential. Orthogonal frequency division multiplexing (OFDM) decomposes the original

Table 3.4 Comparison of physical layer communication technologies for underwater networks

Medium	Range	Data rate	Power efficiency	Environment suitability
Acoustic	up to 20 km	∼kbps	High	Long-range, turbid water
Optical (visible light communication)	1–100 m	Mbps–Gbps	Low	Clear water, line-of-sight required
Magnetic induction	<10 m	kbps	Medium	Turbid/shallow water
RF (surface/6G)	km-scale (above water)	Mbps	High	Aerial links from buoys
Terahertz (6G)	1–10 m (Air/near-surface)	Gbps	Low	Line-of-sight surface nodes

channel into perpendicular subcarriers, each modulated individually. This can be useful in eliminating multipaths and their effects. Chirp spread spectrum (CSP) works best with Doppler shifts and in low signal-to-noise ratio environments. It can be employed across nodes in the depths of oceans. Another rule of thumb is to dynamically switch between binary phase-shift keying (BPSK), quadrature phase-shift keying (QPSK), or higher-order quadrature amplitude modulation (QAM) based on the available signal-to-interference and noise ratio and bit error rate [12]. Forward error correction (FEC) schemes, such as turbo codes and convolutional codes, can be used to mitigate errors that are caused by channel distortions.

3.2.5.4 Potential use of 6G and optical communication for underwater networks

6G has transformative capabilities that can be employed for underwater communication. Low-earth orbit (LEO) satellite constellations like Starlink can act as mediating bodies between oceanic buoys or lighthouses, or any surface-level infrastructure serving as an SDN controller, and the terrestrial base stations. LEO offers global coverage of the topology, providing high data rates at the cost of very low latency due to the mesh topologies designed for the satellite relay systems. 6G also introduces IRS. IRS can reflect and refract signals toward better communication paths. Floating IRS systems can be made on buoys or ships, which can optimize the propagation of the acoustic signal near the surface.

3.2.5.5 AI/ML integration at the physical layer

Convolutional neural networks and recurrent neural networks can be used to model and predict suitable underwater channels. Autoencoders and denoising neural networks can also be used to improve signal strength, signal-to-interference-noise ratio, and fidelity in high-noise scenarios. Also, the choice of the optimal modulation or coding scheme can be made using cognitive adaptation algorithms that utilize reinforcement learning agents.

3.3 Security considerations in underwater networks

The hostile environment, constrained resources, and unique communication mechanisms pose a broad range of security challenges in the design of underwater sensor and AUV networks. The conditions underwater make the topologies susceptible to conventional cybersecurity threats and physical vulnerabilities. The integration of SDN, 6G, and AI-based architectures in underwater networks brings about various methods [13] and reasons for securing each layer of the protocol stack.

3.3.1 Threats to underwater networks

Underwater networks are exposed to numerous threats that can compromise their integrity, confidentiality, availability, and authenticity [14]. Eavesdropping is a critical issue in which adversaries capture high-risk data by passively listening to systems and exploiting privacy. If an interference is introduced intentionally in the medium, a jamming attack can occur that can either disrupt or cause a complete breakdown of data transmission. Spoofing attacks can also occur in underwater sensor networks. Attackers can impersonate as legitimate nodes to disrupt trust mechanisms and data aggregation. Injection attacks can further worsen the situation by allowing retransmission of malicious packets. Routing-based threats, including selective forwarding attacks, blackhole, and wormholes, falsify the routing process, which eventually leads to a drop in packet delivery ratio. One of the very important concerns is the energy depletion attacks. The limited battery life of AUVs can be exploited by flooding them with message requests, which ultimately exhausts the operational lifespan of legitimate nodes in the medium.

3.3.2 Security measures at each layer

A comprehensive and layered approach is required to secure underwater communication networks, primarily due to the use of unconventional media, such as acoustic signals. At the application layer, lightweight end-to-end encryption and strong authentication and authorization protocols should be enforced. DL can be used for intrusion detection by identifying anomalies in incoming traffic. Data integrity and confidentiality are ensured at the transport layer using datagram transport layer security, efficient key exchange mechanisms, and adaptive retransmission strategies. This will resist denial-of-service attacks in underwater conditions prone to latency. Blockchain-enabled routing protocols [15] at the network layer can proactively counter routing attacks. This can be complemented by AI-driven IPSec adaptations and anomaly detection schemes. The MAC or data link layer is designated for encryption and frame integrity. AI can also be employed here to predict and avoid jamming attacks or collision-based attacks. At the physical layer, mechanisms such as signal fingerprinting, secure modulation, and channel-based key generation can safeguard the medium. In conclusion, these strategies work together across the protocol stack to establish a secure and resilient customized communication network for underwater environments.

References

[1] Pottier A, Bouvet PJ, Tomasi B, *et al.* Data rate performance of mobile multiuser MIMO underwater acoustic communication systems. *IEEE Journal of Oceanic Engineering*, 2025;50(2):477–488.

[2] Nkenyereye L, Nkenyereye L, and Ndibanje B. Internet of underwater things: a survey on simulation tools and 5G-based underwater networks. *Electronics*, 2024;13(3):474.

[3] Gupta S and Singh NP. Underwater wireless sensor networks: a review of routing protocols, taxonomy, and future directions. *The Journal of Supercomputing*, 2024;80(4):5163–5196.

[4] Bormann C, Castellani AP, and Shelby Z. CoAP: an application protocol for billions of tiny internet nodes. *IEEE Internet Computing*, 2012;16(2):62–67.

[5] Quincozes S, Emilio T, and Kazienko J. MQTT protocol: fundamentals, tools and future directions. *IEEE Latin America Transactions*, 2019;17(09):1439–1448.

[6] Masiero R, Azad S, Favaro F, *et al.* DESERT underwater: an NS-miracle-based framework to design, simulate, emulate and realize test-beds for underwater network protocols. In *2012 Oceans-Yeosu*. IEEE; 2012. pp. 1–10.

[7] Abbas CJB, Montandon R, Orozco ALS, *et al.* EBVBF: energy balanced vector based forwarding protocol. *IEEE Access*, 2019;7:54273–54284.

[8] Hao K, Shen H, Liu Y, *et al.* Integrating localization and energy-awareness: a novel geographic routing protocol for underwater wireless sensor networks. *Mobile Networks and Applications*, 2018;23:1427–1435.

[9] Syed AA, Ye W, and Heidemann J. T-Lohi: a new class of MAC protocols for underwater acoustic sensor networks. In *IEEE INFOCOM 2008 – The 27th Conference on Computer Communications*. IEEE; 2008, pp. 231–235.

[10] Ahmed F and Cho HS. A time-slotted data gathering medium access control protocol using Q-learning for underwater acoustic sensor networks. *IEEE Access*, 2021;9:48742–48752.

[11] Guo J, Song S, Liu J, *et al.* Exploring applicable scenarios and boundary of MAC protocols: a MAC performance analysis framework for underwater acoustic networks. *IEEE Transactions on Mobile Computing*, 2024;23(12):12717–12730. https://doi.org/10.1109/TMC.2024.3417895

[12] Bharathi PS and Reddy MJB. A security enriched energy efficient data communication over underwater wireless sensor network using AI powered hybrid deep learning procedures. In *2024 9th International Conference on Communication and Electronics Systems (ICCES)*. IEEE; 2024, pp. 819–825.

[13] Adam N, Ali M, Naeem F, *et al.* State-of-the-art security schemes for the internet of underwater things: a holistic survey. *IEEE Open Journal of the Communications Society*, 2024;5:6561–6592.

[14] Yuan Y, Shen X, Sun L, *et al. Modeling cascading failures and invulnerability analysis of underwater acoustic sensor networks based on complex network. IEEE Sensors Journal*, 2024;24(5):6942–6952.

[15] Tomović S, Krivokapić B, Na, *et al.* BEKMP: a blockchain-enabled key management protocol for underwater acoustic sensor networks. *IEEE Access*, 2024;12:74108–74125.

Chapter 4

Integrating SDN and NFV to deploy early warning systems in marine distributed underwater networks (MDUN)

Muhammad Noman Javid[1] and Huma Ghafoor[1]

This chapter takes an in-depth look at how software-defined networking (SDN) and network function virtualization (NFV) are applied in the emerging field of the marine distributed underwater networks (MDUN). The research includes reviewing and categorizing recent approaches that overcome the limitations of existing underwater systems caused by centralized control, long input and output delays, energy needs, and fixed architectures. Covered topics are distributed SDN controller placement, virtualization of edge network functions, piezoelectric energy sources, communication systems using sound, low-earth orbit (LEO) satellites for networking, error-resistant data delivery with erasure-coding, and energy-conscious multipath routing. Our proposed integrated architecture explains what works well in supporting tsunami early warning in real-time and what needs progress in terms of operating on its own for a long time. The taxonomy groups research into categories defined by water depth (underwater, surface, and land), management of controls, use of energy resources, and flexibility in function. Existing issues and possible areas for further study are listed, such as using artificial intelligence for routing, making virtual network functions able to repair themselves, and automating service orchestration for safety during disasters. This chapter provides a solid understanding that helps develop energy-efficient, scalable, and resilient undersea networks.

4.1 Introduction

The internet of underwater things (IoUT) is a transformative technology that is going to shape maritime safety, raise standards of ocean observation, and provide better protection for underwater areas. Advances in smart node sensors, underwater relays, and surface computer systems provide IoUT with new opportunities to quickly discover and react to natural disasters [1,2]. In these applications, early detection of

[1]School of Electrical Engineering and Computer Science (SEECS), National University of Sciences and Technology (NUST), Pakistan

tsunamis is especially important because it provides enough time for people in dangerous areas to take safety measures while avoiding significant financial losses. Yet, working in the ocean creates problems in gathering and sharing information. Systems need to be highly reliable, respond quickly, use little energy, and be able to operate independently when there is no easy maintenance access [3].

Most current underwater IoT systems depend on a centralized architecture, using surface buoys to periodically exchange data between underwater sensors, satellites, and ground stations [4]. Although it works in many ways, this approach still has some major limitations. Feeding all data through a central point creates chances for malfunctions; if these central nodes fail, it can result in many parts of the network losing contact and essential data becoming unavailable in case of an emergency [5]. Additionally, dealing with centralized construction brings up issues with growth and flexibility. On top of this, having remote controllers leads to delays and challenges with synchronization, as the paths between the acoustic devices and the main controller introduce variability in delays.

The mismatch, together with time zone differences and the changing nature of the ocean, limits the effectiveness of flexible, real-time networks. For this reason, this chapter focuses on the integration of SDN and NFV in underwater network architecture. This framework distributes SDN controllers to different points both on the ship and in different layers of the underwater network, making control of the network less centralized. With distributed control, the whole system is not affected if some parts fail, allowing shaken network segments to continue to work during seismic events and similar disasters [6]. The design considers energy efficiency in underwater acoustic communication and makes use of advanced routing protocols suited to the limited bandwidth and slow communication common in subsea channels.

A key difficulty in deep-sea IoUT is finding a way to power the sensors over long periods of time. This problem can be solved by integration, which powers its seabed sensors using piezoelectric methods that transform pressure and movement into electricity [7]. This method helps reduce reliance on batteries, limits costly repairs, and expands the service time of equipment from a few months to possibly years [1,8]. Also, by using erasure coding, the system helps ensure that the data is reliable and improves bandwidth efficiency, which is important due to the small amount of usable acoustic spectrum. Because of this coding, early warning messages are still reliably received, even if a few data packets are missing. Maritime communications to land analytics rely on low-earth orbit (LEO) satellite networks, such as Starlink. Because of their low orbit, LEO systems provide a much shorter delay, between 30 and 50 ms [9], which enables them to supply tsunami updates and seismic data essential to emergency actions.

Furthermore, the architecture uses NFV to automatically deploy and handle network safety systems such as firewalls, intrusion detection, and load balancers on ship-based edge platforms. Since the system uses virtualization, it can quickly change its service and less heavily rely on dedicated hardware when dealing with unexpected changes or disasters [5,6]. As a result, the six pillars build a flexible, strong, and energy-efficient network suitable for underwater systems. With the help of decentralized management, sustainable energy, improved data sharing, and advanced satellite and virtualization, the chapter focuses on minimizing the problems and weaknesses found in traditional underwater sensing networks. Consequently, it allows for the development of warning and monitoring systems that can work reliably in any harsh marine environment.

4.2 Motivation and related work

Conventional tsunami early warning systems (TEWS), including deep-ocean assessment and reporting of tsunamis (DART)-II, use ocean bottom pressure recorders (OBPR) at seabed together with surface buoys to send warning information. When seismic sea waves appear, these OBPRs measure sudden changes in pressure underwater and broadcast readings to nearby floating buoys, which then pass the information to coastal monitoring centers through a satellite network. Despite its usefulness in many tsunami events and its key role in monitoring tsunamis worldwide, this architecture faces important limitations when it comes to scalability, reliability, and responsiveness in regions likely to be affected by tsunamis. A main issue with these systems is that the buoys may not work reliably when physical conditions and operations are poor. As time goes on, strong currents, high tides, and wind may move buoys away from the places where they were positioned to monitor. As a result, the accuracy of the data's geography is reduced, as well as the size of the warning network. Since water is salty and can encourage rust, marine areas are often very harmful to boat equipment. Maintaining such units is very difficult and costly, and it becomes more so the farther away they are. Moreover, each DART-II setup currently costs more than $250,000, and the deployment of a basic yet redundant regional model would take several millions to finance, on top of the costs to operate the network and its connections.

Besides environmental damage, these systems are vulnerable to other types of risks. Continuity of operations is threatened by vandalism, collisions with commercial ships, and disagreements between nations in strategic maritime areas. In addition, these old architectures are very strict and hard to change. They do not have network programmability, so they cannot update sensor data traffic based on new risks or events happening around them. When both storms and seismic events strike at the same time, the system is not smart enough to route data to help with tsunami monitoring. As soon as tsunamis happen in the nearby sea, they have a greater effect on people living along the shoreline, which makes quick alerts necessary. End-to-end communication latency should ideally not exceed 200 ms to ensure that early warnings reach people facing risk in time. Unfortunately, traditional TEWS systems are often unable to provide very low latency. Because there are four main stages in seismic data transmission, delays can occur, and events may fail at any point along the way. Failing to transfer key telemetry quickly in real-time usually cuts down on the system's usefulness for emergencies that happen in a short period. So, even though systems like DART-II have helped in monitoring tsunamis, the framework should be more flexible, efficient with energy, and uncompromising, as outlined in this chapter.

Introducing SDN has changed the way networks are built and maintained, mainly in terrestrial environments where there is sufficient bandwidth and a steady infrastructure. Because of SDN, the control and data planes are separated, making it possible for centralized controllers to regulate the network, put policies into effect, and alter routing based on ongoing needs. Because of this separation, programs can be made more easily, managed better, and adapted to new needs. It was common in traditional networks that every device needed to be individually reconfigured whenever changes were made, which took a lot of time and introduced chances of error. SDN overcomes the problem by introducing a software-based controller that manages traffic for all devices on the network from one location. This permits for quick adaptation, better traffic optimization, and the addition of advanced policies without the need to change the hardware. Due to the problems found in underwater environments, SDN is considered a promising solution for IoUT. These are having a small amount of bandwidth, using sound signals to communicate, not enough energy, and having problems reaching submerged devices. Traditionally designed networks did not handle these limitations very well. For instance, routing algorithms that count on the same level of link quality and regularly refreshed information are unable to operate in changing underwater environments, where noise and mobility affect the links.

Underwater signals are limited by the difficulty of reaching a large distance, slow signal arrival, the possibility of multiple delays, and a high chance of message disruptions. For this reason, existing networks cannot handle sudden changes in the environment or emergencies such as tsunamis. One of the ways in which SDN deals with these issues is by using centralized control that can be instantly reconfigured based on shifting network circumstances. During a sudden underwater earthquake detected with sensors, SDN redirects tsunami-related data to standard data traffic and chooses the most efficient route for the data. Even so, putting SDN into practice in underwater networks is not easy. Centralized control cannot work well due to the high levels of latency and the amount of energy needed for undersea communications. With the low bandwidth and long network delay of the acoustic links, it is almost impossible for the underwater nodes to communicate regularly with a central SDN controller located far away. In addition, most underwater sensor nodes run on limited batteries and lack the ability to communicate often with a main control device. Therefore, underwater deployments require the use of distributed SDN to overcome these issues. In this setup, SDN controllers are embedded in dynamically moving platforms above and below water, such as research ships, powered buoys, and autonomous underwater vehicles (AUVs). They function like small data centers that help with quick control because they are very close to the nodes underwater.

Distances that are closer between devices still use less energy, respond more quickly, and enhance communications. Due to the integrated architecture, each network segment can function well independently if the main coordination system fails momentarily. The idea of integration also enables faster rearrangement when nodes die or topology changes, both of which happen often underwater. In addition to SDN, NFV improves the flexibility of IoUT systems. With NFV, important functions in any network like firewalls, intrusion detection systems, data compressors, and telemetry preprocessors are activated as software on normal computing devices instead of separate physical equipment. This method makes it possible to use virtual network function (VNF) on marine platforms or edge devices because the data is generated or gathered there. In this integrated architecture, having NFV makes it possible to quickly set up new services, including data compression, prioritization, and duplication, if a seismic event happens and it is important to ensure alerts are reliable and timely. With SDN and NFV, networks can be adjusted both in their routing and in their use of services, and this happens remotely. With SDN and NFV, this integrated underwater system can respond and scale quickly. When marine sensors detect irregular findings or seismic changes, resources are distributed, traffic routes are altered, and additional alert measures are applied to ensure that helpful information reaches coastal areas as rapidly as possible.

Because it understands the entire local section of the network, the SDN controller can detect traffic jams and faulty nodes and move data along the most efficient routes. At the same time, NFV lets services be added or changed as needed to ensure data and communication quality is maintained. Because of this connection, SDN and NFV allow networks to adjust on the spot, while traditional submarine networks need special effort to make real-time changes. As a result, our integrated architecture ensures reliability. It is efficient with its power consumption and can be easily programmed, ensuring all the features needed to monitor maritime disasters. The architecture supports fast actions in the event of tsunamis by keeping up with data, restricting data loss, and ensuring that critical alerts reach the intended users. Besides SDN, NFV is also important for improving the adaptability of IoUT systems. NFV takes network services such as firewalls, intrusion detection systems, data compressors, and telemetry preprocessors, and runs them as software programs on common computing platforms. VNFs can now be used directly on the same equipment where data is generated or combined at the edges of marine systems. With our integrated architecture, NFV helps make sure that new services, especially for data compression, prioritization, or duplication, are quick to deploy should there be a need after a seismic event. As a result, networks can now be adjusted in both routing and the services handled, all from a remote control point, without changing the hardware.

SDN and NFV help the integrated underwater network to be flexible and responsive. When there is a change or disturbance seen by ocean sensors, the system acts by readjusting resources, traffic paths, and activating more alerts to ensure data reaches coastal areas promptly. Because the SDN controller oversees the local area, it notices issues such as congestion or problems with nodes, and changes the data path to the least crowded or shortest route. At the same time, NFV facilitates the adaptable deployment of services to keep data safe and communication effective. With the operation of SDN and NFV together, underwater networks can be adapted on the fly; a traditional task that fixed hardware systems find difficult. As a result, our integrated architecture can withstand failures, minimize energy usage, and offer the programmability necessary to monitor maritime disasters. With this architecture, high-level underwater incidents like tsunamis are reliably handled, as there are few delays, data loss is kept to a minimum, and safety alerts go through successfully, no matter how far the system is from any resources. Table 4.1 presents the contributions and limitations of all the techniques discussed in this section.

Table 4.1 Key-related work and limitations

Study	Contributions	Limitations
[1] Mohsan *et al.* (2023)	Comprehensive review of IoUT components, protocols, and challenges.	Does not propose SDN, NFV, or disaster-resilient IoUT architecture.
[2] D. R. K. M *et al.* (2020)	Highlights management issues in centralized underwater IoT networks.	Lacks programmable solutions and decentralized architectures.
[3] Sedaghat and Jahangir (2024)	SDN-based energy-efficient routing for tsunami early warnings.	Focused only on routing; excludes virtualization and acoustic constraints.
[4] Mohammadi *et al.* (2021)	SDN framework to ensure QoS routing in IoUT.	Does not include energy harvesting or multipath reliability strategies.
[5] Saha *et al.* (2020)	Energy-aware SDN/NFV architecture for IoT applications.	Not tailored for underwater environments or acoustic-based systems.
[6] Luo *et al.* (2022)	SDN-based multi-modal sensor network for marine monitoring.	Lacks practical deployment and integration with disaster warning.
[7] Mohammadi (2025)	Proposes multicast traffic engineering for SDN-Fog IoUT.	Focuses on traffic reliability; omits energy and NFV integration.
[8] Ullah *et al.* (2024)	Improve underwater connectivity using decision-based methods and remote sensing.	Does not address SDN/NFV or LEO satellite-based backhaul.
[9] Niknami *et al.* (2023)	Discusses SDN and SDR integration in maritime networks using LEO satellites.	Surface-layer focused; lacks acoustic or underwater device support.

4.3 Integrated architecture overview

The integrated framework brings together various components designed for the demanding and shifting environments common in oceanic and underwater areas (see Figure 4.1). Every part of the architecture handles an important aspect such as picking up energy from below, sensing earthquakes, changing the way data flows, and communicating in real-time through satellites, leading to a reliable system for monitoring natural disasters underwater, detecting seismic events, and delivering early tsunami warnings. The foundation of the system is the seismic-piezoelectric sensor node which is attached directly to the ocean floor. These nodes detect the beginning of undersea tectonic activity by collecting information on small earthquakes, bed movements, and changes in pressure. Inside each sensor node, piezoelectric substances, for instance, lead zirconate titanate (PZT), create an electric charge whenever the node experiences mechanical stress from seismic vibrations. Due to this, energy

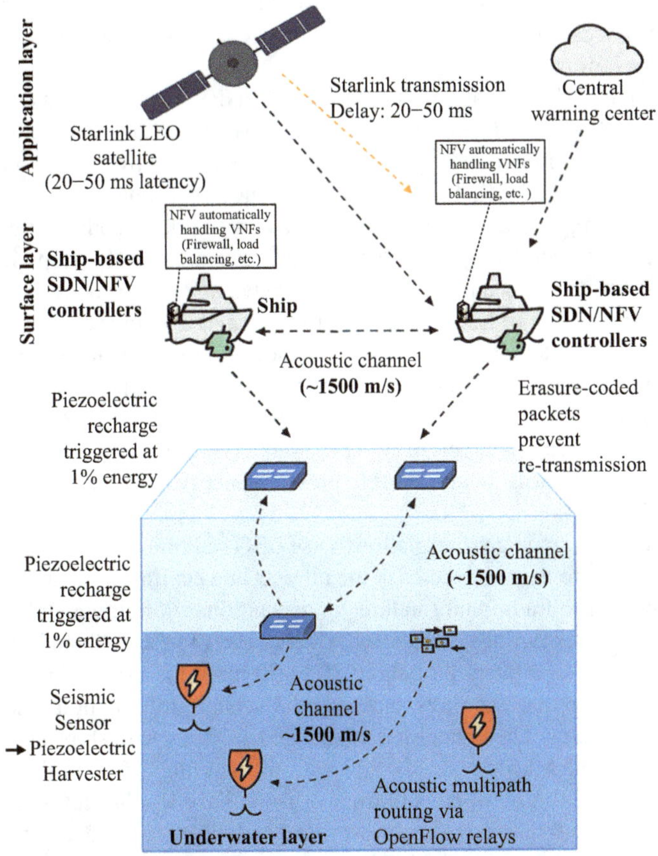

Figure 4.1 Integrated SDN/NFV-based MDUN architecture

can be gathered from underwater motion, so the nodes do not need to be maintained or supplied with power for many years. Since batteries cannot be exchanged easily at deep-sea locations, autonomous action is vital for these deployments. Sensors in these systems have very low-power microcontrollers and monitoring systems that work only in given time periods to conserve power. They only transmit data when a seismic event is detected or found to be abnormal [1,8].

The underwater relay nodes are located above the seabed, covered throughout the water, and play the main role in the underwater communication infrastructure. Being like underwater switches or routers, these nodes are enriched with Open vSwitch (OvS) and OpenFlow support, meaning that they can be programmed to direct network traffic as needed. Sensor nodes send collected data through relay nodes and these nodes then communicate using acoustic modems that send messages through sound waves when doing so underwater. Despite facing problems like slow transmission speeds and sluggish updates, these relay nodes are built to assist in mesh routing, change traffic priorities, and handle traffic where it is generated. Relay nodes keep other nearby SDN controllers informed about links, battery levels, and network structure, so intelligent switching and reliable data transmission can happen during hard circumstances, including signal deterioration, changes in water currents, or some network links failing [2,4].

Due to their surface location, ship-based SDN/NFV controllers function as agile and flexible control platforms at the edge. All ships have strong computing systems that run VNFs stored in containers, including those offered by Docker and the kernel virtual machine (KVM). Such functions include real-time filtering of network data, sending compressed information about the network, encoding signals, looking for intrusions, and choosing routes that use less energy. The ships function as local SDN controllers, managing underwater relays and sensor nodes around them through the application of flow rules, path updates, and creating services as the mission requires. NFV integration also supports handling service changes, upgrades, and movement between ships, without the requirement of updating physical hardware. As an example, during disasters, it is useful because some functions such as maritime hazard analysis or sending emergency signals can be given priority, with less important functions temporarily switched off to preserve energy or move processing power elsewhere [5,6].

Ship-based communications platforms use LEO satellite uplinks like those operated by Starlink to remain connected to the internet and central commands around the world. In contrast to traditional satellites in geostationary orbit, which have a lot of latency and space restrictions, LEO satellites are nearer to the earth and can give fast broadband internet at a latency of about 30 to 50 ms. This fast backhaul system is necessary to ensure that warnings and essential seismic information reach stations on land without delay. These satellites exchange messages with ships using phased-array antennas, which keep a steady link with the satellites' fast motions. Because LEO uplinks cover almost all parts of the world and have low latency, our integrated architecture can send data from the most distant sections of the ocean to mainland planning offices almost immediately [9].

The terrestrial analytics and command centers make up the final and important part of our integrated architecture (see Figure 4.1). Aggregated telemetry and raw or pre-processed data are received by these facilities from LEO satellite links that carry data from ship-based controllers. When data are received, they are sent to strong analytical devices that use methods like seismic inversion filtering technique (SIFT) and the community modeling interface for tsunamis (ComMIT). SIFT helps sort and understand the data from seismic waves to predict where, how, and how badly an earthquake happened. To find out how much the seafloor moved, it uses inverse modeling and predicts the likelihood of a tsunami. Alternatively, ComMIT is a popular tsunami simulation program started by the national oceanic and atmospheric administration (NOAA) to model waves moving through the water and rising onto the shorelines. Integrates digital elevation models (DEMs) of bathymetric and topographic data to simulate impact zones and predict wave heights.

By working together, these computer systems allow command centers to tell the difference between a normal earthquake and a tsunami threat. As soon as an important warning is received, details are sent to the national firewall and warnings are broadcast to both government officials and coastal populations. These centers manage remote problem-solving, regular updates, and planning for maritime solutions to guarantee that the whole system stays healthy, flexible, and responsive. Together, these features help the system operate on its own in ocean environments while collecting dependable data, routing it intelligently, and sending out alerts as an emergency happens. Underwater hardware, mobile ship-based devices, global satellite links, and sophisticated analysis at shore represent the peak of SDN and NFV in supporting critical underwater systems. This design allows our integrated architecture to help during disasters and to continuously monitor the ocean and support various maritime activities.

4.3.1 Architectural layers

The layered architecture of our integrated system features three different but linked layers, each supporting specific tasks in its operation. The structure of these layers follows the traditional SDN model, data control, and application planes, but is designed for use in underwater and marine monitoring, as indicated in Figure 4.1. The underwater layer, like the data plane, is made up of the subsea elements that perform environmental sensing and nearby data transmission. The seismic-piezoelectric sensor nodes and the acoustic OpenFlow-based relay nodes are both included in this challenge. By combining these elements, we can catch seismic movements and turn the data into acoustic waves that travel step by step toward the source. Due to the limitations of underwater acoustics, such as narrow transmission, long delays, and noise, this layer is made to use less energy, deal with errors, and automatically adjust to any changes in the network topology from mobile submarines or the environment. The surface layer acts as the control and NFV plane. The system uses SDN/NFV controllers and methods for inter-ship communication.

Here, distributed SDN controllers manage routing, surveillance, and flow rules throughout the network, with VNFs dealing with processing telemetry in real-time. If the underwater network is overloaded or if anomalies are detected, services like virtual firewalls, telemetry preprocessors, data compressors, and routing optimizers can be automatically increased or decreased as needed. This layer allows the system to adapt to any seismic event or network failure in real-time, making the system strong and reactive [5,6]. The space or terrestrial plane is the name given to the application layer, where Starlink LEO satellites and land-based centers are located for processing and emergency help. In this layer, the data are understood in detail, presented visually, and decisions are made. All input from the SDN/NFV controllers is collected and processed, and the resulting information is shared with the disaster response, government, and impacted groups. Because of satellite technology, data are shared without interruptions, and terrestrial systems help keep complicated analysis and alert delivery up-to-date.

4.4 The integrated SDN- and NFV-based MDUN architecture

Using the integrated architecture, researchers can create energy-saving, durable, and expandable tsunami detection systems in real- time. Divide duties among various operational layers, allowing each to manage a specific task in sensing, communication, control, and alert activities.

4.4.1 Underwater data layer

In the underwater data layer, there are networks of seismic pressure sensors connected to small piezoelectric energy units. These gadgets are located directly at sea level or close to the sea floor to detect signals from pressure changes triggered by earthquakes. Most energy harvesters are made from PZT and convert ocean vibrations or movement in the earth's crust into electricity using the direct piezoelectric effect. The main controller in sensor nodes, such as MSP430 and STM32, handles data processing at or near the sensor's location. Having signal processing abilities and low-power design are the main reasons that microcontroller units are used in this area. The energy of acoustic sensors, which typically lies between 10 and 100 μW, is preserved in large supercapacitors (up to 50 mF), to backup brief but key acoustic communications.

Data from sensor nodes are sent to the relay nodes at different places in the ocean by a network that hops sounds several times. OvS with OpenFlow is used to create these relays, which function as underwater routers that can route traffic with programmable decisions. Hydromea LUMA X-class modems enable acoustic communication by sending data at a rate between 10 and 50 kbps over ranges of around 200 m on low-frequency waves. By using adaptive equalization, error correction coding, and retransmission of data at the MAC layer, the system can avoid multipath fading, Doppler shifts, and high bit errors while guaranteeing correct data transfer to controllers on the surface. Acoustic signals are preferred in the underwater data plane as they are well suited for sending data over long distances in noisy and turbid places. Table 4.2 outlines a comparison of acoustic, optical, and radio-frequency (RF) methods to communicate in underwater environments.

4.4.2 Surface control layer

It is the surface control layer that links the underwater sensor network to the terrestrial facilities used for processing. It relies on maritime vessels that use SDN/NFV control software. Data processing, decision making, and traffic control are managed by these ships, which serve as floating data centers, using containerized VNFs, running through Docker or KVM. The underwater relay nodes are connected to distributed SDN controllers that communicate using OpenFlow rules. They use sensor information, oversee how traffic is routed, and control network states. Besides, the NFV framework, within the system, can deploy major services, such as data compression, error correction, packet classification, and energy-aware scheduling in real-time. Low-latency internet contacts of approximately 50 ms, along with bandwidth up to 100 Mbps, are provided to ships by Starlink terminals from LEO satellites. Because of this setup, ships can transmit important seismic data in real-time to disaster response centers on land without going through the usual data offloading process at ports.

Table 4.2 Comparison of underwater communication channels

Parameter	**Acoustic**	**Optical**	**Radio frequency (RF)**
Operating frequency	10–60 kHz (typically)	400–700 THz (visible spectrum)	10 kHz to 1 MHz
Typical bandwidth	Low (kHz range)	High (MHz–GHz range)	Very low (Hz–kHz range)
Data rate	10–50 kbps (Hydromea LUMA X class modems)	Up to 1 Gbps (in clear water)	1–100 bps (highly attenuated)
Communication range	Long (up to 20 km in ideal conditions; 200–500 m typical hop)	Short (2–10 m)	Very short (<1 m)
Latency	High (propagation delay $\sim$1.5 km/s)	Low (speed of light in water $\sim 2.25\times 10^8$ m/s)	Low, but practically constrained due to signal loss
Medium suitability	Turbid, noisy, and deep-sea environments	Clear, shallow water with minimal scattering	Extremely limited in conductive saltwater
Power consumption	Moderate to high (depending on range and power level)	High (especially at higher data rates)	High, with minimal data throughput
Hardware example	Hydromea LUMA X Acoustic Modem	BlueComm (optical modem)	Experimental short-range RF antennas
Susceptibility to noise	High (affected by ambient noise, marine life, etc.)	Medium (affected by turbidity, bubbles)	High (affected by high conductivity of seawater)
Deployment complexity	Moderate	High (alignment and clear water required)	Experimental and low practicality
Use case suitability	Long-range relay, seismic sensing, low-bandwidth telemetry	High-speed video/data over short distances (AUV docking)	Rare; research focused applications in shallow freshwater

4.4.3 Application layer

The application layer uses air and ground infrastructure to help with data review, tsunami predictions, and notifying people during emergencies. The connection between ships and land data centers is made possible by Starlink and several other LEO systems in the space segment. The earth portion of tsunami monitoring consists of modern centers for analyzing data and using software such as ComMIT and SIFT. Seismic data arriving from ships are used together with bathymetry and topographic models by ComMIT to understand the path of tsunami waves and estimate when they reach the coast. To complete this procedure, SIFT uses the simulated data to build visual maps showing where the risks are highest. In addition to shaping and predicting wave movements, these tools also create useful information for people making decisions. When the risk is calculated, the alerts are shared through many types of communication technologies, enabling multiple agencies to respond and inform the public. Researchers working in the maritime industry should monitor cell broadcast systems, get alert through NOAA, look for the national disaster risk reduction and management council (NDRRMC) messages, and use navigational telex (NAVTEX) and automatic identification system (AIS). Having many types of channels means that all alerts reach both coastal people and boats promptly and safely, which is the goal of our system.

4.5 Energy sustainability

This architecture is especially designed to handle long-term, fully autonomous operation in areas like the deep-sea, since manned support is not easily available or cost-effective. To handle this issue, the system includes different kinds of sustainable energy approach, low-power hardware, and effective communication protocols that help reduce energy use without decreasing the system's reliability. The most important aspect in this approach is a technology called piezoelectric energy harvesting, which helps the underwater nodes generate electricity on their own through mechanical action from the water or earthquakes. Vibrations are transformed by piezoelectric materials into electric energy which is then kept in large storage capacitors. The nodes reduce energy use by going to sleep mode and waking up only when they need to. Furthermore, energy-aware routing with OpenFlow helps SDN controllers on surface ships adjust data paths to prevent nodes experiencing energy shortage. All of these mechanisms make it possible for the underwater network to work for several years, lower maintenance costs, and ensure safer preparation for disasters.

4.5.1 Piezoelectric energy harvesting

It is almost impossible to supply electricity through cables underwater due to the deep oceans, the challenges posed by environmental conditions, and the large distances. The integrated architecture addresses this by using piezoelectric energy harvesting, making electricity from mechanical stress through materials such as PZT. When pressure waves or seismic activity apply pressure, these materials create an electrical charge by the direct piezoelectric effect. On the seafloor, each sensor node has a PZT film that detects underwater vibrations and generates electricity. The power output from wearable micro-electro-mechanical system (MEMS) can vary from 10 to 100 μW, but it can rise to nearly 30 μW in bursts linked to seismic activity. It requires less electrical power than most devices, but can be safely used if stored and buffered up to 50 mF in supercapacitors. After being stored, these energy reservoirs help power quick and intense processes, such as sensing, local computing, or transmitting sound information in packets. As shown in the latest studies [1,8], the existing design allows for a 5 s burst every 10 min, meeting the needs for stable monitoring with an energy use that stays within safe levels. Several bridge nodes in our integrated architecture are equipped with solar panels and floats which are linked to devices underwater. The system can continue to function for five years or more before human help is needed. Eliminating battery changes is available with piezoelectric systems, and this also enables deployments that can survive and scale in dangerous seismic areas. Consequently, the system handles better maintenance, reduces costs, and remains effective in different sea situations.

4.5.2 Duty cycling and recharge strategy

An important way to save energy in this architecture is through duty cycling, which restricts node work to only when it is necessary. With this approach, sensor and relay nodes are regularly asleep, but they come awake when events happen. An onboard sensor picks up any unusual seismic or pressure behavior above the assigned allowable level, triggering the first condition. The second type means that the node is periodically turned on, whether it detects events or not, to send heartbeat packets or system news. Using this mix of event-based and scheduled activity allows a network to be flexible and maintain regular service. When in sleep mode, the system only consumes 2 μW; this way, environmental detection happens without starting communication. Every microcontroller in a node always monitors the voltage of its connected supercapacitor. When the energy in the battery drops below about 10%, the system stops transmitting and only works on recharging itself. During this time, the piezoelectric unit gathers more energy until the capacitor is filled enough to work normally again. Thanks to this logic, you can continue to use critical sensors, as it keeps the data safe by preventing zero storage capacity.

Moreover, ships that function as SDN/NFV controllers monitor live data about the energy status of each node. Thanks to these instant updates, OpenFlow lets them route network traffic differently, avoiding traffic through nodes that have a low battery. The centralized approach spreads the energy load better and does not burden any one relay link too much. In addition, the energy used by acoustic modems during underwater communication is high, so minimizing broadcasts during non-important times greatly increases the time a node can function. By combining adaptive duty cycling, energy-aware routing, and autonomous recharging, our integrated architecture ensures that it remains available and reliable, regardless of environmental or energy-harvesting conditions. It reduces manual activities and prepares systems better for long-lasting use in tough ocean settings where constant monitoring of disasters occurs.

4.6 Fault-tolerant data delivery via erasure coding

The reliability of underwater acoustic communication is often reduced by environmental disturbances, long distance travel times, a limited amount of frequency, and interference caused by multiple paths. This architecture meets these challenges by using erasure coding and dynamically relying on SDN-based routing and NFV for data transmission support. This two-level technique guarantees that the data remain safe and correct even when there are underwater link failures, little available energy, or signal interruptions.

4.6.1 Coding mechanism

Reed–Solomon (RS) and Luby transform (LT) codes are used at the transport layer to help avoid communication failure if a few packets are damaged. With these forward error correction schemes, a data message is split into n packets, so that if k packets from n arrive safely, the original message can be recreated. This relationship between (n, k) changes depending on live information about the network, including the state of the link and the error statistics. With a configuration (6, 4), the system remains able to recover the data even if two packets are lost during transmission. Redundancy matters a lot in underwater situations because chronic Doppler spread and signal scattering may cause some packets to be lost. To build greater resilience, the system sends data using k different routes, preventing many failures from occurring at the same time. Thanks to this method, data are always available and packets are more likely to reach their destination in underwater acoustic channels, where there are long delays and subtle changes.

4.6.2 SDN and NFV integration

Together, SDN and NFV make an adaptable and highly reliable delivery of data possible here. With worldwide network visibility, the SDN controller used in the surface vessel decides on k-disjoint routes and allocates them as needed for erasure-encoded segment deliveries. Telemetry collected live from the network guides the selection of these paths, such as the energy state of every node and the frequency with which link failures occur. In addition, erasure decoding VNFs are deployed and operated by the NFV environment inside the ship. They are set up using light containerization (such as Docker or Linux container daemon (LXD)), which makes it possible to grow and process things swiftly. The decoding is done using local hardware, avoiding the requirement to retransmit misplaced data from those same nodes. Also, real-time updates from underwater relays, looking at packet loss and power use, are always shared with the controller, which responds by reorganizing the packets or changing the way data are encoded. This design lowers the overhead caused by retransmissions and manual updates. With research, networks can continue to function after encountering temporary faults, weak signals, or depleted energy in a few nodes. This means that our integrated architecture is a reliable system for crucial and disaster-resistant operations in the ocean.

4.7 Routing and optimization framework

Routing in underwater networks by balancing the load and finding intelligent routes helps to maintain their trustworthiness and environmental impact. Energy imbalances, delays, and shorter network lives are possible outcomes when using traditional static routing schemes. Meanwhile, the integrated architecture provides optimized, energy-conscious, and delay-sensitive routing through rules managed by SDN and supported by NFV. Due to the modular design, the system can handle shocks from the environment, distributes traffic evenly, and improves the overall bandwidth.

4.7.1 Energy-aware forwarding

Since there is limited energy in the underwater world, routing decisions should consider how much power the sensor and relay nodes have remaining. The routing framework compares routes by computing the average and variability of the nodes' remaining energy. Such routes without many energy factors are removed to keep the load on weak nodes to a minimum. To ensure stability, an interquartile range (IQR) filter is added to delete outlier routes that may seem efficient but depend on nodes that have unreliable energy usage. As a result, this process ensures that every part of the network uses equal resources, which helps the network operate longer. Based on current energy data, the SDN controller updates the routes to ensure even network traffic and reduce the chance of failures.

4.7.2 Multipath optimization

Because acoustic links are easily disrupted by fading, using only one way to communicate is risky. To solve this issue, our integrated architecture applies the depth-first search (DFS) and breadth-first search (BFS) methods to find several disjoint or redundant paths that connect nodes. These paths are arranged in priority order, with the stability index calculating how successful their transmissions have been over time. Routing decisions consider only paths that do not come from nodes in recharge mode or display a high error rate. When there are seismic quakes or heavy traffic, the SDN controller means data is quickly rerouted to less crowded paths. As a result, the network is better protected from failures and has an improved overall packet delivery ratio, even when services are disrupted.

4.7.3 Controller and NFV optimization

Our integrated architecture increases its overall energy efficiency by decreasing control plane traffic and appropriately distributing smart network functions using NFV. Each data packet's control headers are made to be very small, to keep the acoustic bandwidth intact. NFV is essential for supporting functional virtualization, with VNFs grouped as critical (such as tsunami alert generation), routine (such as sensor telemetry reporting), and support (such as diagnostics and maintenance). At regular intervals, the resource manager VNF tracks, computes memory and bandwidth demand, and shifts resources as needed. In such incidents, the VNF guarantees that critical tasks have priority over ordinary tasks. Also, service function chaining (SFC) makes use of SDN rules to arrange traffic so that it passes through a series of VNFs such as compression, filtering, and duplication before being delivered to ground stations or notified to the appropriate authorities. As a result, data are transmitted with less repetition, the link is used more efficiently, and significant tasks are shifted from underwater devices to those on the surface. This allows the integrated architecture to maintain high throughput and remain reliable by using only a small amount of energy, making it perfect for reliable maritime monitoring and quick reaction systems.

4.8 Advantages of integrated architecture

Our integrated architecture provides many useful features to overcome unique obstacles in underwater networks and pressing needs for resilience in emergencies. By combining distributed control, energy harvesting, the use of satellites for communication, and modular service orchestration, this architecture improves reliability, ensures long-term sustainability, and allows for larger networks than previous models.

4.8.1 Key advantages

A major characteristic of integrated architecture is its distributed control plane. In contrast to a single-controller approach, this architecture has several SDN controllers working onboard ships, buoys, and autonomous surface vessels. All of these controllers handle local segments in their areas and update, share, and exchange topology, flow, and telemetry information with their peers. The network resilience is greatly improved using this distributed method. A system crash like hardware failure, an earthquake, or a ship accident hits only the part that is down and the rest keeps operating as always. If a tsunami wrecks parts of the network underwater or moves a buoy out of water, the other operating parts will still collect and relay disaster warnings. As a result of this mechanism, no single failure of a component prevents the whole network from operating; this is important for emergency response systems. Additionally, in the case of underwater earthquakes or movement of equipment, the decentralized setup makes it possible for the architecture to continue to work safely and effectively. If a controller detects a change, it dynamically updates the data paths, redirects messages, and changes the control among different nodes around it. This ability to fix problems is necessary for continuous operations in rapidly changing and risky undersea conditions. Table 4.3 summarizes all key advantages.

4.8.2 Sustainability through piezoelectric energy harvesting

In situations where environmental monitoring, seismology, or data transmission need underwater nodes, they are often placed in very difficult locations like the ocean bottom or regions where plates of the earth's crust are meeting. In such cases, keeping up with routine maintenance, and most importantly, changing the battery, would not be smart or cost-effective. A piezoelectric energy scraping method is used in the sensor nodes of our architecture to address this problem. This method makes use of the energy found in changes in ocean currents and tectonic processes. Due to piezoelectricity, these materials make it possible for nodes to produce their own energy. After some time, these nodes run independently for many years, which cuts out the

Table 4.3 Key advantages of our integrated architecture

Features	Description	Benefits
Distributed SDN control	Controllers deployed on ships, buoys, and surface vessels coordinate network functions locally and share state information with peers	Enhances fault tolerance and ensures continued operation during partial network failures
Piezoelectric energy harvesting	Sensor nodes harness mechanical energy from ocean currents and tectonic activity to generate power	Enables long-term, maintenance-free operation of deep-sea nodes
Low-latency LEO satellite communication	Starlink and similar LEO satellites offer rapid data relay between maritime controllers and ground centers	Ensures near-real-time delivery of disaster alerts and sensor data
SDN- and NFV-based scalability	Automatic discovery and integration of new network elements via southbound protocols and lightweight virtualization of services	Simplifies horizontal scaling and supports modular service deployment
Dynamic resource allocation	VNFs like the resource manager monitor and re-prioritize functions based on disaster or system conditions	Maintains critical services like tsunami alerts under constrained resources
Service function chaining (SFC)	Orchestrates packet processing (e.g., encoding, duplication, and security) before transmission	Optimizes data handling and reduces bandwidth consumption
Platform independence via NFV	VNFs is deployable on general purpose hardware or virtualized platforms using docker or other containers	Lowers infrastructure costs and improves flexibility during deployment and recovery operations

need and cost of battery servicing [1,8]. Not only does this advancement help the nodes work longer, it also boosts safety. Battery replacement in deep-sea missions can put divers or remotely operating vehicles in danger. As a result of getting rid of this requirement, the system leaves out those dangerous steps and guarantees that important underwater environments are continuously monitored and communicated through data.

4.8.3 Low-latency global access via LEO satellites

It is especially important for underwater applications in disaster response that data are sent quickly and accurately. Because traditional geostationary satellites take so long to deliver information, they are not suitable for time-critical notices. Starlink and other LEO satellite constellations make it possible for our architecture to deal with this problem. The latency values for LEO satellites fall between 30 and 50 ms because they orbit at heights ranging from 300 to 550 km. Due to this low-latency communication, data and emergency messages get to ground-based disaster centers almost right away [9]. In addition, maritime SDNs rely on LEO satellites to provide a superfast link between controllers and global data centers and cloud-based analytics systems. As a result, information is processed right away, displayed visually, and shared among teams across the world, ensuring key alerts reach everyone and situations are handled promptly.

4.8.4 Scalability and modularity in network design

Our architecture benefits from an easily scalable and modular system that allows the smooth addition of new devices, vessels, and sensors without much reorganization. To make this system flexible, SDN and NFV are used in combination. If new nodes or links appear, they are found by the controller through southbound protocols, such as OpenFlow and the network configuration protocol (NETCONF), and included on its map of the network. After that, the controller designs the best routes and assigns the resources to them. Because of this, the system can expand horizontally by adding more sensors or surface buoys as needed, making it useful for ongoing ocean studies or for growing the area being observed. For NFV, services like packet filtering, compression, intrusion detection, and video streaming are handled by VNFs. Their platform-independent nature means that they can be arranged, moved, or ended whenever needed using light virtualization tools such as Docker. Consequently, service expansion is managed with software instead of requiring hardware replacements, making it more efficient and cost-effective [5].

4.8.5 Dynamic service reconfiguration for disaster support

Reconfiguring service chains and adjusting resource allocation after a disaster greatly help in both seismic and maritime emergencies. The integrated architecture incorporates a VNF resource manager that monitors system resources regularly, including CPU, memory, power, and manages the allocation of resources to ensure good service delivery. In the event of an earthquake, the system could put priority on virtual functions that help with tsunami detection, studying deeper waters, or finding survivors. At the same time, tasks including environmental monitoring or keeping archives may be put on pause or moved. Using priority scheduling, the network will not disrupt emergency workflows even when there is high traffic on the network [5,6]. With the help of service function chaining (SFC) in the SDN controller, data packets are processed at each step of their sequence (including encoding, security checks, and duplication) and are either sent by satellite or kept for future look-up. This control guarantees that each piece of data is used efficiently, and that the data transfer consumes little bandwidth. Overall, our integrated architecture uses a well-adapted, dependable, and future-ready method for underwater data networks. Using distributed SDN control, environmentally friendly power harvesting, fast satellite access, and adaptable virtualization, it meets the demanding needs of current marine and disaster response units. In three important applications, marine research, seismic surveillance, and defense, this framework makes it simple to have underwater connections that are never interrupted and uses less power.

4.9 Future enhancements

4.9.1 AI-powered routing

If machine learning and artificial intelligence (AI) are used in next-generation integrated systems, the systems could even better adapt. Using earlier data on packet loss, energy used, and current weather, these algorithms can predict possible problems with link speed and routing failures. Alternatively, real-time bathymetry can deliver details to AI, so that sonar avoids noisy areas and other hazards. Currently, these enhancements are missing, but they are following the latest trends in smart ocean networking.

4.9.2 Self-healing VNFs

Self-healing VNFs are also recommended as a useful change. If an issue occurs with a VNF, AI technology can highlight it and help our system relocate it to a better location. That is why services would not stumble when there is a powerful earthquake or failure of a major network node. Moreover, when activity from seismic sensors is expected to go up, predictive VNF scaling can be used to add new resources before it happens [5]. Because of such progress, the system would respond more promptly and independently, and thus be more useful for taking care of numerous tasks, for instance, disaster rescue, exploration of the sea, and navy duties.

4.10 Conclusion

Our integrated architecture introduces new ways to increase stability and intelligence in underwater network systems. Since traditional underwater networks face low bandwidth, high delays, low energy, and difficulty maintaining reliability underwater, our architecture gives a balanced, customizable, and expandable solution to these problems. The central idea of this framework relies on placing SDN controllers on both ships and edge processors, which helps prevent service failures and lets the system operate unaffected by such issues. In contrast to centralized structures that may stop working because of damage to a main controller, our architecture segments can still function independently if they lose contact with the control center. Real-time routing and resource allocation happen at every vessel and control node, letting the network to respond to any changes in the water and task goals. When deep-sea deployments make physical support difficult, energy sustainability is achieved by merging new hardware and software ideas. Using piezoelectric energy welded with low-power microcontrollers and regulated energy use, underwater devices can do their job for years without any attention. Such nodes turn oceanic roughness and major movements of the earth's crust into electricity, which they later store in powerful supercapacitors. When adaptive sleep-wake cycles and power-efficient protocols are used, the device can keep operating and transmit only important readings while using very little energy.

Early tsunami warnings and seismic activity detection rely greatly on reliable data delivery. With erasure-coding techniques, this becomes much stronger. With Reed-Solomon and Luby Transform codes, the system can still retrieve all parts of a message even if packets go missing. Combining these approaches with k-disjoint routing implemented through SDN helps packets be delivered on several routes, reducing the impact of a single point of failure. The SDN controller sees the entire network and reads data from all devices, regularly enhancing routing decisions by taking links quality, power, and environment into account. As a result, connecting with Starlink and similar networks keeps the gap between detection of underwater activities and surface or land response time very brief. Because of this, LEO satellites can end urgent national disaster warnings in about 30–50 ms. The speed at which an early warning can be delivered matters a lot with tsunamis; every second can save thousands of lives. Our architecture stands out because of its flexible and virtualized design powered by NFV. Lighter versions of the seismic alerting, telemetry logging, and link monitoring software are configured as VNFs on skiffs and other maritime devices. As the mission changes, these VNFs can be set up, upgraded, or shut down independently of any underlying hardware. As an example, following an earthquake, the network could introduce VNFs designed for monitoring or navigation assistance instead of those for seismics, again without having to change the hardware.

Due to these new arrangements, both resource savings and the ability to reuse the same infrastructure for various missions are possible. This new approach to research and development supports innovations for the future. With AI and machine learning now paired with the SDN controller, later versions of this architecture could handle self-fixing networks, set optimal energy levels automatically by watching usage patterns, and automate mission planning. Besides, if different sea vessels could work together, using distributed algorithms to divide the sensing, message sharing, and computing duties, it could improve the overall safety, accuracy, and energy distribution of the fleet. Basically, our integrated architecture framework combines smart technologies and multiple aspects to monitor the oceans. Because it manages energy use, provides reliable data delivery, offers flexible use, and ensures network reliability, it is the perfect architecture for disaster alerting, marine research, and long-run environmental monitoring. The use of distributed SDN control, service modularity with NFV, energy harvesting tools, and proven routing methods in our architecture help create future underwater networks that can work independently, change, and adapt, yet be reliable under tough and unpredictable underwater conditions.

References

[1] S. A. H. Mohsan, Y. Li, M. Sadiq, J. Liang, and M. A. Khan, "Recent advances, future trends, applications and challenges of internet of underwater things (IoUT): A comprehensive review," *Journal of Marine Science and Engineering*, vol. 11, no. 1, p. 124, 2023. https://www.mdpi.com/2077-1312/11/1/124.

[2] D. R. K. M. J. Lee, E. Ko, S.-Y. Shin, J.-I. Namgung, S.-H. Yum, and S.-H. Park, "Underwater network management system in internet of underwater things: Open challenges, benefits, and feasible solution," *Electronics*, vol. 9, no. 7, p. 1142, 2020. https://www.mdpi.com/2079-9292/9/7/1142.

[3] S. Sedaghat and A. H. Jahangir, "Statistical analysis and energy-efficient routing for tsunami early alarming in internet of underwater things using SDN infrastructure," *IEEE Transactions on Green Communications and Networking*, vol. 9, no. 2, pp. 561–573, 2025.

[4] R. Mohammadi, A. Nazari, M. Nassiri, and M. Conti, "An SDN-based framework for QoS routing in internet of underwater things," *Telecommunications Systems*, vol. 78, no. 2, pp. 253–266, 2021. https://doi.org/10.1007/s11235-021-00812-y.

[5] D. Saha, M. Shojaee, M. Baddeley, and I. Haque, "An energy-aware SDN/NFV architecture for the internet of things," in *2020 IFIP Networking Conference (Networking)*, IEEE: Paris, France. 2020, pp. 604–608.

[6] H. Luo, X. Wang, Z. Xu, C. Liu, and J.-S. Pan, "A software-defined multi-modal wireless sensor network for ocean monitoring," *International Journal of Distributed Sensor Networks*, vol. 18, no. 1, 2022, pp. 1–19.

[7] R. Mohammadi, "Multicast reliable traffic engineering technique for SDN-fog based IoUT," *Scientific Reports*, vol. 15, no. 9069, 2025, pp. 1–9.

[8] I. Ullah, F. Ali, A. Sharafian, A. Ali, H. M. Y. Naeem, and X. Bai, "Optimizing underwater connectivity through multi-attribute decision-making for underwater IoT deployments using remote sensing technologies," *Frontiers in Marine Science*, vol. 11, p. 1468481, 2024. https://www.frontiersin.org/journals/marine-science/articles/10.3389/fmars.2024.1468481.

[9] N. Niknami, A. Srinivasan, K. St. Germain, and J. Wu, "Maritime communications—current state and the future potential with SDN and SDR," *Network*, vol. 3, no. 4, pp. 563–584, 2023. https://www.mdpi.com/2673-8732/3/4/25.

Chapter 5

NOMA-based underwater acoustic communication

Arooj Fatima[1], Muhammad Sufyan Haider[1], Muddassir Sadiq[1] and Salman Ghafoor[1]

Underwater acoustic communication (UAC) faces challenges such as limited bandwidth, multi-path fading, and Doppler shifts. This chapter explores non-orthogonal multiple access (NOMA) as a solution to improve spectral efficiency and user fairness. By enabling simultaneous multi-user access through power-domain multiplexing and successive interference cancellation (SIC), NOMA outperforms traditional orthogonal multiple access (OMA) schemes. The integration of NOMA with technologies such as filter bank multi-carrier (FBMC), orthogonal frequency division multiplexing (OFDM), full-duplex relays, deep reinforcement learning, and underwater optical wireless communication is examined. Despite the simulation and experimental advances that have been made in the literature, practical deployment requires adaptive, energy-efficient, and scalable solutions to address environmental variability and mobility.

5.1 Introduction

Military operations, oil pipeline monitoring, and underwater exploration are just a few of the critical applications that rely on underwater wireless communication. Due to the extreme absorption of electromagnetic wavelengths by water, underwater communication mostly uses auditory signals, in contrast to land or air environments, where radio frequencies function well. However, there are a number of obstacles that UAC must overcome, including long propagation delays, substantial multipath effects, Doppler shifts brought on by movement, and restricted bandwidth. To avoid these, experts explore methods to increase data transmission speeds without sacrificing dependability [1]. Compared to terrestrial wireless networks, bandwidth is severely constrained in underwater environments. Two types of users, natural and artificial, share the same acoustic band. Traditional multiple access methods that assign users to different times or frequencies, such as time division multiple access (TDMA)

[1]School of Electrical Engineering and Computer Science (SEECS), National University of Sciences and Technology (NUST), Pakistan

or frequency division multiple access (FDMA), waste valuable resources. In addition, OFDM, which effectively counteracts multipath fading by splitting the bandwidth into orthogonal subcarriers, is frequently used in older systems to get around these issues.

The utilization of multiple-input, multiple-output (MIMO) systems, which use several hydrophones to increase capacity and reliability, is another development. But in order to handle inter-symbol interference, OFDM needs a cyclic prefix (CP), which lowers spectral efficiency, a significant disadvantage in underwater channels with limited bandwidth. Additionally, when bandwidth is limited, conventional OMA approaches such as TDMA or FDMA inefficiently assign different resources to each user. NOMA is used, since these techniques do not fully use the potential of simultaneous multi-user transmission. The advent of NOMA marked a sea change. By dividing users by power levels, NOMA enables several users to share the same time and frequency, in contrast to orthogonal systems that provide distinct resources to each user. More power is granted to users with weaker channel conditions and less to those with stronger ones. Each user's signal is sequentially decoded at the receiver using a technique, SIC. This change is revolutionary. All of a sudden, more users may be served without requiring extra bandwidth.

Fairness improves as weaker users are no longer marginalized, and data throughput rises. NOMA provides an essential step forward for underwater habitats, where resources are limited and demands are increasing. These sophisticated communication infrastructures are making it possible to establish smarter, quicker, and more reliable links below the sea surface, whether they are used to guide submersibles on research missions, monitor deep-sea infrastructure, or coordinate autonomous fleets. For a long time, the deep-sea has remained elusive. But now that these advances have been made, researchers are finally beginning to pay attention and react. In this chapter, our focus is on solving the underwater bandwidth problem using NOMA by discussing various existing techniques in the literature and exploring different solutions that open up opportunities for young researchers and scientists in this field. The remainder of the chapter is arranged in the following manner. Section 5.2 discusses related work, Section 5.3 provides new solutions to bandwidth challenges in underwater networks, and Section 5.4 concludes the chapter.

5.2 Related work

The literature has offered solutions to the bandwidth problem that involve integrating NOMA with various technologies. Bocus *et al.* [2] discussed the combination of FBMC and OFDM systems. These two technologies were developed to handle the challenging underwater environment while effectively allocating available bandwidth to numerous users. Additionally, NOMA was used to further enhance the system performance. In comparison to TDMA and FDMA, NOMA, in conjunction with OFDM or FBMC, allows for the simultaneous service of more users, greatly boosting data speeds and spectral efficiency. Real-time activities, such as live video transmission from remotely operated underwater vehicles (ROVs) back to surface ships, depend on this. Although OFDM is effective, it is not the only waveform that could be modulated by multiple carriers; therefore, a more sophisticated technique, FBMC is becoming more suitable for underwater conditions. Unlike OFDM, FBMC eliminates the need for a cyclic prefix by tightly localizing the signal in both time and frequency using carefully designed filters on each subcarrier, which leads to higher spectral efficiency, a crucial advantage when every bit per second counts. Hence, the researchers are able to achieve even higher performance by merging FBMC and NOMA.

FBMC filters increase resilience to Doppler and multipath effects while lowering interference. NOMA ensures that several users can use the same frequencies at the same time without interfering with the signals of each other. This combination provides high data speeds, low error probability, and a higher overall utilization of the acoustic spectrum, even in fast-changing and frequency-dispersive underwater channels. Furthermore, this enhancement is not only beneficial, but also necessary for situations such as real-time video transmission from underwater vehicles to a ship on the surface. Although waveform improvements addressed efficiency, a more comprehensive architectural change is required to scale up communication across large networks, such as those comprising many sensors, robots, and control buoys. Hierarchical full-duplex NOMA was developed to address this problem in Makled *et al.* [3]. Suppose that a surface buoy concurrently sends control signals to robotic arms below and sensor nodes on the seafloor provide data upward. These uplink and downlink transmissions are used to alternate. By enabling simultaneous transmission and reception on the same device, over the same frequency, full-duplex technology alters the rules of the game. Since each node's transmission could interfere with its own reception, integrating full duplex into a NOMA-based underwater network-needed intelligent interference handling.

However, the system could thrive with efficient SIC and relay nodes that facilitate communication between obstructed or remote devices. In their configuration, a relay node was placed between the surface buoy and the sensor nodes, acting as an intelligent bridge. In order to preserve synchronization and lessen the load on each device's battery, it decoded the incoming signals and passed them in real-time. The entire network had a hierarchical structure, with the buoy serving as the central controller, the relay serving as the intermediate tier, and the sensor nodes and robotic arms forming the lowest tier. The efficiency of their hierarchical full-duplex NOMA system was astounding. They balanced power consumption between nodes, reduced latency, and increased data throughput. Their model continuously outperformed previous systems, such as conventional half-duplex or non-relay networks, even in situations where interference cancellation was not perfect. According to them, underwater communication advances beyond its previous limitations by combining FBMC with NOMA and expanding it into a full-duplex relay-assisted hierarchy. Despite constant interference from water, these systems provide reliable, efficient, and simultaneous communication between several users. More than just a technical advancement, this represents a significant advancement in understanding and approaching the underwater environment.

Similarly, the integration of an underwater terrestrial uplink system that combines SIC based on quality of service (QoS) and NOMA was built on previous developments in underwater communication [4]. By prioritizing customers according to their service needs, the QoS-based strategy ensured that crucial data was transmitted on time, in contrast to the classical SIC that decodes signals based only on channel strength. To serve several users over the same spectrum without compromising efficiency or fairness, the underwater nodes in their scheme interacted acoustically with a surface gateway, which then transmitted data via radio to a terrestrial base station. Compared to traditional systems, their architecture optimizes energy consumption, reduces the likelihood of outages, and increases spectral efficiency. In areas with limited bandwidth, their system allows for more intelligent resource management by balancing user demands and channel conditions. However, issues such as real-time adaptation, underwater mobility, and interference management remain. Future research could include artificial intelligence (AI) to dynamically optimize decoding and power allocation.

5.3 Recent advancements in NOMA-based UAC networks

To increase the intelligence and adaptability of underwater acoustic networks, digital twin architecture has been recently studied. The research in Song *et al.* [5] proposed pairing each physical acoustic network with a virtual twin that mimics both local nodes and the general state of the network. At the local level, a resource allocation module received real-time performance parameters (such as traffic load or connection quality) and quickly adjusted transmission power and channel assignments. The data of every local twin was integrated into a worldwide collaborative learning system. The dynamic network slicing functionality of this twin-based design allowed the network to form virtual slices or resource pools dedicated to specific missions or applications. By predicting user needs and channel fluctuations with the help of the digital twin, the network offered more timely and resilient resource allocation. The twin structure showed to improve multi-task scheduling flexibility and dramatically speed up AI controller training (using simulated data). In complex underwater settings, their study led to lower latency and more efficient spectrum utilization.

Another important advancement in optimizing power and spectrum consumption in NOMA-based UAC networks is deep reinforcement learning (DRL) [6]. Because the underwater acoustic spectrum is so small, recent designs allowed several users to share parts of the spectrum (partial spectrum sharing) under regulated interference. A joint optimization problem that optimized throughput or fairness was used to define the resource allocation issue, which determined how much power each node should use and which sub-band each node should occupy. To tackle this complex, non-convex problem, researchers developed hybrid techniques that combine classical optimization with data-driven DRL agents. The model-based component handled the structured aspects of the problem (such as satisfying power or fairness criteria), while a deep learning agent interacted with the environment continuously. By adjusting the broadcast powers and channel assignments, the agent successfully learned to avoid excessive collisions through trial and error. As nodes adopted policies that maximized the minimum data rate among users and reduced inter-user interference, their technique gradually increased spectral efficiency.

However, the DRL-driven method is more durable and efficient than fixed allocation strategies by effectively enabling the NOMA-based UAC network to reconstruct itself autonomously and adaptively in response to fading, migrating nodes, or unique traffic patterns. Improvements have also been made by combining the ideas of NOMA with underwater optical wireless communication (UOWC), which offers much more bandwidth. These systems use the rate-splitting multiple access (RSMA) technique [7], in which each user's message is split into a common portion that is decoded by all users and a private portion that is decoded by a single user. The transmitted precoders (beamforming weights) needed to be constructed to maximize the long-term (ergodic) sum rate in the presence of optical channel turbulence. The ergodic capacity under realistic underwater optics models was recently approximated analytically. These closed-form rate formulas are used to formulate the precoder design as a convex optimization problem. They enable the RSMA-UOWC system to precisely manage interference between user broadcasts while utilizing the wide optical spectrum.

Likewise, to increase the performance of NOMA networks underwater, a cooperative communication structure was developed that included an unmanned aerial vehicle (UAV) and a buoy-based decode and forward (DF) relay with energy harvesting capabilities [8]. The relay used a power splitting mechanism, conserving the remaining energy for future transmissions and decoding the incoming data with a fraction of the received signal power. In their system, the UAV served as a mobile source, transmitting signals to the surface relay modulated with quadrature pulse shift keying (QPSK). The relay then used on-off keying (OOK)-modified acoustic signals to provide data to two underwater sensor controllers. The UAV, relay, and underwater sensors communicated using the shadowing fading distribution. This distribution accounted for multi-path propagation and shadowing in an underwater environment. The moment generating function (MGF) and Meijer-G functions were utilized to create analytical equations for the average bit error rate (BER) and sum rate, which were used as theoretical reference points. The analysis discovered that consumers near the relay had consistently lower BER due to better channel conditions, but distant users had higher error rates.

They evaluated the results in multiple scenarios using various parameters, such as the severity of fading, signal-to-noise ratio (SNR) in the source-to-relay connection, and power values. Increasing power values enhanced the sum rate by directing more power to the decoding process. However, this resulted in less stored energy for forwarding, which raised the BER. Lower power preserves more energy for transmission, but reduces the reliability of the decoding. This trade-off highlights the importance of flexible power allocation mechanisms in power-splitting (PS)-based energy harvesting devices. The adoption of NOMA allowed for efficient sharing of spectral resources among several users, enhancing both spectral efficiency and fairness, critical characteristics in the context of restricted undersea bandwidth. SIC was used at the receiver end to distinguish overlapping signals based on power differences and functioned well, especially under asymmetric channel conditions. However, the suggested system failed to account for a number of real-world issues, including node failures, underwater component mobility, and environmental fluctuation.

Incorporating digital twin architectures and DRL may provide real-time decision-making and effective resource management. Furthermore, their model assumes a perfect SIC, which may not reflect actual performance; including the impacts of a faulty SIC provides a more realistic evaluation.

5.3.1 Uplink and downlink NOMA for UAC networks

Finally, employing cooperative or multi-relay schemes can help address scalability issues, and in some situations, UOWC can be introduced to enhance data transmission rates. A UAV-assisted uplink framework was presented for marine IoT systems to overcome the difficulties caused by high path loss and restricted connectivity in maritime communication networks [9]. Through the use of several buffer-aided UAVs acting as DF relays, this design allowed reliable communication between far-shore base stations and maritime sources, including buoys and unmanned surface vehicles (USVs). NOMA was used to control uplink transmissions, increasing spectral efficiency and allowing simultaneous transmission from marine nodes. The UAVs' decoding methods were directed by service-specific rate requirements and instantaneous channel state information (CSI), rather than fixed decoding algorithms. This adaptive technique improved the performance of the SIC, allowing for better separation of overlapping signals. The buffers of the UAVs in their model allowed for variable scheduling and opportunistic relay selection, dramatically improving connection reliability and reducing the likelihood of communication failures.

The communication process consisted of two phases: UAV-to-destination forwarding and source-to-UAV transmission. UAVs dynamically structure SIC signals based on individual channel gains after receiving signals from marine sources transmitting at equal power during the uplink phase. The UAVs retained the decoded packets for later forwarding after dividing them into sub-buffers, one for each source, to ensure user equity. The model included probabilistic line-of-sight conditions based on the altitude of the UAV and spatial configuration, as well as accounting for small-scale fading and large-scale route loss. Their simulation findings showed that NOMA marine IoT systems regularly beat OMA in terms of average sum-rate, particularly at low-to-moderate transmit power levels. Increasing the number of UAVs improved channel diversity, which in turn boosted SIC performance. However, at greater power levels, NOMA's performance advantage faded as OMA benefited from less interference. They provide future directions for exploring hybrid NOMA or OMA systems that can adjust their access management based on changing traffic patterns and SNR levels. Using machine learning to handle tasks such as picking the best UAVs, predicting data rates, or managing buffer space can help boost performance while reducing the need for constant system communication.

For larger-scale maritime IoT setups, moving toward more flexible and open network designs, such as network function virtualization (NFV) and open radio access network (O-RAN), can make it easier to scale, manage resources more efficiently, and respond quickly to real-time changes. As underwater communication systems become more crowded and complex, researchers look for more adaptive ways to decode multiple overlapping signals. Instead, the system proposed in Zuberi *et al.* [10] learns to recognize combined user signals and decode them directly without intermediate steps. This is achieved through offline training using synthetic underwater data that mimics real-world signal interference, user distances, and varied power allocations. Once trained, the model no longer needs updated channel information during use. Their study reported promising results. The model successfully handled two-user and three-user scenarios using different modulations, without any increase in processing delay. This scalability is essential for underwater environments, where multiple devices, such as AUVs or sensor nodes, often operate together. Its ability to function without per-user decoding also means fewer delays in real-time deployments. This marks a notable shift from conventional receiver design toward flexible and more learning-driven systems suitable for evolving underwater conditions.

Despite strong simulation results, their study's biggest limitation lies in its exclusive reliance on synthetic data. The underwater conditions in the real world are far less predictable, influenced by currents, temperature gradients, and physical movement. These can significantly distort the signals and reduce the accuracy of the model. A model trained on idealized or static conditions may not perform well when exposed to fluctuating noise or hardware-level distortions. The model depends on large, labeled datasets and substantial processing power, resources not always available in typical underwater setups. Once deployed, it performs efficiently, but retraining the model for a new environment could require access to lab-grade infrastructure. This raises questions about long-term usability, especially in dynamic or mobile networks. They can have addressed this by proposing incremental learning techniques or showing how much performance drops when channel conditions deviate from the training set. Without such adaptability, the receiver risks becoming ineffective in unpredictable field conditions. Flexibility is key in underwater systems, and static models often fall short when real-time adjustment is needed. Improving adaptability could make this receiver more resistant to underwater variability. Their model has great potential, but its success depends on how well it manages unexpected changes. Future designs should incorporate fallback mechanisms, such as rule-based decoding, to recover data if the neural network fails.

Another possibility is to combine this type of receiver with smarter network control tools. This can allow the system to react not only to signal quality but also to operational needs. Their study opens new doors for underwater networks, but the technology must evolve beyond simulation. Field testing in real-world settings is essential prior to adoption. In addition, the performance of their model under time-varying conditions, such as node movement or interference from external sources, needs further analysis. Designing a version that adapts on the fly, without full retraining, can be more useful for long-term deployments. Battery-powered underwater nodes especially benefit from systems that combine low power with high adaptability. Including energy-efficient decoding strategies can improve the practicality of the model in actual field missions. Their system shows strong decoding capabilities, but its practical integration into full underwater networks remains uncertain. One gap in their study is how the receiver handles mobility. Nodes, such as AUVs or drifting sensors, change position constantly, which alters the channel characteristics. If the model is trained on stationary data, its accuracy may degrade in mobile scenarios. Moreover, the system focuses only on downlink communication. Uplink transmissions, where multiple nodes send data to a base, often involve more timing uncertainty and noise. Including those cases can make the approach more complete.

Space and power constraints in underwater devices make it difficult to deploy large neural models. Although inference is fast, the model might need compression or edge-level optimization to fit onboard. They did not discuss these requirements. Exploring lighter architectures can help solve the issue discussed above. In addition, combining the model with swarm-based systems or collaborative learning across nodes could allow local adaptation without central training. Such techniques can help networks learn and respond collectively, creating a more intelligent and resilient underwater communication system suitable for larger deployments. A study in Li *et al.* [11] designed a real-time underwater communication system that sends data to two users at the same time. Rather than mixing signals with analog circuits, they controlled the light-emitting diodes (LEDs) digitally. This approach allows them to adjust which groups of LEDs are turned on or off depending on which user the data are for. They separated the user signals by controlling how strong the light was for each. This meant that stronger light was used for users with weaker channels, which helped balance things out. The hardware included a driver circuit that quickly removed excess charge from the LEDs so that the lights could be switched on and off without delay.

This is important because slow switching causes errors in signal clarity. Their driver made the LED response cleaner and faster. However, the system was designed for only two users. Adding more users could make it harder to properly manage power levels. That might lead to overlap or confusion between signals. A more flexible system is needed to handle more users in future applications. One possible fix can be to use real-time feedback to automatically adjust the power for each user. Another issue they faced is that LEDs usually have a very narrow bandwidth, which limits how fast they can send data. To handle this, they added a simple equalizer after the receiver that cleaned up the signals. Instead of using a separate equalizer for every LED group, they used one equalizer for all. This made the design easier to build and cheaper as well. The equalizer helped improve signal quality so much that they were able to achieve data rates of 40 Mbps per user. That is a big improvement compared to older systems that often had to compromise between speed and clarity. The system handled this speed well in their testing environment. But real underwater conditions are harder to predict. The water could suddenly become cloudy or full of particles, which could affect the efficiency of the equalizer. Since the current equalizer is fixed, it might not be able to deal with those changes. A better approach can be to use a smart equalizer that changes its settings based on real-time feedback from the received signals.

That can help to make the system more dependable when used outside the laboratory. Unlike many studies that stop at simulation, this one was tested in a real water tank. They added turbulence and particles to the water to make it more like the ocean. Even under those conditions, both users were able to receive the data clearly, and the error rates remained within the acceptable range. They noticed that giving the weaker user more power, around a 2-to-1 ratio, helped balance the performance for both. This shows that power control plays an important role in the reliability of the system. Testing a setup like this made their findings more believable and practical. However, the tank was only one meter long, which does not reflect the long distances often needed in real underwater tasks. For example, if this were used to connect a sensor on the seafloor to a boat above, the system would need to work across a much bigger distance. The current design might not be able to handle that without changes. In the future, the system could be improved by adding signal relays to extend its range. Another option could be combining it with another type of communication system when light can't reach far enough.

One of the strengths of their system is that both users receive their data at the same time, instead of waiting their turn like in TDMA-based systems. This saves time and uses the available bandwidth better. The custom-built driver helps keep the signals sharp and reduces delay from LED switching problems. Their system runs smoothly during testing, and the signals stays clear. But even though the setup works well in fixed conditions, it may not do as well when the users are moving or when the environment keeps changing. The power levels given to each user are fixed during the test, which might not be helpful if something changes suddenly. A system that adjusts power based on the actual situation can likely perform better. Also, the authors did not include any kind of error correction in the system. That is something underwater systems usually need because of the noise and unpredictable conditions. Including an error correction method can help avoid lost or damaged data. It could make the system much more reliable. These small upgrades can help the system move beyond the laboratory and into real underwater missions.

5.4 Conclusion

NOMA has become vital for marine applications, yet underwater environments impose constraints, limited bandwidth, long delays, Doppler shifts, and high multi-path, which hamper communication reliability. Traditional schemes like TDMA or FDMA fail under such conditions due to static resource allocation and poor spectral usage. OFDM improves multi-path resistance but sacrifices efficiency via its cyclic prefix, which is costly in bandwidth-starved channels. In contrast, NOMA enhances spectral use by enabling power-domain multiplexing, letting multiple users share time-frequency resources. Through SIC, it favors user fairness and throughput, especially when combined with multi-carrier methods like FBMC, which removes the cyclic prefix and improves Doppler and multi-path resilience. Hierarchical full-duplex NOMA networks further allow bidirectional communication using buoys, relays, and underwater nodes, achieving reduced latency and energy-efficient coordination. Moreover, DRL empowers adaptive power spectrum allocation, balancing efficiency and fairness. NOMA's fusion with UOWC and RSMA boosts capacity in optical turbulence using refined precoding. Energy-harvesting DF relays and UAVs support wider coverage via power splitting and modulation schemes tailored to underwater links. Learning-based decoders trained on synthetic data enable scalable, low-latency communication but struggle with adaptability and high computational demands. Likewise, LED-based NOMA systems require adaptive feedback to handle environmental variability. Despite promising laboratory results, real-world deployment still faces challenges like environmental uncertainty, energy limits, and scalability. Future progress depends on smart, energy-efficient, and adaptive architectures.

References

[1] J. Yao, L. Zhou, T. Wu, *et al.*, "Exploring fairness for FAS-assisted communication systems: From NOMA to OMA," *IEEE Transactions on Wireless Communications*, vol. 24, no. 4, pp. 3433–3449, 2025.

[2] M. J. Bocus, D. Agrafiotis, and A. Doufexi, "Non-orthogonal multiple access (NOMA) for underwater acoustic communication," in *2018 IEEE 88th Vehicular Technology Conference (VTC-Fall)*, Chicago, IL, USA. 2018, pp. 1–5.

[3] E. A. Makled, A. Yadav, O. A. Dobre, and R. D. Haynes, "Hierarchical full-duplex underwater acoustic network: A NOMA approach," in *OCEANS 2018 MTS/IEEE Charleston*, Charleston, SC, USA, 2018, pp. 1–6.

[4] N. Sharma, A. Girdher, and A. Bansal, "Uplink NOMA with QoS-based SIC for mixed underwater-terrestrial communication system," *IEEE Transactions on Vehicular Technology*, vol. 73, no. 5, pp. 6470–6485, 2024.

[5] S. Song, B. Huangfu, J. Guo, J. Liu, J. Cui, and X. Shen, "A digital twin-based intelligent network architecture for underwater acoustic sensor networks," *IEEE Transactions on Mobile Computing*, vol. 24, no. 9, pp. 8196–8213, 2025. https://doi.org/10.1109/TMC.2025.3555640

[6] Y. Zhang, M. Li, X. Feng, X. Han, and M. Jia, "Resources allocation for underwater acoustic soft frequency reuse network based on multi-agent deep reinforcement learning," *Physical Communication*, vol. 67, p. 102487, 2024. https://www.sciencedirect.com/science/article/pii/S1874490724002052.

[7] F. Xing, F. Tong, X. Liu, Z. Wang, and V. C. M. Leung, "Precoding optimization for rate splitting enabled internet of underwater things over optical wireless underwater turbulent channels," *IEEE Internet of Things Journal*, vol. 12, no. 14, pp. 26782–26798, 2025. https://doi.org/10.1109/JIOT.2025.3563505

[8] K. Mittal, A. Gupta, and P. Garg, "Energy harvesting in NOMA based UAV-acoustic system," In: *2023 IEEE 20th India Council International Conference (INDICON)*, IEEE; Hyderabad, India. pp. 578–583, 2023. https://doi.org/10.1109/INDICON59947.2023.10440918

[9] N. Nomikos, A. Giannopoulos, P. Trakadas, and G. K. Karagiannidis, "Uplink NOMA for UAV-aided maritime internet-of-things," in *2023 19th International Conference on the Design of Reliable Communication Networks (DRCN)*, 2023, pp. 1–6.

[10] H. H. Zuberi, S. Liu, M. Bilal, *et al.*, "Deep-neural-network-based receiver design for downlink non-orthogonal multiple-access underwater acoustic communication," *Journal of Marine Science and Engineering*, vol. 11, no. 11, p. 2184, 2023. https://www.mdpi.com/2077-1312/11/11/2184.

[11] X. Li, L. Gui, Y. Xia, *et al.*, "NOMA assisted real-time downlink UWOC system using a software-configurable driver under different channel conditions," *Journal of Lightwave Technology*, vol. 42, no. 14, pp. 4860–4873, 2024.

Chapter 6

Reconfigurable intelligent surfaces in challenging underwater environment

Mohammad Ahsan Atteque[1], *Syed Muhammad Azeem ul Hassan*[1] *and Sajjad Hussain*[1]

Underwater links demand kilometer-scale reach yet suffer a sharp bandwidth ceiling, driving the search for high-rate yet low-cost alternatives. Acoustic reconfigurable intelligent surfaces (RIS) rearrange this trade-off by using piezo actuated reflector arrays to sculpt and, when active, amplify wave fronts, thereby boosting signal-to-noise ratio and extending coverage across complex three-dimensional water columns. The chapter follows the evolution of these surfaces from proof-of-concept dual-element tiles with centimeter-level angle-tracking loops to network-scale designs that embed light-weight control logic. The study then synthesizes depth-aware placement heuristics, duty-cycled "wake-on-need" operation, and cost-driven tile sizing to show how coverage, power, and complexity can be balanced for practical deployment. Next, this chapter explores future additions to the field by mapping these foundations onto emerging 6G class scenarios, autonomous underwater vehicle (AUV)-assisted relays, and single-access point localization. Our goal is to highlight a technological path toward resilient, high-speed acoustic networks that can self-adapt to ocean dynamics.

6.1 Introduction

Underwater communications are very important for a multitude of applications in the ocean. Due to the high challenges in this environment, there is a need to focus on long communication ranges and high data rates, which are needed to share information over long distances to cover large expanses of ocean space. Simultaneously, high data rates (beyond 1 Mb/s) and long ranges (beyond 1 km) for underwater communications can hardly be achieved with existing technologies [1]. Existing technologies offer the following benefits to underwater communication systems.

- Acoustic communication: With narrow bandwidth and highly marked multipath fading, it is the only medium that can achieve long ranges (several kilometers)

[1]School of Electrical Engineering and Computer Science (SEECS), National University of Sciences and Technology (NUST), Pakistan

today. However, this condition yields extremely low data rates (i.e., in kilobits per second).

- Optical communication: It is good for high data rates, but is by its very nature short-range (measured in hundreds of meters). It is also susceptible to turbidity and scattering.
- Electromagnetic (EM) communication: It uses medium- or low-frequency signals where the frequency of signals used to penetrate and move in water for communication shows reliability. However, both the data rate and communication range are restricted because of low carrier frequencies and high water absorption.

Among all these, acoustic transmission is preferable due to its lower attenuation and wider propagation range compared to optical and EM waves. Additionally, acoustic waves can travel much further away without experiencing significant degradation. But the primary obstacle related to acoustic transmission is its limited bandwidth, which leads to low data rates. Hence, the main objective of this chapter is to determine how substantially it increases the data rates in long-range underwater acoustic systems by reducing the cost and complexity of the underwater network. The acoustic reconfigurable intelligent surface (RIS)-based system seems to be a viable solution that can achieve high data rates and provide long-range underwater communications.

6.2 What is RIS and how does it impact underwater networks?

An RIS surface consists of many micro-reconfigurable units that can intelligently reflect an incident wave toward a desired receiver. RIS, through controlled beamforming and EM amplification of some reflected signals, creates strong re-reflected pathways, thus boosting the receiver's signal-to-noise ratio (SNR). Acoustic RIS can significantly increase point-to-point data rates, decrease inter-user interference, enhance overall network throughput, and serve as an underwater infrastructure beamforming for all devices, even low-cost sensors or small robots, without changes to the end-user devices. Although there is EM-based RIS available in the literature [2] for terrestrial networks, acoustic RIS works on different physical principles (mechanical waves versus EM waves) and faces its own underwater challenges:

1. Hardware design: New hardware based on acoustic principles must be obtained, as terrestrial EM designs (such as patch antennas) are not suitable.
2. Wide-band beamforming: Underwater acoustics is extremely wide-band about the carrier frequency, and entirely the reverse is seen in terrestrial narrow-band systems. This calls for new beamforming strategies that can accommodate the dispersion of signals.
3. Network coverage and operation: An underwater RIS must cover extensive three-dimensional volumes (possibly thousands of meters deep), in contrast to a terrestrial RIS that is mostly two-dimensional or height-limited. Moreover, long and variable delays in acoustic propagation require intelligent lightweight operation protocols.

4. Environmental effects: The underwater RIS moves under the effects of waves and turbulence. Having a static background requires terrestrial installations. The high-level challenge is offset by motion effects.

In order to overcome the above-mentioned issues, an array of reconfigurable piezoelectric reflectors can be used. To meet the needs of 3D coverage, the RIS may assume a planar (n × n array) or one of several other forms, such as spheres or polyhedral. Each unit (reflector) has one piezo-element connected to a metal head and tail. It responds to incident acoustic pressure on the head microcontroller, which adjusts the resonant impedance (for example, by a variable capacitor) of the control circuit of the piezoelectric element. Changing the impedance tunes the unit's elasticity, which in turn modifies the phase of the reflected acoustic wave. An off-resonant circuit induces a phase shift, while a resonant circuit does not. Mechanical components (masses) are represented as inductors, acoustic force as a voltage source, velocity as a current, and piezoelectric effects through the transformer and compliance. The water interaction is represented as a combination of resistance and inductance. A passive unit reflects only. An active unit can include a negative resistance element (reactive) to amplify the reflected signal, thus providing gain. Figure 6.1 shows an IRS-based acoustic communication where different underwater nodes can enhance their performance using these surface elements.

Sun *et al.* [1] addressed these research challenges by proposing a framework that was organized into four components. The first component showed the RIS-assisted network architecture, where RISs were meant for infrastructure deployment in such a 3D underwater environment. The second comprised the broadband beamforming design whose objective was to mitigate the dispersion effects inherent in underwater acoustics. The third component focused on the ocean movement mitigation strategy, which dealt with the impact of RIS motion (rotation primarily) by currents, waves, etc. As distance changes within a measurement packet time were negligible compared to distances measured in kilometers, rotation became increasingly dominant in affecting the beam alignment, that is, angle of arrival (AoA) or angle of departure (AoD). The fourth was the light-weight operation protocol, which was designed to enable RIS benefits to standard single-input single-output (SISO) devices without any changes to existing network protocols. The RIS operated as a decoupled infrastructure element.

RIS passively listened to the channel and was associated with the transmitter-receiver pairs whose pilot signal arrived first (assuming that the MAC was handled by the network). They used a Bellhop ray tracer model to simulate the long underwater channel. Their 3 × 3 RIS simulation results showed that by varying the capacitor values, the reflection direction can be controlled. Hence, according to their scheme, RIS provided the largest increase in data rate over the same range of the order of a kilometer. The data rate increased with an increase in the size of the RIS but it increased complexity and cost of the entire network. Likewise, [3] proposed a hardware that began with a tile sandwiching piezo-electric rings between metal end caps. An equivalent circuit description unified its electrical, mechanical, and acoustic behavior and exposed a tunable impedance element in series with the piezo stack. In its passive form, the tile offered only a modest acoustic return; the addition of a small reflection

Figure 6.1 The RIS-based underwater acoustic communications

mode amplifier boosted the same transfer characteristic with a few milli-watts of extra drive per element on a surface of 15 × 15.

Whether passive or powered, the impedance of every tile selected from a pre-computed table allowed a central controller to impose any desired phase map across the whole array. The active tiles behaved as reflection mode amplifiers, so their gain boosted the desired echo together with the local acoustic noise. Using the element radiation pattern, they obtained a closed-form expression for the electrical power that a single piston must draw and introduced a directive factor that captured how strongly it coupled into the water column. Summing those requirements over the entire surface, they imposed an array-level ceiling of only a few watts for several hundred reflectors; in the 15 × 15 demonstrator, this came to a few tens of milli-watts per tile, well within the reach of small batteries or energy harvesters. They contributed the first tunable acoustic surface, a dispersion-canceling beamformer suited to 20% wide bands, and an angle-tracking protocol that was transparent to existing networks. Overall, by blending a piezo-electric circuit model with wide-band array theory and a lightweight control loop, it is possible to convert kilometers of unreliable water into a managed medium. Hence, from this discussion it is clear that to reduce the complexities of RIS, it is necessary to introduce refined control algorithms.

6.3 How control algorithms can reduce network cost and complexity?

Control algorithms are intelligent computational methods dealing with real-time adjustment of RIS. These algorithms perform phase shifts, beam steering, and signal optimization according to environmental dynamics or requirements. They are rule-based or driven by optimization or machine learning to minimize signal distortion and maximize performance. Their purpose is to make the RIS partly autonomous with reduced manual control and a lower reliance on expensive network infrastructure. Control algorithms play a major role in minimizing the complexity of RIS-assisted networks because they intelligently and adaptively configure actual RIS elements without the need for continuous human intervention and the cost of expensive hardware upgrades. For example, underwater RIS should quickly adapt to rapid signal fading and multipath due to water disturbance, variations in salinity, and temperature differences. Control methods such as deep learning could pre-train models offline to control real-time configurations, saving processing time and overhead in the network involved in the configurations [4]. Thus, with optimized delay spread and RIS phase shifts, such algorithms increase the perception of the signal without implementing large installations of physical infrastructure and at the same time decreasing both operational complexity and requirement in system size. Control algorithms make RIS capable of operating autonomously in industrial and disaster environments, accommodating rapid changes that could introduce obstacles, such as debris, equipment, or damaged buildings.

Rather than deploying additional antennas or relay stations, algorithms deploy real-time environmental data to adaptively alter the RIS's signal steering or boosting configuration. This adaptability minimizes the requirement for supporting components and permits the design of a network with fewer, but smarter, RIS units. Similarly in the case of active RIS, low-power amplifiers fitted with intelligent control are capable of boosting signals without using full transmitter chains, thereby ensuring less cost and power requirements while keeping the coverage and reliability intact. The control logic embedded in RIS hardware, primarily when integrated with machine learning, provides a scalable and cost-effective way to deploy systems. The logic adjusts and optimizes phases, energies, and beamforms in conjunction with established systems such as the internet of underwater things. According to simulation studies conducted in [5,6], having control algorithms allows RIS to adapt to current signal conditions in real time, achieving very high boosts (up to 20 dB) without requiring extra transmission power or infrastructural expansion. This intelligent management converts passive surfaces into programmable nodes, reducing the physical and monetary footprints of advanced wireless communication networks. The research gap therefore lies in creating a tunable acoustic reflector, a beamforming strategy immune to frequency squint, and a protocol that hides all this complexity from existing modems.

6.4 Existing solutions in the literature and future directions

Zhao *et al.* [7] offered a blend of theory, optimization, and prototyping. First, it developed depth-aware, closed-form ray acoustic expressions that allowed designers to map the exact footprint of every convergence and shadow band in both the V-shaped deep-sea duct and the near-surface waveguide. Building on that analytical lens, positioning an acoustic RIS (aRIS) right on the natural sound channel axis in deep water or anchoring it on the seabed below the surface waveguide in shallow water maximized lateral reach and lifted simulated coverage from a modest baseline to virtually blanket performance, without demanding extra source power. They showed that daisy-chained aRIS panels configured as multi-hop relays can push connectivity far beyond the first shadow zone, while a hybrid sequential quadratic programming or genetic algorithm routine systematically finds the smallest tile count for any depth and aperture combination, keeping hardware lean. On the hardware side, their work tackled the full-duplex constraint of piezoelectric transducers by pairing two elements per cell, one to absorb, one to reradiate, so the array followed a first-absorb and then radiate logic. A small two-unit prototype demonstrated a beamforming gain that closely matched theory in pool trials, validating both the dual-element architecture and the fine-grained phase-shifting network that drove it.

To keep links stable under real ocean motion, centimeter scale ultra-short baseline (USBL) tracking and micro-electro-mechanical systems (MEMS) gyroscopes fed a live phase correction loop that trimmed vertical, horizontal and rotational pointing error by almost two orders of magnitude, maintaining coherent beams under sub-meter drift and only a few degrees of rotation. A complementary cost model then traded the panel area against the tile count, giving designers a clear view of life-cycle economics. In our view, merging these analytically derived placement rules, dual element tiles and adaptive self-calibration with RIS enabled network architecture we have already discussed could be a possible solution and a future direction which can lead to practical deployment. A promising next step is to embed the closed form coverage metric within the machine learning controller so that an autonomous fleet can predict drift induced coverage gaps, reposition buoyant RIS units accordingly, and activate only those tiles whose predicted link margins fall below a target threshold, thereby sustaining near perfect coverage with minimal energy overhead. Now we shift the RIS conversation from simply how a surface should shape a wavefront to when it should even wake up. Continuous operation, Ndjiongue *et al.* [8] argued, wastes energy because every extra bounce through a passive panel inserts extra fading, training overhead, and processing delay.

Instead, according to their perspective, why not use an on-demand (OnD) model in which a surface remains in deep sleep until the direct path falters, then helps only long enough for the blockage to clear. Inserting a single OnD panel can move the received signal comfortably above outage and multiply the usable throughput, yet because the device is duty cycled, its power budget stays a small fraction of a permanently active wall. Although the numerical gains depended on the context, the principle was cleanly mapped to acoustic, radio frequency, and hybrid underwater optical links, where pilot signaling and multiplicative fading were even more punitive. By weaving this trigger-based philosophy into an RIS-enabled network, they allowed tiles to hibernate until a line-of- sight alarm fired, then faded back to silence once healthy margins returned. Now, this idea can be broadened beyond static walls. Mobile or temporary panels can ride drones, AUVs, or shoreline fixtures and check in only when local traffic spikes or new obstacles appear. Element counts can scale on the fly; if a single board saturates, a sibling is summoned, while lightweight wake-up packets, Markov queues, or compact neural networks decide whether the extra hop is worth the added latency. The multi-cell coordination, edge-hosted inference, joint energy information transfer, and sub-millisecond commissioning protocols, all of which dovetail with their own goals, can be considered as future directions.

Folding these insights into a novel architecture can let an autonomous fleet sense fading, reposition buoyant panels, and activate only the tiles whose link margins drop below target, maintaining nearly seamless coverage with minimal energy overhead. In our view, adopting an OnD mindset could be transformative for underwater RIS deployments where battery swaps or tethered power are expensive. Acoustic or hybrid acoustic optical panels could monitor link margin through low rate control pings, wake only when seasonal thermocline, turbidity spikes or vessel traffic degrade the channel, and sleep otherwise. Coupling the machine learning phase profile optimizer with such wake-sleep logic not only extend mission endurance but also free scarce onboard energy for active relays in truly unavoidable shadow zones, turning these into possible solutions into field ready practice. In Hossain *et al.* [9], an evolved wireless communication system designed for high-speed transmission was proposed to ensure data security and efficiency within the 6G system. Their system appended unmanned aerial vehicles (UAVs) and RISs to wireless networks to increase signal quality and coverage. Frequency-domain spectrum shaping (FDSS) and discrete cosine transform (DCT) spreading techniques minimized the peak-to-average power ratio with bandwidth conservation.

The transmission employed massive multiple-input multiple-output (MIMO) with orthogonal frequency division multiplexing (OFDM) to accommodate multiple users concurrently and a special data encryption method utilizing the two-dimensional cosine-sine interleaved chaotic map (2D-CSIM) scheme to secure the physical layer of transmitted data. This made it difficult to decode the intercepted signals by increasing the confidentiality of the signal. Block diagonalization (BD) precoding was used in interference reduction between users. On the receiver side, techniques like Cholesky decomposition-based zero forcing (CD-ZF) and lattice reduction-based minimum mean square error (LR-MMSE) helped eliminate signal impurities, making the received message clearer and more precise. Their simulation results showed excellent performance with low error rates at high speed. Thus, they contributed to less stress on the hardware. It processed every user information through the system, encryption, modulation, compression by DCT, and shaping the frequency signal before sending it. Null subcarriers were introduced to avoid an increase in the dispersion of adjacent channels and assured good spectrum efficiency. The information was relayed through a direct phasing axis confirmatory system and were using drones switched on RIS panels to bounce signals in controlled manners up to users hardly reached by straight-line transmission.

The system allowed for a dual-path that greatly increased the reliability in challenging environments, especially cities with tall buildings. Demodulated, decoded, and decrypted, the signal on the receiving end created the original audio after very few errors. Repeat-and-accumulate coding, low-order quadrature amplitude modulation (QAM), and LR-MMSE detection applied in the system produced the best performance. It maintained very low bit error rates, even for SNR above 27 dB. Compared with already existing techniques such as OFDM, orthogonal time frequency space, and orthogonal chirp division multiplexing, this design was more efficient in terms of productivity, signal strength, and error handling. The complexity analysis proves that all this can be done with acceptable processing power, which makes it commercially viable in 6G and beyond. The authors finally said that their system could offer strong, secure, and flexible communication in modern wireless environments. In our opinion, their method of wireless communication underwater bears a new fruit of ease, especially where UAV robotics integration is in combination with 6G architectures. The issues of signal leakage, interference, and high peak-to-average power ratio levels are mitigated entirely. The addition of a 2D-CSIM encryption method greatly increases the encryption and security of the system, a very important dimension in communication.

The use of robust and layered design strategies using techniques such as BD precoding and signal detection also adds to the security of the proposed system. However, again, the architecture of the system is quite heavy and is greatly limited by the power and size of the computational equipment used, which is a big factor in funding research. Using AI models of scalability, including lightweight cryptographic alternatives, definitely helps increase the accuracy and the performance of the UAV. The 2D-CSIM encryption method, while effective, has computational limitations. Using lighter algorithms, such as artificial noise (AN) injection, to stop eavesdroppers. Secure beamforming using RIS-configuration to not let signals be scattered and only be sent to the intended user is just one of the solutions assisting in the solution of heavy algorithms and low computational power. Overall, the methodology, emphasis on integrating security and feasibility, show promise for future refinements in intelligent and adaptive 6G systems. Pu *et al.* [10] presented a pulse model for indoor localization, which integrated RIS and generative artificial intelligence (GAI) for accurate positioning in a case where only a single access point (AP) was present. Conventional indoor positioning systems, such as GPS, wireless local area network (WLAN), and Bluetooth, encounter immense constraints in non-line-of-sight situations and usually require more than one AP, making them less suited for a smaller space or isolated scenario.

To this end, they introduced a RIS-assisted model that establishes virtual line-of-sight links via multiple reflective elements, essentially mimicking a multi-AP environment. The heart of the solution lay in processing the noisy received signal strength (RSS) data employing a variational autoencoder-convolutional neural network (VAE-CNN) model. This hybrid neural architecture simultaneously reduced the noise present and extracted relatively identifiable features from the signal data. By representing the target's actual position as a sparse vector, they could reformulate the localization problem as a sparse recovery problem. To further deal with the computational challenges presented by high-dimensional matrices arising from numerous reflective elements, a semi-tensor product (STP)-based recovery approach was then implemented. This distinctively reduced both memory requirements and latency for localization, while increasing accuracy and robustness. Through simulations, the system demonstrated robust performance indoors while choosing fewer reflective elements based on RSS strength. In contrast to traditional methods, such as K nearest neighbors (KNN) and weighted KNN, the proposed technique achieved superior localization accuracy and generalization with denoised RSS data using the VAE-CNN model. Using performance evaluation metrics such as mean square error (MSE) and peak signal noise ratio to compare generative denoising and generic filtering methods such as Gaussian and median filtering, they established the greater merit of generative denoising.

The system was also tested against various compressed sensing (CS) techniques employing different measurement matrix sizes, where it was observed that any selection of fewer reflective elements by the proposed STP-based approach not only maintained the accuracy of the predictions, but also outperformed CS with respect to confidence probability and error range. Notwithstanding the successes, the authors acknowledged challenges, such as RIS configuration variability, and proposed some future directions toward dynamically adapting the recovery model. In a more general sense, this work filled a major gap in indoor positioning systems and provided a basis for putting AI-based signal processing into practice in real-life localization applications in 6G underwater networks. The proposal of a system to tackle indoor localization using a single access point via RIS and generative AI was a solution that takes the emphasis away from an extensive infrastructure setup, alongside introducing easier access to smart environments previously tested in homes, warehouses, and small-scale setups. The use of VAE-CNN, while fascinating, introduces complex problems of training and instability alongside the restrictiveness of computational power, which is not a new issue. Using small computers and equipment, while an important reason for logistics and costs, heavily restricts the effective use of the system.

Extensive training, newer research, time, and manpower are not easy-to-come-by resources. The integration of sparse recovery with STP is another high point, reducing the complexity of the architecture while retaining accuracy, introducing a new challenge of lack of interpretability and adaptability. Their STP system greatly solves the issue related to the lack of hardware. Overall, their work presents not only a new algorithm but also a practical framework fit for deployment in an ever-evolving AI-powered 6G networks landscape. RIS communication systems arrive with their own challenges, and with these challenges we are introduced to a multitude of solutions. Solutions to the problems mentioned above that we have discussed so far include covering underwater shadow zones using RIS, using 6G networking alongside RIS systems, and indoor localization with a single access point via GAI. These solutions pave new ways to tackle the hardships of underwater communication, while noting that newer algorithms can be added to these solutions to make them more effective. Technologies like adding small AI models to add autonomy while ruling out human error, adding encryption and security by utilizing the laws of physics underwater. These newer solutions readily solve the problem of computational power. What if we add optical panels to the mix? That surely helps to monitor communication. More powerful wireless communication systems using the ever-evolving 6G networking technology. With many more solutions to follow and to add convenience, in our opinion this system is very versatile for new scientists and engineers to get creative and find the right answers to the right questions.

6.5 Conclusions

Thus far, we have established that in an underwater environment, traditional acoustic links offer long reach but sacrifice bandwidth, while optical and EM paths reverse that trade off; the core challenge is therefore to raise data rates over kilometer-scale distances, and acoustic RIS emerges as the most promising avenue. Acoustic RIS, built from piezo-controlled reflectors, can steer and amplify wavefronts to lift SNR, extend coverage, and reduce interference, yet must overcome hurdles such as broadband beamforming, 3D deployment, and motion-induced misalignment. Adaptive control algorithms, whether rule-based or learning-driven, autonomously retune phase maps, energy budgets, and beam directions in real time, cutting hardware overhead, trimming latency, and making the network self-healing in dynamic water columns. Next, we surveyed emerging solutions: depth-aware placement and dual element tile architectures that tighten link budgets, on-demand duty cycling that slashes energy consumption, and machine learning controllers that fuse coverage analytics with wake–sleep logic to enable lean, scalable, and resilient underwater connectivity. Looking ahead, the convergence of underwater acoustic RIS with 6G class technologies, such as UAVs equipped with RIS panels and AI-driven single AP localization, heralds resilient, high-speed links that can adapt autonomously to ocean dynamics. Collectively, these insights chart a practical path toward high-rate, low-cost acoustic networks.

References

[1] Z. Sun, H. Guo, and I. F. Akyildiz, "High-data-rate long-range underwater communications via acoustic reconfigurable intelligent surfaces," *IEEE Communications Magazine*, vol. 60, no. 10, pp. 96–102, 2022.

[2] P. Ramezani, B. Lyu, and A. Jamalipour, "Toward RIS-enhanced integrated terrestrial/non-terrestrial connectivity in 6G," *IEEE Network*, vol. 37, no. 3, pp. 178–185, 2023.

[3] H. Wang, Z. Sun, H. Guo, P. Wang, and I. F. Akyildiz, "Designing acoustic reconfigurable intelligent surface for underwater communications," *IEEE Transactions on Wireless Communications*, vol. 22, no. 12, pp. 8934–8948, 2023.

[4] M. A. E. A. Marshall, "A review of reconfigurable intelligent surfaces and their application to machine learning-assisted underwater communications," *Journal of Ocean Technology*, vol. 19, no. 4, pp. 52–84, 2024.

[5] A. K. P, S. P. N, T. Parveen, and A. Priya, "Challenges encountered with reconfigurable intelligent surfaces (RIS) in underwater communication," in *2024 Third International Conference on Electrical, Electronics, Information and Communication Technologies (ICEEICT)*, 2024, pp. 1–5.

[6] S. Kisseleff, S. Chatzinotas, and B. Ottersten, "Reconfigurable intelligent surfaces in challenging environments: Underwater, underground, industrial and disaster," *IEEE Access*, vol. 9, pp. 150214–150233 2021.

[7] L. Zhao, J. Tan, J. Wang, I. F. Akyildiz, and Z. Sun, "Covering underwater shadow zones using acoustic reconfigurable intelligent surfaces," *IEEE Conference on Computer Communications*, pp. 1–10, 2025. https://doi.org/10.1109/INFOCOM55648.2025.11044529

[8] A. R. Ndjiongue, O. A. Dobre, and H. Shin, "On the concept of on-demand RIS-assisted 6G networks: applications and the way forward," *Communications Magazine*, vol. 63, no. 5, p. 174–181, 2025. https://doi.org/10.1109/MCOM.001.2400249

[9] M. N. Hossain, A. Kaniz, S. T. Kamal, S. E. Ullah, and T. Shimamura, "Multiuser FDSS-based DCT-spread massive MIMO-OFDM system for secure RIS-assisted UAV-enabled networks," *IEEE Access*, vol. 13, pp. 35254–35269, 2025.

[10] Q. Pu, W. Yu, X. Lan, M. Zhou, and X. Yang, "A novel RIS-aided indoor localization in single access point scenarios via generative AI," *IEEE Internet of Things Journal*, vol. 12, no. 10, pp. 13500–13510, 2025.

Chapter 7
Routing strategies for 6G underwater communication networks

Syed Muhammad Jameel[1], *Parsa Rukhsar*[1] *and Syed Ali Hassan*[1]

Underwater communication networks (UCNs) are important both for natural and artificial underwater users. The main objective of researchers in this field is to design networks that can support various underwater applications while maintaining the safety of marine life. UCNs are crucial for these applications, including environmental monitoring, oceanographic exploration, the safety of marine mammals, and defense operations. For communication among underwater sensor nodes, routing is essential. The collection of data from the ocean floor and the transmission to the surface depend on safe and stable routes. The objective of this chapter is to examine routing protocols that have been designed in the literature specifically for underwater acoustic communication (UAC) and to discuss the potential applications of these protocols in the changing environment of sixth-generation (6G) networks. Our categorization of these protocols includes localization-based, localization-free, and cooperative routing protocols. We conclude by discussing artificial intelligence (AI)-based routing schemes and cell search techniques that can enhance the performance of UCNs. Hence, by combining foundational concepts with advanced techniques, the chapter provides a holistic view of current advances and future directions in underwater communication systems.

7.1 Introduction

Covering more than 70% of the Earth's surface, the world's oceans remain one of the most enigmatic and underexplored environments on the planet. Beneath these vast waters lies a wealth of untapped natural resources, ecological knowledge, and strategic infrastructure critical to global sustainability and security. The demand for effective underwater communication systems has never been greater, driven by applications in marine environmental monitoring, offshore oil and gas operations, submarine detection, oceanographic data collection, and maritime security [1,2].

[1]School of Electrical Engineering and Computer Science (SEECS), National University of Sciences and Technology (NUST), Pakistan

However, despite decades of research, the underwater domain continues to pose formidable barriers to reliable wireless communication, hindering our ability to unlock its full potential. Traditional terrestrial wireless technologies are ill-suited for the aquatic medium due to the unique physical characteristics of underwater channels. Radio frequency (RF) signals, which are ubiquitous in terrestrial networks, suffer from rapid attenuation in seawater, limiting their range to only a few meters unless extremely low frequencies are used at the expense of the data rate. Optical communication offers high data rates and low latency. Still, it is severely constrained by turbidity, absorption, and the need for line-of-sight (LoS) conditions, making it feasible only for short-range, clear-water applications [3].

Acoustic communication, on the contrary, remains the most viable option for medium- to long-range communication underwater; however, it comes with significant limitations, such as low bandwidth, long propagation delays, and vulnerability to Doppler spread and ambient noise [4]. These constraints require specialized protocols and architectures tailored for the underwater environment. UCNs, consisting of sensor nodes, autonomous underwater vehicles (AUVs), and surface relay stations, have emerged as a primary solution for enabling persistent and scalable underwater monitoring. However, the effectiveness of UCNs is closely coupled with the robustness of their communication infrastructure and, in particular, the efficiency of their routing strategies. Challenges such as three-dimensional node mobility, high packet loss, limited energy budgets, and intermittent connectivity further complicate the design of routing protocols [5]. As we enter the era of 6G wireless networks, there is growing momentum toward expanding connectivity by integrating space, air, terrestrial, and ocean networks. 6G envisions ultrareliable low-latency communication (URLLC), massive machine-type communication (mMTC), and enhanced mobile broadband (eMBB) capabilities, features that, if adapted to underwater networks, could revolutionize marine operations [6].

The convergence of 6G with underwater communications introduces new architectural paradigms, enabling seamless end-to-end connectivity from the seabed to space [7]. Despite this ambitious vision, underwater networks remain among the most neglected infrastructures in the 6G ecosystem. Therefore, in this chapter, we focus on the ocean environment of the 6G architecture to enhance communication among sensor nodes. A recent survey classifies underwater routing protocols into categories such as location-based, energy-aware, data-centric, and delay-tolerant protocols, each developed to address specific limitations of acoustic or hybrid channels [8]. However, many of these protocols are still in the early experimental stages or have been tailored to narrow use cases, lacking the scalability and adaptability needed for wide-area, heterogeneous deployments, envisioned in future 6G-enabled networks. To overcome these bottlenecks, researchers are now exploring the use of AI, machine learning (ML), and reinforcement learning (RL) to dynamically enhance routing decisions [9]. These intelligent protocols can adapt to environmental variability, optimize energy consumption, and mitigate packet loss, offering a promising direction for the next generation of underwater networks. Moreover, AUVs guided by AI-based path planning algorithms are increasingly used to complement static sensor networks, enhancing spatial coverage and data reliability in dynamic oceanographic missions [10].

7.1.1 Advancements in underwater communication mediums

Acoustic waves have long been the dominant medium for underwater communication, offering the advantage of long-distance propagation with moderate energy consumption [11]. However, acoustic waves are susceptible to multipath effects, Doppler shifts, and environmental noise, which can degrade signal quality and introduce high bit error rates (BER). Recent advances in cognitive acoustic communication have introduced AI-driven dynamic spectrum detection, which enables underwater networks to adaptively select the least congested frequencies and optimize transmission paths in real time [12]. RF communication in underwater environments has traditionally been limited due to high signal absorption in seawater. However, emerging research on very low-frequency (VLF) and extremely low-frequency (ELF) RF signals suggests that long-range underwater RF transmission is possible with specialized antennas and power-efficient modulation techniques [13]. Although RF remains unsuitable for deep-sea applications, it shows promise for short-range, high-data-rate underwater communications in freshwater lakes and coastal areas.

Optical waves offer an alternative to high-speed, short-range underwater communication, particularly in clear waters where blue-green spectrum transmission is effective [14]. Advances in adaptive optical transceivers allow real-time adjustments to transmission power based on water clarity, scattering effects, and turbulence, improving data rates and reducing power consumption. Despite its LOS requirement, optical communication is gaining traction in applications such as underwater networks and submarine-to-submarine data exchange [15]. The integration of hybrid communication models that combine acoustic, RF, and optical technologies is a promising approach to enhance underwater 6G networks [16]. By leveraging the strengths of each medium, future underwater communication systems can achieve higher data rates, lower latency, and improved energy efficiency, making them viable for large-scale deployment in scientific, industrial, and defense applications. In addition, AI-based routing schemes and cell search techniques optimize routing decisions and selection of communication cells or regions, respectively, in complex and highly dynamic underwater networks, ensuring efficient data routing.

7.1.2 Architecture of underwater communication networks

The 6G underwater architecture ranges from one-dimensional (1D) to four-dimensional (4D) structures, as can be seen in Figure 7.1, which involves seamless communication from underwater nodes to space, considering air and terrestrial users as relays. In the 1D structure, the data travel from the floor upward, that is, to the surface of the ocean. The depth in the vertical direction is the only parameter this architecture considers. The two-dimensional (2D) structure involves horizontal and vertical communications, which is complex than 1D communication networks. The main challenge in 2D networks arises from node mobility and uneven node distribution, which can lead to coverage gaps or routing voids [17]. However, the 3D underwater model is more complex, leading to a volumetric structure. To achieve volumetric monitoring of the aquatic environment, 3D-UCNs deploy sensor nodes at various depths, forming a 3D mesh topology. This configuration allows for comprehensive

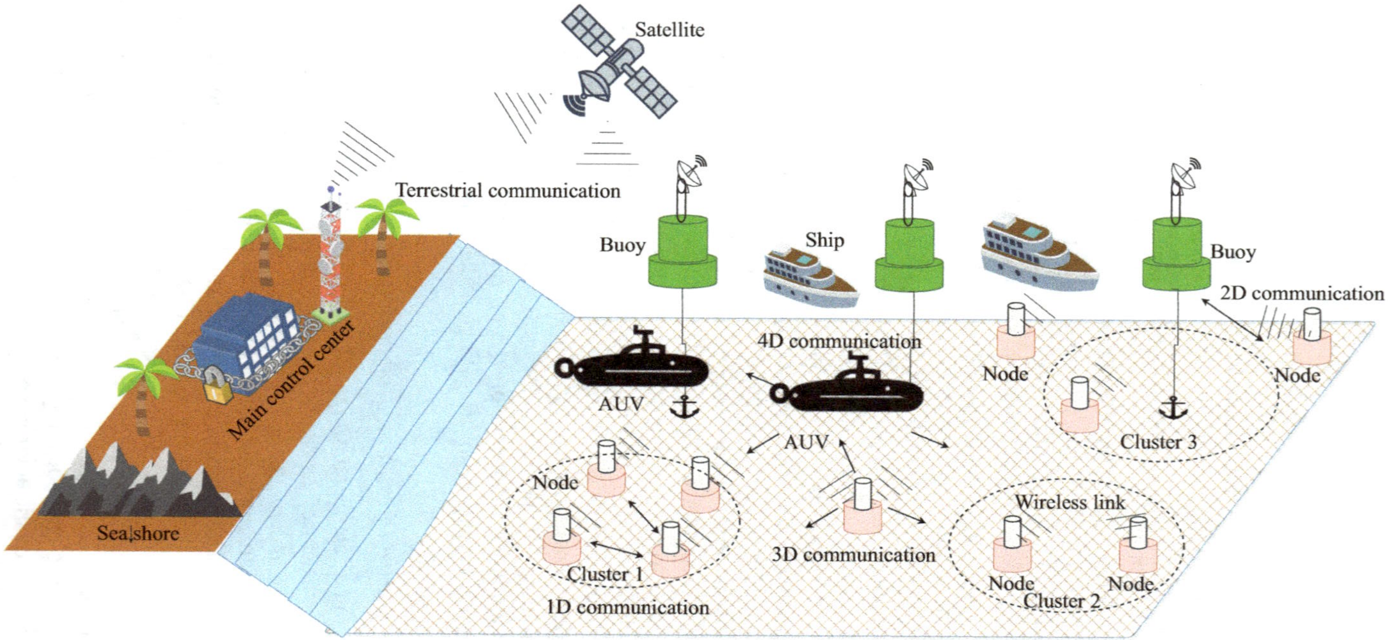

Figure 7.1 Underwater communication architecture

environmental sensing, such as oceanographic data collection, chemical pollution monitoring, and military surveillance. Nodes in a 3D architecture face challenges related to node localization, mobility due to ocean currents, and communication delay variations across depth levels. Addressing these challenges often requires clustering protocols and depth-aware routing mechanisms to optimize data delivery and energy efficiency. The 4D-UCN represents the most advanced architectural paradigm, incorporating the temporal dimension by integrating mobile elements such as AUVs, ROVs, and floating relay nodes. These nodes navigate within the water body, collecting and relaying information while dynamically updating the network topology. This network architecture is particularly suited for adaptive surveillance, sampling, disaster response, and target tracking in expansive and unknown marine territories. Routing in 4D-UCNs demands mobility-aware, predictive, and AI-augmented routing protocols to maintain performance under continuous topology changes [18].

7.1.3 Key challenges in designing routing protocols for 6G underwater networks

The design of efficient routing protocols for underwater 6G communication presents numerous challenges due to the harsh and unpredictable nature of the underwater environment, as shown in Figure 7.2. The underwater domain is significantly different from terrestrial and aerial environments, requiring specialized routing strategies that can handle dynamic topology changes, high propagation delays, limited bandwidth, and energy constraints [19]. These factors complicate the design of robust and efficient routing mechanisms that ensure reliable data delivery while optimizing network resources.

7.1.3.1 Latency and propagation delay

One of the most defining characteristics of underwater communication is its inherently high propagation delay. Unlike terrestrial systems that rely on electromagnetic waves traveling near light speed, underwater networks predominantly use acoustic waves, which propagate at approximately 1500 m/s, five orders of magnitude slower. This delay becomes further complicated by factors such as temperature gradients, salinity, and depth pressure variability, which cause non-uniform propagation speeds and introduce variable latencies. For routing protocols, this means that link state updates, synchronous communication, and feedback mechanisms become computationally expensive and often impractical. Routing strategies must, therefore, be designed with delay tolerance and asynchronous behavior in mind, particularly for critical applications like disaster response or AUV collaboration.

7.1.3.2 Bandwidth constraints and spectral scarcity

UAC offers extremely limited bandwidth, typically constrained to a few kilohertz depending on depth and range. As the frequency of the signal increases, the attenuation increases exponentially, making high-bandwidth channels impractical for long-distance communication. Routing protocols must therefore incorporate bandwidth-aware path selection, often preferring shorter, multi-hop paths over fewer long-range

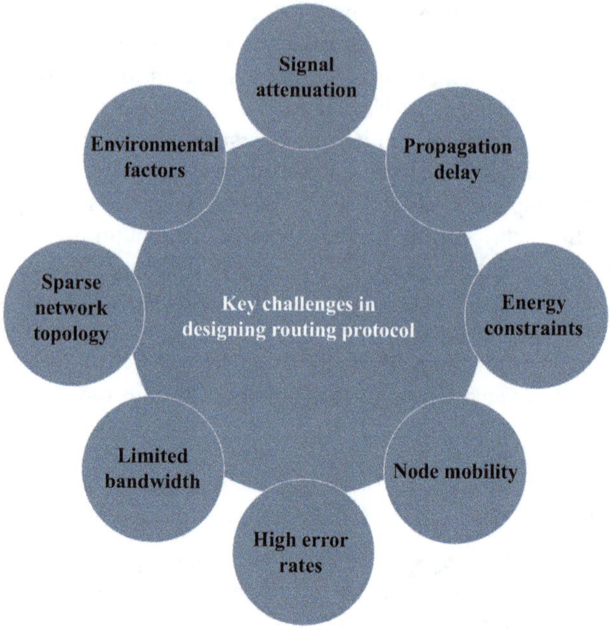

Figure 7.2 Overview of challenges in underwater communication routing

links to balance spectral efficiency. Furthermore, cross-layer optimizations are essential to prioritize routing decisions that minimize channel contention, optimize spectral reuse, and adapt transmission rates dynamically based on ambient noise levels.

7.1.3.3 Energy constraints and power asymmetry

Energy efficiency is perhaps the most critical design constraint in UCNs. Most sensor nodes are powered by non-rechargeable batteries and are often deployed in inaccessible or hazardous environments, rendering energy replenishment either cost-prohibitive or impossible. Efficient routing requires strategies such as residual energy-aware forwarder selection, duty cycle communication, and even simultaneous wireless information and power transfer (SWIPT) to harvest energy from transmission signals. Routing algorithms must be designed to extend the lifetime of the network while balancing the load distribution, avoiding overburdening critical relays, and ensuring graceful degradation of network performance over time.

7.1.3.4 Node mobility and topological dynamics

Unlike terrestrial sensor networks, underwater nodes are subject to constant motion due to ocean currents, waves, and deployment drift. This induces frequent topological changes, rendering static routing tables and deterministic paths obsolete in a matter of seconds. Mobility-aware routing must therefore rely on predictive models, including Kalman filters, recurrent neural networks (RNNs), or long-short-term memory (LSTM) architectures, to forecast node trajectories and proactively adapt routing decisions. Furthermore, protocols should enable on-demand path reconstruction with minimal signaling overhead to remain robust in dynamic environments.

7.1.3.5 Localization inaccuracy

Traditional localization mechanisms, such as global positioning system (GPS), are ineffective underwater due to the poor penetration of RF signals in saline water. Although the acoustic range and dead reckoning provide partial alternatives, they suffer from drift, environmental variability, and energy overhead. Routing strategies must be designed to function without absolute localization (e.g., depth-based or hop-count-based routing) or include confidence-aware localization modules that quantify positional uncertainty and adjust routing logic accordingly. Hybrid protocols may also benefit from partial anchor node localization supported by surface buoys.

7.1.3.6 Channel unreliability and interference

The underwater acoustic channel is plagued by multipath fading, Doppler spread, and ambient noise from marine life, vessels, and seismic activity. This results in a high BER and unpredictable link quality. Protocols must integrate channel-aware metrics such as signal-to-noise ratio (SNR), link quality estimation (LQE), and time-of-arrival (ToA) predictions into their routing decisions. Adaptive modulation and opportunistic forwarding, where multiple candidates compete to forward a packet based on channel states, can further mitigate unreliable links.

7.1.3.7 Scalability and routing overhead

As UCNs expand to support large-scale ocean monitoring missions, routing protocols must maintain scalability without incurring excessive control overhead. Traditional flooding or broadcast-based discovery mechanisms become unsustainable due to energy and bandwidth constraints. 6G-oriented routing strategies must employ hierarchical clustering, geographical region, or graph compression techniques to minimize routing overhead. Moreover, leveraging edge intelligence for local decision-making reduces the need for frequent global updates.

7.1.3.8 Time synchronization and coordination

Underwater networks struggle with precise time synchronization due to propagation delay and acoustic channel variability. However, synchronization is crucial for collision avoidance, MAC layer coordination, and cooperative routing schemes. Routing algorithms must therefore accommodate clock drift and adopt event-triggered communication models or synchronization-free scheduling to maintain robustness. Some recent solutions incorporate probabilistic synchronization models to improve coordination without incurring high synchronization costs. In general, the design of routing protocols for 6G underwater networks requires addressing these fundamental challenges while balancing trade-offs between reliability, energy efficiency, scalability, and security. Future research must focus on developing adaptive and intelligent routing mechanisms that take advantage of AI, ML, and hybrid communication techniques to overcome the inherent limitations of underwater environments.

In this chapter, we have presented a comprehensive taxonomy of routing strategies designed to meet the multifaceted challenges of underwater communication within the emerging paradigm of 6G networks. These strategies are classified into three core categories: localization-based routing, localization-free routing, and cooperative routing. Each offering a distinct solution framework adapted to the constraints of the underwater acoustic environment. Localization-based protocols, such as vector-based forwarding (VBF) and hop-by-hop (HH)-VBF, use positional or directional information to make geographically informed forwarding decisions, enabling structured data flows in sparse or mobile networks. In contrast, localization-free protocols, exemplified by depth-based routing (DBR) and energy-efficient (EE)-DBR, operate without explicit location information, instead using depth or energy metrics to achieve lightweight and energy-aware routing, a crucial factor in battery-limited underwater deployments.

Cooperative routing protocols such as cooperative energy-efficient optimal relay selection (Co-EEORS) and channel-aware cooperative routing (CACR) are based on collaborative node behavior, enabling redundancy and link reliability through relay diversity and channel awareness. Beyond these foundational strategies, the chapter also introduces cell search techniques and AI-based protocols as advanced enablers of underwater 6G. These emerging approaches further extend network adaptability, latency optimization, and energy efficiency through intelligent synchronization and autonomous decision-making. As depicted in Figure 7.3, this holistic classification not only organizes the current protocol landscape but also lays the groundwork for integrating 6G features such as AI-native routing, low-latency synchronization, and cross-domain network coordination. Together, these strategies offer a scalable and intelligent framework to support the ultra-reliable, energy-efficient, and

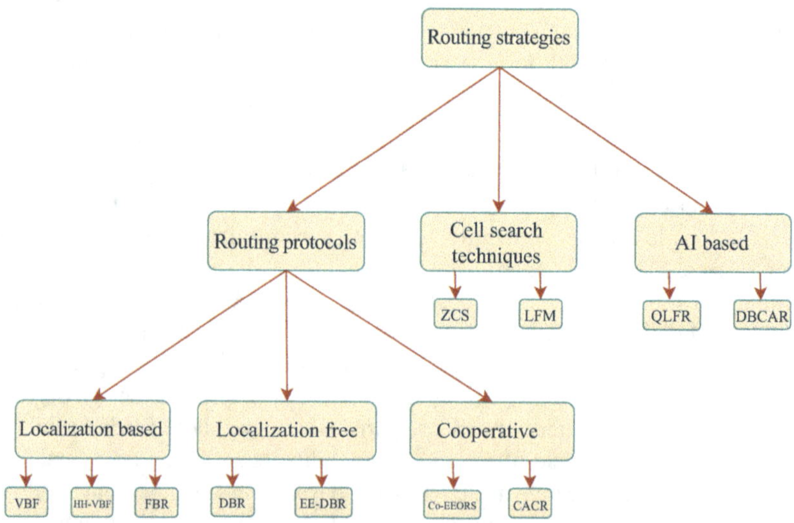

Figure 7.3 Routing strategies for underwater communication

autonomous communication capabilities envisioned in next-generation underwater communication systems.

7.2 Routing protocols for 6G underwater communication

Routing determines the optimal path for data transmission from a source to a destination node in a network. It involves selecting intermediate nodes that relay the data and ensuring that it reaches the destination reliably and efficiently. Routing in underwater networks requires specialized protocols designed to mitigate the challenges discussed previously while maintaining high reliability, low latency, and energy efficiency. With the advent of 6G underwater networks, where the scale of communication increases dramatically to support autonomous vehicle coordination, environmental monitoring, and large-scale sensor deployments, efficient routing plays a central role in ensuring network stability and performance. Thus, routing strategies must be able to address both the technical and environmental constraints unique to underwater communications. The section provides a detailed explanation of each category of underwater routing protocols.

7.2.1 Localization-based routing in underwater communication networks

Localization-based routing is a fundamental approach in UCNs where nodes rely on their positional information to make routing decisions. Like terrestrial networks, where GPS facilitates precise localization, underwater environments lack reliable global positioning due to the rapid attenuation of RF signals [20]. As a result, localization-based routing protocols in UCNs primarily depend on acoustic signals, which exhibit the unique challenges discussed earlier. Localization-based routing protocols typically function by leveraging geographic forwarding mechanisms, where nodes determine the best forwarding path based on their relative positions. These protocols can be classified into distributed and centralized localization schemes. Distributed localization approaches enable sensor nodes to estimate their positions using limited communication with neighboring nodes, whereas centralized schemes rely on surface buoys or anchor nodes to establish global positioning references. The efficiency of localization-based routing is heavily influenced by network topology, node density, and energy constraints, making it crucial to design protocols that can adapt to underwater environmental dynamics [20]. The section focuses on localization-based protocols that optimize paths for UCNs.

7.2.1.1 Vector-based forwarding (VBF)

VBF is capable of scalable, robust, and energy-efficient packet forwarding through the use of a vector-based approach. A forwarding vector is maintained between a source and a destination node, ensuring that the most optimal nodes participate in communications. This promotes less redundant information, network longevity, and high data delivery in such a complex environment [21]. Geometric forwarding is the basis of VBF, which calculates the perpendicular distance of all neighboring nodes in the communication range of the sender node. In order to send a packet, the sender selects a relay node to the farthest node on the perpendicular line. This selection is done by the sender that calculates the perpendicular distance, and if the relay node lies within a predefined threshold distance, it qualifies as a candidate forwarder. Once a relay node is selected, it becomes the new transmitter and repeats the same process, progressively relaying the packet to the final destination. This is how packets from the ocean floor reach the surface of the ocean. This distributed hop-by-hop decision-making enables the protocol to dynamically adjust to varying network densities and topology changes caused by node mobility or environmental factors. VBF demonstrates strong resilience, path reliability with a high packet delivery ratio, stable communication by reducing unnecessary retransmissions, energy efficiency by decreasing energy consumption, and increases the lifetime of the routing link, thus improving overall network performance in the ocean environment.

7.2.1.2 Hop-by-hop vector-based forwarding (HH-VBF) protocol

The HH design addresses some of the key challenges faced by VBF, particularly in sparse networks where data delivery may be hindered due to inadequate nodes within the predefined virtual pipe. In HH-VBF, each node that receives a packet computes the vector from its current position to the destination sink. This vector forms the local routing pipe for the node. This per-hop vector approach allows for greater flexibility in selecting forwarding nodes and ensures that packets can reach the sink even in sparse areas where a global vector might fail. HH-VBF works by continuously adjusting the routing decision at each hop. The nodes forward packets based on local vectors instead of a fixed global path, ensuring that each hop has its own independent routing pipe. In doing so, HH-VBF improves packet delivery rates in sparse networks and reduces sensitivity to the radius of the routing pipe. The self-adaptation algorithm of HH-VBF is different from VBF, which allows nodes to evaluate their desirability to forward packets based on their proximity to the routing vector. The desirability factor is computed like VBF, where the node calculates its distance to the vector and uses this value to assess its suitability as a forwarder.

However, HH-VBF improves upon VBF by modifying the desirability factor and incorporating redundancy control. Nodes with a smaller desirability factor will forward the packet first, ensuring that the most optimal forwarding node transmits data to the sink. This self-adaptation reduces the redundancy of packet forwarding and ensures efficient use of network resources. HH-VBF provides several advantages over its predecessor, VBF. The primary benefit of the protocol is its ability to improve packet delivery rates in sparse networks. Since it defines a per-hop routing vector, it can adaptively find alternative forwarding paths even in regions where nodes are sparsely distributed. This results in improved data delivery in environments with low node density. The self-adaptation algorithm reduces redundant packet forwarding, conserving energy, and extending the lifetime of battery-powered nodes. The ability to control redundancy makes it more energy-efficient, especially in large-scale underwater networks where energy conservation is critical.

7.2.1.3 Focused beam routing (FBR) protocol

The FBR protocol is also a location-based routing technique designed to enhance energy efficiency and reliable data delivery in UAC networks. FBR employs a beamforming inspired transmission mechanism, where packets are forwarded only through a well-defined conical region directed toward the destination. By leveraging localized power control and dynamic relay selection, FBR minimizes unnecessary transmissions, conserves node energy, and adapts to the unique challenges of the underwater communication medium. The FBR protocol operates on the basis of a multi-hop transmission strategy, where data packets are progressively relayed through intermediate nodes until they reach the final destination. The core idea behind FBR is to restrict forwarding nodes to those that fall within a focused transmission beam, effectively reducing energy wastage due to redundant transmissions. The protocol is initiated when a source node needs to send data to a destination node. The source node first transmits a request-to-send (RTS) control message at the lowest transmission power level, which includes the geographic coordinates of both the source and destination nodes. Neighboring nodes receiving this RTS compute their relative position to the source-destination axis. Only nodes that fall within a predetermined transmission cone are considered candidate forwarders. If no candidate forwarders respond, the source node incrementally increases the transmission power until a node is found within the beam. This adaptive power control ensures that the nodes only use the minimum necessary energy for forwarding, thereby optimizing power consumption. To enhance efficiency, FBR incorporates a relay selection mechanism that prioritizes nodes closer to the destination while ensuring that only the most energy-efficient paths are chosen. When multiple candidate forwarders exist within the transmission cone, the protocol selects the relay based on the following:

- *Proximity to the source-destination axis:* Nodes that are closely aligned with the optimal routing vector are preferred.
- *Transmission power constraints:* Nodes that can be reached with lower power levels are favored to conserve energy.
- *Avoiding redundant paths:* If a node hears that another forwarder has already taken the transmission role, it suppresses its response to minimize redundant forwarding.

The protocol dynamically adjusts the beam direction and transmission power to accommodate varying node densities. If a relay node is not found after reaching the maximum power level, the source node changes the transmission cone to explore alternative paths. This mechanism prevents routing voids and ensures robust data delivery even in sparse networks, and provides the following advantages.

1. *Energy efficiency:* By restricting the nodes to a focused transmission region and employing adaptive power control, the FBR minimizes energy consumption and extends the lifetime of the battery-powered underwater nodes.
2. *Reduced network congestion:* Unlike flood-based protocols, FBR significantly reduces redundant transmissions, leading to lower channel contention and minimized collisions.
3. *Scalability and robustness:* The distributed nature of FBR enables scalability in large-scale UCNs while maintaining reliable packet delivery in dynamic underwater environments. The adaptive relay selection method of the protocol ensures resilience to changes in mobility and topology.

7.2.2 Localization free protocol

Localization-free routing protocols are crucial for UCNs, where traditional localization techniques such as GPS are ineffective due to the inability of electromagnetic signals to propagate efficiently underwater [22]. Nodes do not rely on absolute location information but instead use relative information, such as the depth of the node or the number of hops to the destination, to make routing decisions. These protocols are often preferred in underwater networks, as they are less resource-intensive compared to location-based protocols, which require significant energy to continuously update and maintain location data, often relying on control messages exchanged among nodes. The section focuses on localization-free protocols that optimize routing decisions for UCNs.

7.2.2.1 Depth-based routing (DBR) protocol

DBR uses a greedy routing algorithm in which the source node selects the relay node based on the depth of all neighboring nodes. As the target of the sender is to deliver the packet to the surface of the ocean, it sends packets to nodes located at the lowest depth that are far away from the source node within its communication range. At each selected relay, the protocol is repeated until the packet arrives at its destination [23]. If a node at the higher depth receives a packet, it discards it, because the destination is at the surface of the ocean. In the literature, there have been many protocols that are the improved version of the DBR. DBR enhances network performance by allowing reliability and stability in sparse underwater networks, energy-efficient routing decisions, collisions-free routing, and channel efficiency in cognitive underwater routing schemes.

7.2.2.2 Energy-efficient depth-based routing (EEDBR)

EEDBR is a sender-based protocol that operates in two phases: the knowledge acquisition phase and the data forwarding phase [24]. During the knowledge acquisition phase, each sensor node broadcasts periodic hello packets that contain its current depth and residual energy. Neighboring nodes store this information, specifically focusing on neighbors that are at a shallower depth. This information allows each node to locally maintain a ranked list of forwarding candidates based on depth and energy. In the data-forwarding phase, a source node transmits data by including a list of eligible forwarding nodes at a lower depth, ordered by their residual energy. The top node in the list (that is, the one with the highest energy) forwards the packet immediately, while others apply a hold time before forwarding. This ensures that only one node transmits and others cancel their transmissions if they hear the same packet, thus reducing redundancy. The core of EEDBR's energy-aware mechanism lies in its adaptive hold time formula, which ensures that nodes with higher residual energy are prioritized to forward packets first. The priority term ensures that even when multiple nodes have identical residual energy, they do not forward at the same time. This value is determined by the index of the node in the ordered forwarding list. This strategy ensures that high-energy nodes take precedence in forwarding decisions, while low-energy nodes suppress their transmissions, promoting balanced energy usage across the network.

Energy suppression can affect the packet delivery ratio in critical applications; EEDBR introduces a delivery ratio-sensitive suppression mechanism. This is accomplished by dynamically adjusting the suppression probability based on feedback from the sink node, thus balancing delivery reliability and energy conservation based on real-time network conditions. Upon receiving a packet, each node checks its presence in the forwarding candidate list. If the node is not on the list, the packet is dropped immediately to reduce redundancy. When the node is on the list but not the highest ranked, it computes a holding time based on its residual energy and priority index. During this holding interval, if the same packet is overheard, it cancels its transmission to conserve energy. Otherwise, it proceeds with forwarding once the timer expires. The nodes at the top of the list, those with the highest energy, are allowed to transmit the packet without delay, ensuring that the most capable node performs the forwarding task. The balance between localized decision making and global network performance adaptation exemplifies the next generation of intelligent routing design for underwater acoustic sensor networks. Moreover, the sender-based selection mechanism is more effective than receiver-based approaches, as it allows fine-grained control over forwarding candidates. The suppression feedback loop enables it to adapt to both energy-sensitive and reliability-critical scenarios, making it suitable for various underwater applications.

7.2.3 Cooperative routing

Cooperative routing is a paradigm in UAC in which the nodes collaborate to enhance data transmission performance by sharing their communication resources [25]. In multi-hop forwarding, where data is simply relayed from node to node along a predefined path, cooperative routing exploits the broadcast nature of the underwater acoustic channel by allowing multiple relay nodes to jointly assist in forwarding packets. This spatial diversity technique, known as cooperative diversity, significantly improves link reliability, energy efficiency, and overall network throughput, which are critical to the performance of energy constrained and delay-sensitive UCN. The concept of cooperative routing has evolved into a powerful non-cross-layer strategy that integrates the mechanisms of the physical layer with the routing decisions of the network layer [26]. In a typical cooperative transmission model, in addition to the direct communication link between a transmitter and a receiver, multiple nearby nodes, referred to as relays, participate in forwarding the data. Each relay node independently receives the signal and then re-transmits it to the destination, where the receiver either selects the best quality signal or performs signal combining to enhance decoding accuracy. This dual path approach mitigates underwater channel impairments such as multi-path fading, high latency, and signal attenuation, which are especially pronounced in underwater environments. Relay node selection is critical to the performance of the cooperative routing protocol. An optimal set of relays can significantly increase diversity gain and reliable data delivery, but excessive use of relays can also introduce network overhead and interference, particularly in densely deployed networks. Therefore, many modern cooperative routing protocols incorporate relay selection strategies that balance performance and energy consumption. These strategies often consider factors such as residual energy, channel quality, and node proximity to destination. Cooperative routing protocols have gained increasing relevance in 6G underwater networks, where reliability, adaptability, and energy efficiency must be jointly optimized across layers of the protocol stack.

7.2.3.1 Cooperative energy-efficient optimal relay selection protocol (Co-EEORS)

The Co-EEORS protocol introduces an advanced cooperative routing mechanism that integrates energy efficiency, optimal node selection, and link quality awareness to ensure reliable data delivery [27]. This protocol builds on the limitations of previous cooperative schemes by avoiding the reliance on full location coordinates and synchronization while still achieving high packet delivery, network lifetime, and resilience to underwater acoustic channel impairments. At its core, Co-EEORS is designed to enhance routing decisions through a dual-parameter node selection strategy. Unlike conventional schemes that rely on implicit broadcast or multiple relays, Co-EEORS explicitly identifies a single relay and destination node for each transmission. This avoids unnecessary contention and interference by suppressing redundant forwarding from other neighbors. Co-EEORS leverages a decode-and-forward strategy. The source node transmits data to both the relay and the destination node. If the effective SNR exceeds a predefined threshold, the destination node confirms successful reception by sending an acknowledgment (ACK) to the source. Otherwise, it requests retransmission, thus enhancing reliability even under poor channel conditions. Through the integration of energy-aware node selection, cooperative signal forwarding, and adaptive SNR-based reception, Co-EEORS provides a resilient and scalable solution for underwater routing in future 6G underwater networks.

7.2.3.2 Channel-aware cooperative routing (CACR) protocol

The CACR protocol introduces a dynamic and adaptive routing architecture for UCN. Recognizing the limitations imposed by underwater acoustic channels, CACR implements a cross-layer framework that tightly integrates the physical and network layers [28]. The protocol aims to optimize routing decisions by assessing both channel quality and topological position, thus improving energy efficiency, packet delivery ratio, and latency. When a source node has data to transmit, it broadcasts the RTS packet. Neighbor nodes respond with clear-to-send (CTS) only if they meet dual conditions: (i) their SNR exceeds a quality threshold, and (ii) their hop count is less than or equal to the source. This ensures that data are forwarded over high-quality and progressive links. From the responding nodes, the source selects the routing relays based on either maximum SNR or minimum ToA, depending on whether throughput or latency is prioritized. Simultaneously, cooperative relays are selected from the intersection of nodes that are neighbors of both the source and the routing relays, with preference given to nodes having the lowest ToA to the routing relays, minimizing delay spread [28]. Once these relays are selected, the source transmits the data to both nodes. The routing relays perform joint signal combining using the original and cooperative copies, improving signal fidelity through diversity gain. If the routing relay successfully decodes the packet, it broadcasts an ACK, halts further transmission from the cooperative relay. In case of failure, a negative ACK is sent, triggering retransmission either by relay or source, depending on the number of previous attempts. This process ensures that retransmission is minimal and only occurs when strictly necessary, conserving energy and reducing medium access contention. As we discussed throughout the chapter, underwater communication presents unique challenges; localization-free routing protocols, such as DBR, and cooperative routing protocols such as Co-EEORS provide fundamental solutions for improving network performance under the constraints of dynamic topologies, energy limitations, and environmental factors such as signal attenuation and long propagation delays. However, to further optimize routing and ensure robust communication in these complex environments, techniques that can enhance resource management and routing adaptability are needed.

7.3 Cell search techniques

The routing protocols discussed in the previous section may fail in high-state underwater environment when network becomes sparse due to ocean currents and nodes find it difficult to deliver data upward. To overcome these issues, cell search techniques offer promising solutions by defining cell-based regions to ensure optimal routing paths in highly challenging underwater environment. The section discusses two state-of-the-art routing schemes that falls under this category.

7.3.1 ZCS-based cell search technique (ZCS-CST)

Historically, Zadoff–Chu sequences (ZCS) have been extensively used in LTE and other wireless systems for synchronization due to their efficient signal detection and low cross-correlation properties. In the context of UAC, the ZCS-CST works by first using the linear frequency modulation (LFM) waveform for coarse synchronization. The sequence is robust in identifying carrier frequency offsets (CFO) and Doppler shifts, which makes it highly suitable for systems like UAC where signal degradation due to Doppler shifts and multi-path effects is significant [29]. ZCS-based synchronization signals are then transmitted for fine synchronization. The fine synchronization steps ensure that the system can overcome the Doppler-induced frequency shifts that typically occur in underwater environments. The LFM waveform is first used for coarse timing acquisition for preamble detection. The LFM waveform is ideal for this task due to its Doppler insensitive nature, ensuring reliability in timing acquisition even under mobility. The received signal is correlated with an LFM signal to perform coarse CFO estimation and initial synchronization. The Doppler shift is compensated using the resampling process with the estimated Doppler scale factor. After synchronization, cell searching is performed to estimate the cell identity (CID) and fine-tuned timing. Channel estimation is carried out using the least squares (LS) method with pilot signals, which are interleaved with data symbols [29].

7.3.2 LFM-based cell search technique (LFM-CST)

The LFM-CST is designed for underwater environments in which Doppler shift becomes a significant challenge. LFM-CST integrates signal detection and cell searching into a single, joint process, allowing for a more efficient approach in dynamic networks [29]. LFM waveforms are inherently Doppler insensitive, making them ideal for systems where mobility is needed, such as that of AUVs, thus introduce unpredictable frequency shifts. LFM waveforms have historically been used in radar communication systems for applications where mobility is a concern because of their ability to provide accurate synchronization even with significant motion-induced frequency changes. In the UAC system, LFM-CST uses LFM waveforms for both preamble detection and cell search. The process begins by using LFM signals to estimate the coarse synchronization parameters, followed by cell ID detection in the same step. This eliminates the need for a separate synchronization stage, making it faster and more efficient, especially in mobile environments. The technique compensates for the Doppler shift by estimating the frequency offset and adjusting the received signal accordingly [29]. Coarse synchronization is obtained through the correlation of the received signal with the LFM signal. The Doppler shift is compensated through the frequency offset estimation process.

7.4 AI-based routing protocols

The section discusses the intelligent routing protocols for underwater environment. These are called "intelligent" because they adapt according to the environment, thus trying to overcome all the existing issues in the previously discussed routing methods.

7.4.1 Q-learning-based localization-free anypath routing protocol (QLFR)

In QLFR, each node in the network calculates its Q-value based on two primary reward functions: one related to depth and the other related to residual energy. The depth-based reward encourages forwarding toward nodes with smaller depths, indicating proximity to the sink. The energy-based reward ensures that nodes with higher residual energy are prioritized for data transmission, preventing nodes from overloading and failing prematurely. The Q value for a node is computed using the Q-learning algorithm, which updates the node's state action pair over time to maximize the long-term reward. When a node needs to forward a packet, it consults its neighbors and computes the Q-values for each candidate. The node then forms a priority list based on these values, which is included in the packet header. The nodes on the priority list are ranked in descending order of their Q-values. This allows the system to make globally optimal routing decisions that take into account the overall energy consumption of the network and the delay incurred during data transmission.

7.4.2 Depth-based channel-aware routing (DBCAR) protocol

DBCAR is a combination of DBR and channel-aware routing (CAR). As mentioned earlier, there are various modifications of DBR in the literature. One of those is to improve channel availability of underwater sensor nodes because of their limited bandwidth in UAC networks. Therefore, DBCAR considers both greedy forwarding- and channel availability-based communication. Only those relays are selected in this protocol, which have a common channel features with the source node. DBCAR thus improves the packet delivery ratio, latency, energy efficiency, number of hops, channel parameters (such as SNR, BER), and network overhead by taking intelligent decisions in forwarding packets to the destination.

7.5 Future strategies for 6G underwater communication

The advent of 6G communication introduces a transformative phase for UCNs, demanding not just incremental improvements, but a comprehensive reimagining of the communication architecture. Building upon previously discussed routing strategies, this section delves into forward-looking solutions designed to overcome the multifaceted challenges of the underwater domain. These strategies integrate emerging paradigms, sustainable energy mechanisms, and inter-domain cross connectivity to meet the high-reliability, ultralow latency, and energy efficiency benchmarks of 6G underwater networks.

7.5.1 Integration of non-terrestrial networks (NTNs)

Future underwater communication infrastructures will increasingly rely on the integration of NTNs, which include satellite constellations, high-altitude platforms (HAPs), UAVs, and AUVs. This vertical extension allows seamless global connectivity between subsea nodes and terrestrial cloud systems, highlight the pivotal role of AUVs as mobile relay agents that bridge gaps between isolated underwater clusters and long-range communication backbones. These AUVs operate dynamically, performing optimized path maneuvers to maintain LoS acoustic or optical links with sensor nodes and uplinking data to satellite gateways, enabling persistent data collection in vast marine territories [30]. From a systems engineering perspective, NTNs enhance the scalability and redundancy of UCNs by incorporating spatial and temporal diversity into the communication channel. This multi-layered approach supports hybrid routing architectures where data hop vertically across media (acoustic, RF, optical) and horizontally across domains (oceanic, aerial, and space). Inclusion of high-altitude and satellite links mitigates geographic isolation, reduces latency through selective relay engagement, and supports network slicing for application-specific quality of service guarantees [30].

7.5.2 AI-native networks and cognitive intelligence

6G underwater networks must inherently possess the ability to adapt to volatile channel characteristics, mobile topologies, and energy limitations. Native AI routing protocols that embed ML models directly into network nodes offer the computational agility needed for autonomous decision-making. The cognitive intelligence routing protocol (CIRP) integrates distributed agents capable of learning spatiotemporal patterns of link degradation, node failures, and propagation anomalies. These agents leverage RL frameworks to evaluate and refine routing policies based on live environmental feedback, optimizing path reliability, and minimizing retransmission costs [31]. Beyond reactive adaptation, cognitive intelligence enables predictive analytics through LSTM models and federated learning. Such models can anticipate node drift patterns caused by underwater currents, forecast acoustic link disruptions, and preselect optimal relays based on residual energy and mobility vectors [31].

7.5.3 Sustainable energy harvesting and SWIPT

Energy autonomy is a cornerstone of resilient 6G underwater networks. Traditional battery-powered nodes, limited by deployment depth and accessibility, are insufficient for long-term missions. Categorize underwater energy harvesting into mechanical, thermal, chemical, and radiative sources. Mechanical systems utilize piezoelectric elements to convert waves or turbulence vibrations into electrical energy [32]. Thermal systems exploit temperature gradients via thermoelectric generators (TEGs), often deployed near hydrothermal vents. SWIPT emerges as a dual function solution, allowing nodes to harvest energy while receiving data through the same acoustic or RF waveform. The feasibility of SWIPT in salt water environments using resonant magnetic coupling achieves optimal alignment for mid-range power delivery. This hybrid energy paradigm supports persistent deployments of AUVs and underwater sensor relays, significantly extending operational lifetimes.

7.5.4 Intelligent reflecting surfaces (IRS)

IRS represent a paradigm shift in underwater signal management, enabling environmental reconfiguration to control wave propagation. An IRS comprises a matrix of tunable acoustic or electromagnetic meta-elements capable of altering the amplitude, phase, and direction of incident signals. However, IRSs are deployed on mobile platforms, like AUVs or tethered to seabed nodes, enhance signal focusing and reduce multipath interference. The implementation of IRS in underwater networks introduces a new degree of freedom in the control of the physical layer. It enables passive relay architectures that do not consume active power during reflection, thus conserving energy. IRS-aided communication improves channel reciprocity and link robustness under Doppler spread, which is a significant impairment in underwater networks.

7.5.5 Digital twins and predictive simulation

Digital twin technology in underwater communication involves the creation of high-fidelity virtual replicas of physical UCN deployments. These twins are continuously synchronized with real-world systems using telemetry data, sensor feedback, and environmental metrics. Digital twins can simulate network behavior under varying configurations, enabling predictive maintenance, performance benchmarking, and pre-deployment validation of routing protocols. From a systems perspective, digital twins support adaptive control through closed-loop feedback mechanisms. Anomalies such as excessive energy consumption, packet drops, or routing voids detected in the virtual model can trigger reconfiguration commands in the physical network.

These simulations aid in designing energy efficient, QoS compliant routing policies tailored to dynamic ocean conditions.

7.5.6 Secure and decentralized coordination via blockchain

Security and data integrity are vital in underwater communication networks, especially when involving multi-agency collaboration or sensitive data. Blockchain technology provides a decentralized and tamper-proof framework to ensure trust among nodes without relying on centralized infrastructure. Integrating blockchain with cognitive routing protocols maintains immutable logs of route history, energy contributions, and transmission records. Technically, blockchain implementations in UCNs require lightweight consensus mechanisms such as proof-of-elapsed time (PoET) or practical Byzantine fault tolerance (PBFT) to accommodate limited bandwidth and energy. Smart contracts can automate policy enforcement for route selection, access control, and node authentication.

7.6 Conclusion

In this chapter, we explored a range of routing strategies designed to address the unique challenges of underwater communication within the context of emerging 6G communication technologies. We delved into localization-based, localization-free, and cooperative routing protocols, each offered different advantages and techniques to enhance the efficiency and reliability of data transmission in underwater environments. The chapter then discussed artificial intelligence-based routing schemes and cell search techniques that enhance performance of underwater networks. Finally, future strategies for 6G underwater environments were discussed which may include the use of quantum communication, the incorporation of artificial intelligence and optimization models, and the integration of various environments like terrestrial, space, air with underwater networks.

References

[1] Junhai Luo, Yanping Chen, Man Wu, and Yang Yang, "A survey of routing protocols for underwater wireless sensor networks," *IEEE Communications Surveys & Tutorials*, vol. 23, no. 1, pp. 137–160, 2021.

[2] Khalid Awan, Peer Azmat Shah, Khalid Iqbal, Saira Gilani, Waqas Ahmad, and Yunyoung Nam, "Underwater wireless sensor networks: A review of recent issues and challenges," *Wireless Communications and Mobile Computing*, vol. 2019, pp. 1–20, 2019.

[3] Ian Akyildiz, Dario Pompili, and Tommaso Melodia, "Underwater acoustic sensor networks: Research challenges," *Ad Hoc Networks*, vol. 3, pp. 257–279, 2005.

[4] Emad Felemban, Faisal Shaikh, Umair Qureshi, Adil Sheikh, and Saad Qaisar, "Underwater sensor network applications: A comprehensive survey,"

International Journal of Distributed Sensor Networks, vol. 2015, pp. 1–14, 2015.

[5] Hashim Khan, Syed Ali Hassan, and Haejoon Jung, "On underwater wireless sensor networks routing protocols: a review," *IEEE Sensors Journal*, vol. 20, no. 18, pp. 10371–10386, 2020.

[6] Gurkan Tuna and Vehbi Gungor, "A survey on deployment techniques, localization algorithms, and research challenges for underwater acoustic sensor networks," *International Journal of Communication Systems*, vol. 30, pp. e3350, 2017.

[7] Camila Gussen, Paulo Diniz, Marcello Campos, Wallace Martins, Felipe Costa, and Jonathan Gois, "A survey of underwater wireless communication technologies," *Journal of Communication and Information Systems*, vol. 31, pp. 242–255, 2016.

[8] Zaihan Jiang, "Underwater acoustic networks–issues and solutions," *International journal of intelligent control and systems*, vol. 13, no.3, pp. 152–161, 2008.

[9] Hemani Kaushal and Georges Kaddoum, "Underwater optical wireless communication," *IEEE Access*, vol. 4, pp. 1518–1547, 2016.

[10] John Heidemann, Milica Stojanovic, and Michele Zorzi, "Underwater sensor networks: applications, advances and challenges," *Philosophical transactions. Series A, Mathematical, Physical, and Engineering Sciences*, vol. 370, pp. 158–175, 2012.

[11] Salva Climent, Antonio Sanchez, Juan Capella, Nirvana Meratnia, and Juan Serrano Martín, "Underwater acoustic wireless sensor networks: Advances and future trends in physical, mac and routing layers," *Sensors (Basel, Switzerland)*, vol. 14, pp. 795–833, 2014.

[12] Mustafa Shakir, Mahmood Khan, Izhar ul Haq, and Shahzad Malik, "Design of underwater sensor networks for water quality monitoring and surveillance," *World Applied Sciences Journal*, vol. 17, no. 11, pp. 1441–1444, 2012.

[13] Yu Chen, Wei Yan Pan, Huai Yun Peng, and Hong Qi Zhang, "The ELF/VLF field at the depth of submarine excited by satellite electropult," in *Proceedings of the 9th International Symposium on Antennas, Propagation and EM Theory*, 2010, pp. 505–508.

[14] Kavati Mamatha, K.B.N.S.K. Chaitanya, Shashank Kumar, and A. Arockia Bazil Raj, "Underwater wireless optical communication – a review," in *2021 International Conference on Smart Generation Computing, Communication and Networking (SMART GENCON)*, 2021, pp. 1–5.

[15] Zhaoquan Zeng, Shu Fu, Huihui Zhang, Yuhan Dong, and Julian Cheng, "A survey of underwater optical wireless communications," *IEEE Communications Surveys & Tutorials*, vol. 19, no. 1, pp. 204–238, 2017.

[16] Nasir Saeed, Abdulkadir Çelik, Tareq Al-Naffouri, and Mohamed-Slim Alouini, "Underwater optical wireless communications, networking, and localization: A survey," *Ad Hoc Networks*, vol. 94, pp. 101935, 2019.

[17] Raja Jurdak, Antonio Ruzzelli, and Gregory O'Hare, "Design considerations for deploying underwater sensor networks," *SENSORCOMM '07:*

Proceedings of the 2007 International Conference on Sensor Technologies and Applications, IEEE Computer Society, USA. 2007, pp. 227–232.

[18] Vinayak Khajuria and Manjot Kaur, "Underwater wireless sensor network: architecture, applications and challenges," in *2018 2nd International Conference on Trends in Electronics and Informatics (ICOEI)*, 2018, pp. 939–944.

[19] Guang Yang, Lie Dai, Guannan Si, Shuxin Wang, and Shouqiang Wang, "Challenges and security issues in underwater wireless sensor networks," *Procedia Computer Science*, vol. 147, pp. 210–216, 2019, 2018 International Conference on Identification, Information and Knowledge in the Internet of Things.

[20] Vijay Chandrasekhar, Winston KG Seah, Yoo Sang Choo, and How Voon Ee, "Localization in underwater sensor networks: survey and challenges," in *Proceedings of the 1st International Workshop on Underwater Networks*, New York, NY, 2006, WUWNet 2006, pp. 33–40, Association for Computing Machinery.

[21] Peng Xie, Jun-Hong Cui, and Li Lao, "VBF: vector-based forwarding protocol for underwater sensor networks," In: Boavida Fernando, Plagemann Thomas, Stiller Burkhard, Westphal Cedric and Monteiro Edmundo, (eds.). *Networking Technologies, Services, and Protocols; Performance of Computer and Communication Networks; Mobile and Wireless Communications Systems*, Springer, Berlin, Heidelberg. 2006, vol. 3976, pp. 1216–1221. https://doi.org/10.1007/11753810_111

[22] Muhammad Khalid, Zahid Ullah, Naveed Ahmad, Awais Adnan, Waqar Khalid, and Ahsan Ashfaq, "Comparison of localization free routing protocols in underwater wireless sensor networks," *International Journal of Advanced Computer Science and Applications(IJACSA)*, vol. 8, pp. 408–414, 2017.

[23] Hai Yan, Zhijie Jerry Shi and Jun-Hong Cui, "DBR: Depth-Based Routing for Underwater Sensor Networks", In: Das Amitabha, Pung Hung Keng, Lee Francis Bu Sung, Wong Lawrence Wai Choong (eds.). *NETWORKING 2008 Ad Hoc and Sensor Networks, Wireless Networks, Next Generation Internet*, Springer, Berlin, Heidelberg. pp. 72–86, 2008. https://doi.org/10.1007/978-3-540-79549-0_7.

[24] Abdul Wahid, Sungwon Lee, Hong-Jong Jeong, and Dongkyun Kim, *EEDBR: energy-efficient depth-based routing protocol for Underwater wireless sensor networks*, In: Kim, Th., Adeli, H., Robles, R.J., Balitanas, M. (eds) *Advanced Computer Science and Information Technology AST*, 2011. Communications in Computer and Information Science, Springer, Berlin, Heidelberg, vol. 195, pp. 223–234, 2011. https://doi.org/10.1007/978-3-642-24267-0_27.

[25] Seyed Mohammad Ghoreyshi, Alireza Shahrabi, and Tuleen Boutaleb, "A novel cooperative opportunistic routing scheme for underwater sensor networks," *Sensors (Basel, Switzerland)*, vol. 16, 2016.

[26] Muhammad Shoaib Khan, Andrea Petroni, and Mauro Biagi, "Cooperative communication based protocols for underwater wireless sensors networks: A review," *Sensors (Basel, Switzerland)*, vol. 24, 2024.

[27] Anwar Khan, Ihsan Ali, Atiq Ur Rahman, Muhammad Imran, Fazal-E-Amin, and Hasan Mahmood, "Co-EEORS: cooperative energy efficient optimal relay selection protocol for underwater wireless sensor networks," *IEEE Access*, vol. 6, pp. 28777–28789, 2018.

[28] Hoa Tran-Dang and Dong-Seong Kim, "Channel-aware cooperative routing in underwater acoustic sensor networks," *Journal of Communications and Networks*, vol. 21, no. 1, pp. 33–44, 2019.

[29] Muhammad Asim, Rothna Pec, Tae Ho Im, and Yong Soo Cho, "Cell search techniques for underwater acoustic cellular systems," *IEEE Access*, vol. 7, pp. 106019–106033, 2019.

[30] Boyu Diao, Chao Li, Qi Wang, Zhulin An, and Yongjun Xu, "A channel-aware routing protocol with nearest neighbor regression for underwater sensor networks," *ArXiv*, vol. abs/2108.05057, 2021. https://api.semanticscholar.org/CorpusID:236976128

[31] Kiran Saleem, Lei Wang, and Ahmad Almogren, "Cognitive intelligence routing protocol for disaster management and underwater communication system," *Scientific Reports*, vol. 15, pp. 10004. https://doi.org/10.1038/s41598-025-94561-6, 2025.

[32] Olumide Alamu, Thomas O. Olwal, and Karim Djouani, "Energy harvesting techniques for sustainable underwater wireless communication networks: A review," *e-Prime – Advances in Electrical Engineering, Electronics and Energy*, vol. 5, pp. 1–19, 2023. https://doi.org/10.1016/j.prime.2023.100265

Chapter 8

Energy-efficient routing protocols in underwater acoustic networks

Shazib Ali[1], Malik Noraiz Masood Khan[1], Afnan Mirza[1] and Menaa Nawaz[2]

The underwater networks are very important for maintaining balanced ocean life, helping scientists do research, and finding valuable natural resources. However, stable communication underwater is hard because of the limits of underwater acoustic networks (UANs). These include slow signal travel, limited data transfer, and weak signals. These problems make it difficult for the network to work well, save energy, and remain reliable. This chapter looks at the challenge of sending data in a way that uses less energy in UANs. It starts by explaining common methods of data transfer and why they do not work well in deep-sea environments. The chapter then discusses the growing use of moving smart machines, such as flying drones, underwater robots, and self-driving underwater vehicles, as powerful tools to improve communication underwater. These smart and mobile machines can change routes when needed, making data sharing faster, using less energy, and better adjusting to changes in the ocean. Adding them to the UAN creates a new type of network that focuses on energy saving, lasting longer, and performing better. With their intelligent movement and control, these machines offer new ways to watch the ocean, discover new things, and better manage sea resources. In the end, this chapter shares useful ideas on how these new tools can help build stronger and smarter underwater networks ready for the future of ocean exploration and care.

8.1 Introduction

The world's oceans, comprising approximately 96.5% of the Earth's water, are a vital part of the planet's ecosystem. Exploring and monitoring oceans are crucial for many purposes, including studying climate, managing resources, protecting the environment, and preventing disasters. However, the immense size and complexity of the underwater environment have historically impeded comprehensive investigation,

[1]School of Electrical Engineering and Computer Science (SEECS), National University of Sciences and Technology (NUST), Pakistan
[2]Department of Biomedical Engineering, Air University, Pakistan

largely due to limitations in underwater communication technologies. UANs have become a crucial solution to overcome these challenges. UANs help collect underwater data quickly and reliably by connecting sensor devices spread across different areas. These networks offer great possibilities for a variety of applications [1], such as monitoring marine life, checking water conditions, studying ocean events, identifying and monitoring pollutants (such as oil spills and chemical waste), improving security, monitoring coastal erosion, aiding in disaster response, and helping locate and harvest underwater resources (such as minerals and hydrocarbons).

A conventional UAN architecture comprises the following key components. Sensor nodes are deployed throughout the underwater environment to collect data about various parameters of interest. In underwater sensor networks, sink nodes hold a crucial role by gathering information from the surrounding nodes deployed in the environment. These nodes might either stay in one place or move around, taking the form of autonomous underwater vehicles (AUVs). They act as crucial bridges in the communication framework, facilitating the efficient transfer of the data collected by the sensors to the base station. The presence of mobile sinks, such as AUVs, can significantly improve how well the network adjusts to changing conditions, especially in the constantly shifting of underwater spaces. The data is forwarded to the base station from the sink nodes usually present on the surface or in a remote location, where it gets analyzed and stored. Data transmission typically follows a multi-hop routing approach where information is relayed through intermediate sensors to reach the sink.

In UANs, energy consumption differs significantly from terrestrial wireless sensor networks. Energy is mainly expended on data processing and radio communication in terrestrial systems, whereas UANs face some additional challenges due to the uniqueness of underwater acoustic channels, such as high signal attenuation, limited bandwidth, and propagation delays. These challenges require specialized approaches in the design of communication protocols and the development of network architectures to ensure efficient and reliable underwater operations. The main parameters that need to be optimized in such challenging environments to maintain stable communication between sensor nodes are propagation delay, energy efficiency, bandwidth, signal transmission and reception, speed, and node localization. These parameters are important when nodes in the deep ocean try to send data to the surface of the ocean, thereby establishing an optimal routing path. Acoustic waves, which serve as the primary medium for underwater communication, exhibit a considerably slower propagation speed compared to radio waves in the air. This result leads to significant propagation delays, which adversely affect the efficiency of real-time communication in underwater networks.

Sensor nodes, including AUVs in UANs, are generally powered by batteries with constrained energy capacities. Due to the challenges associated with accessing the underwater environment, battery replacement or recharging is highly impractical, requiring energy-efficient operation strategies to extend the overall useful life of the network. The bandwidth available for underwater acoustic communication is severely limited, thereby restricting data transmission rates and constraining the overall network capacity. Sound signals in water lose strength and get messed up because of things such as absorption, scattering, and bouncing off surfaces. This makes it harder for signals to travel far and causes more problems in communication. Water currents and various environmental factors can induce the mobility of sensor nodes, leading to a dynamic and time-varying network topology. Consequently, routing protocols must be designed to adapt to these changes to maintain reliable network connectivity. In addition, the rapid attenuation of radio waves in aquatic environments significantly limits the effectiveness of the global positioning system (GPS) for the localization of nodes, making it challenging to accurately determine the position of underwater sensor nodes.

This chapter addresses the critical challenges of energy-efficient routing in UANs and provides possible solutions that can open the door for researchers to build energy-efficient, stable underwater communications. The primary focus of this chapter is to answer this research question, guiding this investigation. How can energy-efficient routing protocols be effectively designed and optimized to extend the lifetime of the network, improve data delivery reliability, and adapt to the dynamic and harsh conditions of underwater environments while utilizing acoustic bands? The rest of this chapter is outlined below. Section 1.2 discusses various schemes in the literature that provide energy-efficient underwater routing protocols. Section 1.3 discusses recent advances to improve network connectivity through energy-based routing protocols, their challenges and possible solutions, and Section 1.4 concludes the chapter.

8.2 Related work

Several survey papers have previously examined routing protocols for UANs; however, these studies exhibit certain limitations. Given the complexity of UANs and the critical need for energy efficiency, Khan *et al.* [2] proposed a routing protocol that implements a different approach to tackle this challenge. They developed a method called the energy-efficient cluster routing protocol based on arithmetic progression (EECRAP). The method was built on making the network structure more efficient. EECRAP mainly focused on solving the problem of uneven energy use among the network sensor nodes, often leading to the network stopping work earlier than expected. Using a special way of grouping sensors based on arithmetic progression, EECRAP helps spread energy use more evenly. This means that the network can last longer and work more reliably. The method not only prevents some sensors from running out of power too quickly but also makes sure that data are sent more smoothly in the underwater environment, where saving energy is very important.

The fundamental mechanism of their protocol was based on a unique division of the underwater environment. Rather than utilizing uniform layers, EECRAP organized the network into asymmetric horizontal layers, with the layer thickness progressively increasing with depth according to an arithmetic progression. Their design considered that different depths in the water require different amounts of energy due to the way acoustic signals travel. Within these layers, groups of sensors, or clusters, were formed. The size of each cluster varied, and it was carefully decided based on several factors, such as how much energy is left in the sensors, how deep they are, and where they are located within the overall layered structure. Their approach helps balance energy use across the network, preventing areas from using too much energy too quickly and making the network last longer. Likewise, Gul *et al.* [3] introduced an approach, the energy-efficient regional-based cooperative routing (EERBCR) protocol, which used mobile sink nodes.

They recognized that sensor nodes use energy even when idle and do not do anything useful. To address this issue, EERBCR divides the network into different regions. Mobile sink nodes, which act as data collection points, follow specific paths and visit each region one by one, helping to save energy and improve efficiency. Sensor nodes only become active when a mobile sink node is nearby, which means that they remain in low-power sleep mode when not in use. This reduces energy waste because the nodes do not constantly consume power. The mobile sink nodes move through the network, gathering data only when they come near sensing devices. This technique enhances system efficiency by allowing sensing devices to transmit data only when necessary, which reduces overall energy consumption and lengthens the operational time of the network power sources, thereby increasing network stability and endurance. The EERBCR protocol aims to extend the life of the network to continuous data delivery by addressing transmission delays and energy savings.

EERBCR achieves this by managing the node operation times and optimizing the movement patterns of data collection points, aiming to maintain a uniform energy consumption throughout the UAN. Another approach designed to improve energy efficiency, called energy-efficient guiding-network-based routing (EEGNBR) [4], introduced a digital guidance framework to direct information packets to surface receivers. Within this system, each sensing device received a transmission count that indicated the lowest number of transmissions required to reach a data collection node. These transmission count details were distributed across the entire network using targeted management messages. The resulting dissemination process effectively constructed a logical topology, enabling optimized and energy-efficient routing paths in the underwater environment. A notable advancement introduced by the EEGNBR protocol was its concurrent forwarding mechanism.

Unlike conventional routing strategies that utilize a waiting-based approach, where intermediate nodes delay packet transmission for a predefined interval, resulting in increased latency, EEGNBR enabled multiple neighboring nodes to engage in packet forwarding simultaneously. This parallel transmission strategy effectively minimized forwarding delays and improved the overall reliability of data delivery within the underwater network. The performance of EEGNBR was further optimized by its integrated delay model, which quantifies the end-to-end delay; the packet retention delay had a substantial impact on conventional protocols. The mechanism of EEGNBR concurrent forwarding significantly mitigated this factor, thereby reducing overall latency and improving data delivery efficiency. Their simulations demonstrated that EEGNBR outperforms state-of-the-art protocols in terms of network delay. Although achieving this reduction in delay, EEGNBR maintained comparable or even better performance in energy consumption and packet delivery ratio.

8.3 Recent advancements in energy-efficient routing protocols

This section aims to address underwater acoustic limitations by providing a focused, up-to-date, and comprehensive analysis of energy-efficient routing protocols for

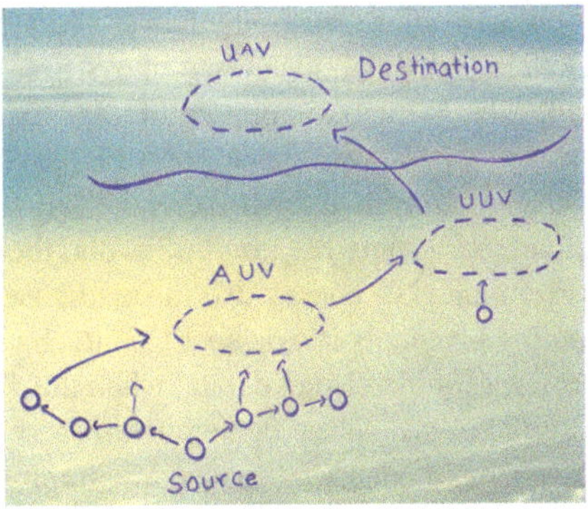

Figure 8.1 The energy-efficient underwater acoustic communications

UANs. Addressing the challenges of UAN routing requires the development and implementation of innovative solutions. We classify energy-efficient protocols into various categories: unmanned aerial vehicles (UAV)-based protocols, unmanned underwater vehicles (UUV)-based protocols, and AUVs-based protocols, and discuss possible solutions in each category below. Figure 8.1 shows that integrating all three can lead to a more stable and effective underwater routing communication.

8.3.1 How do UAVs contribute to the development of energy-efficient routing protocols?

UAVs play an important role in the development of energy-efficient routing protocols in ocean monitoring networks, utilizing the data collection process. The use of UAVs allows for the strategic management of energy consumption by adjusting the UAV's trajectory, the transmit power of underwater sensors and surface buoys, and the time required for data transmissions. UAV-based routing protocols are revolutionized by introducing a dynamic and adaptable layer to data collection, a significant shift from traditional static methods. Static routing methods often face energy problems because they do not use energy evenly across all sensor nodes. This becomes a bigger issue in ocean environments, where sensors are spread out and connections can weaken. UAVs help solve this by allowing flexible planning of data collection routes. By adjusting their flight path, UAVs can get closer to sensors, which reduces the distance data have to travel and reduces the energy used by sensors and buoys. This smart movement helps spread energy use more evenly, stops important sensors from running out of power too soon, and increases the total working time of the whole system. In addition, UAV-based routing can deal with challenges in the ocean, such as weak signals and interference, by choosing better communication routes and adjusting how data is sent. The measures aimed to maximize energy efficiency by balancing the energy used for data transmission against the amount of data collected, while also addressing constraints such as the required data transfer rate and the limited energy available for the system and the energy budgets [5].

Using UAVs helps create routing methods that use less energy throughout the network, allowing underwater sensors and surface buoys to last longer because they have limited power. In addition to the basic benefit of adjusting flight routes to lower energy used in data transmission, UAVs also support the use of more advanced and smarter routing methods. These improved methods take full advantage of the movement of the UAV to respond to live network conditions and the specific status of each sensor node. This helps in managing energy use more precisely and efficiently. For example, with UAVs, routing protocols can be designed to change important settings, such as how much power a sensor uses to send data, when the data should be sent, and which route the data should follow. These decisions are based on key information, such as how much energy remains in each sensor, how important or urgent the data is, and the current quality of the communication signals. Imagine a situation where the UAV organizes data collection so that sensors only send information when the UAV is in the best position to receive it, helping to avoid energy waste from failed transmissions. In another case, the UAV might change its path to reach sensors that still have more energy or are storing important data that need to be collected quickly. This kind of flexible and smart coordination brings about a major improvement in the creation of energy-efficient routing systems, helping ocean monitoring networks run more smoothly and last longer.

8.3.2 How do UUVs contribute to the development of energy-efficient routing protocols?

UUVs are becoming a key part of UANs, not just because they can collect data but mainly because they help create energy-saving ways to send information. The underwater environment is often changing and difficult to predict, making it difficult to maintain steady communication. Old-style fixed routing methods do not adjust well to these changes, leading to wasted energy due to repeated transmissions, unnecessary message sharing, and uneven use of network nodes. UUVs help solve these problems by acting as smart, moving messengers that help carry and send data, allowing the design of more efficient and long-lasting communication systems. The spatial–temporal graph routing (STGR) protocol [6], a smart and reliable routing method made especially for networks with multiple UUVs. UUVs helped save energy by regularly updating and using an STG, which showed both the current and the expected state of the network. Their graph included changing details such as which nodes are nearby, how good the communication links are, how long those links might last, and how much energy is left in nearby nodes. Each UUV created and updated its own STG using information from its own sensors and messages from nearby nodes.

This futuristic map helped UUVs predict when connections appear or disappear and make smart choices about the best time and target to send data, while saving energy. One of the main advantages of UUVs is their ability to support smart and flexible routing using various machine learning techniques (such as Q-learning). In the STGR method, each UUV worked like a smart agent that learned from its surroundings through a trial-and-error approach, just like reinforcement learning. The surroundings, described by the STG, gave a clear picture of how the network changed over time and space, including how the nodes moved, how stable the connections were, and the energy levels of different nodes. The Q-learning system helped each UUV judge different options for passing on data, using a scoring method that included important factors such as how reliable a connection is, how much energy is used, how long communication takes, how energy is shared across the network, and the Doppler effect. This helps UUVs choose the best paths for sending data successfully, while also saving energy and using network resources more fairly. Another important way UUVs help save energy is by holding onto the data for a short time and waiting for a better chance to send it.

Unlike fixed nodes, which have to send data as soon as they connect with another node, UUVs can guess when better connections happen based on their known or expected movement paths. In certain scenarios, a UUV may delay the transmission of data until it moves closer to a more suitable node that has more energy or offers a stronger and more stable communication link. This strategy allows UUVs to conserve energy by avoiding repeated transmissions, which are expensive in underwater networks due to limitations of acoustic communication, and prevents the system from relying on links that may be weak or short-lived. The STGR took this further by the Doppler effect during routing decisions, which was especially useful for multiple UUVs in underwater networks. Fast-moving UUVs can shift the frequencies of the acoustic signal, leading to Doppler shifts, which may result in data loss and the need for retransmission. UUVs can estimate how much a link may be affected by the Doppler effect and avoid such connections by using updated information about their speed and direction. Thus, selecting nodes that are moving too fast in relation to them. This makes the overall system more efficient by helping the network avoid failed transmission that causes energy loss.

Another way of improving the energy use of UUVs is by balancing the workload across the network. In the STGR approach, the UUVs monitored both their energy and that of the nearby nodes. This allowed them to choose smarter paths to send data without overloading a single node. By spreading communication tasks, the network reduced the risk that nodes died early and improved the useful life of the system. UUVs having more energy or a better position could take on more workload, letting the system adjust and stay stable as the conditions change. Additionally, they cut down on extra communication costs by avoiding the typical method of sending updates to all nearby nodes, which could result in energy waste due to signal overlaps and repeated receptions. The STGR method shared status and movement updates more selectively instead. UUVs sent control messages to the nodes that needed them or when specific events occurred. This reduced network congestion and helped conserve energy. This communication was made possible because UUVs could move on purpose and rely on their sensors to be aware of their surroundings and update their route as needed. The simulations from their study showed how valuable UUVs are in saving energy.

In various mobility conditions, such as consistent paths, route deviations, and organized routing paths, the STGR protocol outperformed earlier routing schemes such as vector-based forwarding (VBF), channel-aware reinforcement learning-based multipath adaptive routing (CARMA), and Q-learning-based delay-tolerant routing

(QDTR). STGR achieved significantly higher success rates in data transmission, decreased average delay, improved energy distribution between nodes, and minimized total energy usage. These benefits were most noticeable when the UUVs were moving quickly or had unstable connections, where their ability to predict and adjust gave them a clear edge. UUVs play a key role in the establishment of energy-efficient routing techniques that enable smart, flexible, and environment-sensitive communication. With features like real-time movement, space-time mapping, learning-based decision-making, awareness of Doppler effects, and smart energy sharing, UUVs serve as both data carriers and smart network helpers. Adding them to systems such as the STGR protocol represents a big change in underwater networking, turning fixed and weak sensor systems into moving, strong, and long-lasting communication networks. Their ability to move is not just a physical feature; it helps drive smarter networking that greatly improves energy use in underwater settings.

8.3.3 How do AUVs contribute to the development of energy-efficient routing protocols?

AUVs play an important and diverse role in the creation of energy-saving routing methods, especially for underwater emergency communication networks (UECNs). In Huang and Wang [7], the researchers presented a strong system design that takes advantage of the movement, flexibility, and smart capabilities of AUVs. This setup helps isolated underwater sensor nodes (I-USNs) send urgent information reliably while conserving the limited energy of the network. Using AUVs helps improve the overall lifespan, quick response time, and balanced energy usage of the network, particularly in emergencies where standard communication paths are no longer effective. One of the main functions of AUVs is to help sensor nodes that have lost connection to the main surface unit or other relay nodes. These I-USNs can waste a lot of energy trying to send data unsuccessfully or may not be able to. The system organized all nodes into three emergency response modes (ERMs), direct communication with the surface, communication through another relay node, or communication with the help of an AUV. The choice to send an AUV is not random; it is based on a method that combines a smart search strategy and learning from past actions (reinforcement learning) to find which nodes need AUV help the most. This smart planning makes sure AUVs are used only when truly needed, saving energy and allowing them to work longer.

A key advantage of the system is its use of different communication methods, such as acoustic, optical, and radio frequency links. Each method has its own energy use, range, and reliability. The system actively chooses the best link for each message based on current conditions such as distance, signal loss, and noise levels. AUVs improve this procedure by facilitating nodes located in weak signal regions to transmit data through the most optimal techniques. The innovative blending of AUV movement and various communication channels helps minimize energy consumption not only along the routing path but also in the act of transmitting data, an element commonly ignored by traditional routing protocols. To avoid inefficient or excessive utilization of AUV resources, the system employs a progressive grouping approach, mainly a customized version of the K-means algorithm, to group dispersed nodes based on their geographical positions. Each group is allocated a specific AUV that travels to a carefully chosen stopping location, minimizing both travel and waiting periods. In addition to selecting AUV destinations, the system accurately manages their speed and movement paths through a strong mathematical model known as a convex optimization framework. This method effectively reduces both the data collection time and the energy consumption during movement. The system also considers limitations such as the AUV's energy storage capacity and response time requirements.

The energy usage framework is detailed and includes elements such as buoyancy regulation, movement power, course adjustments, staying in place, and the functioning of onboard electronic hardware. To illustrate, it calculates the energy required to endure high underwater pressure, adjust the direction toward a sensor node, and remain steady while receiving data. By precisely adjusting the route and speed of each AUV, the network enhances efficiency while keeping within the limits of energy capacity. An important feature of the system is its use of adaptive learning strategies, especially Q-learning and deep Q-networks, to identify the most efficient relay points. When sensor nodes can send data through adjacent nodes, these data-driven methods help determine the best power-efficient and consistent routes. If an appropriate relay is not available, an AUV is deployed. This smart decision approach prevents unnecessary AUV deployment and allows the routing method to continuously adapt to shifting scenarios. The system consistently improves by learning over time, enabling it to make increasingly improved routing choices and maintain long-term energy efficiency. An advanced approach, such as an adaptive clustering-based multiobjective evolutionary algorithm (AC-MEA), is employed to achieve an efficient equilibrium among the multiple objectives. The system decides the right number of AUVs to use, what roles they should play, and how much power the I-USNs should use to send data.

In this way, it avoids using too many AUVs, which would waste energy, or too few, which could cause delays. Instead, it finds well-balanced solutions where every choice helps save energy and still delivers important data on time. The suggested strategies were tested using simulations with different network sizes (ranging from 50 to 300 nodes), various node placements, and channel conditions. The results showed that the use of AUVs greatly improves the success rate of data delivery, especially for nodes that are far away or cut off. Each node also uses less energy compared to other methods that do not use AUV or rely on one-to-one communication. In addition, the time it takes to collect data from disconnected nodes is much shorter, making the system reliable even in urgent or disrupted situations. In this advanced UECN system, AUVs are not just simple tools but smart, energy-aware helpers that work closely with other parts of the system to improve communication under limited resources. Their ability to move allows the network layout to change as needed and their use of multiple communication links gives flexible ways to send data. Careful planning of their use helps to avoid waste of energy. Using reinforcement learning to make decisions and advanced algorithms to improve performance helps AUVs become an active part of the communication system. In the end, AUVs help transform underwater communication networks from fixed high-energy systems to flexible, efficient, and reliable ones that can handle challenging and changing underwater environments.

8.4 Conclusion

The progression of UANs is largely motivated by the need to overcome the inherent difficulties of underwater communication, particularly when the goal is to achieve energy efficiency and reliable data exchange across a range of important applications. Conventional communication methods struggle under the unique constraints of the underwater environment, including slow signal propagation, restricted bandwidth, significant signal loss, and constantly changing network structures. These issues are compounded by the limited energy reserves of underwater sensor nodes and the impracticality of recharging or replacing their batteries, which makes energy conservation a critical design concern. To mitigate these challenges, recent research has increasingly turned to the development of advanced routing protocols that incorporate mobile platforms such as UAVs, UUVs, and AUVs. These mobile entities introduce dynamic capabilities into the network, enabling smarter data collection, transmission, and system coordination, surpassing the rigid constraints of fixed sensor nodes. UAV-assisted methods offer greater adaptability by optimizing aerial routes and adjusting transmission settings, which in turn balances energy use among the network components and extends operational longevity. UUV-based systems, such as those using STGR, capitalize on movement intelligence, predictive mapping, and machine learning to make smarter routing decisions, opportunistically forward data, and reduce issues like the Doppler effect that hamper acoustic communications. Similarly, AUVs play a vital role in high-stakes scenarios, such as emergency communication, by utilizing hybrid communication strategies, AI-guided navigation, optimized movement control, and smart relay mechanisms. These strategies result in higher data delivery rates and more efficient energy use throughout the network. Together, these advancements mark a significant shift in the way UANs are designed and deployed. Incorporating mobile agents introduces enhanced flexibility, intelligence, and efficiency, allowing systems to adapt to the harsh and unpredictable underwater environment. By effectively harnessing the unique functions of UAVs, UUVs, and AUVs, researchers are creating more resilient, intelligent, and energy-conscious underwater communication systems. This growing body of work has strong potential to expand the capabilities of UANs in scientific, industrial, and environmental domains, ultimately leading to deeper and more efficient exploration of underwater ecosystems.

References

[1] S. Khisa and S. Moh, "Survey on recent advancements in energy-efficient routing protocols for underwater wireless sensor networks," *IEEE Access*, vol. 9, pp. 55, 045–55, 062, 2021.

[2] M. U. Khan, P. Otero, and M. Aamir, "An energy efficient clustering routing protocol based on arithmetic progression for underwater acoustic sensor networks," *IEEE Sensors Journal*, vol. 24, no. 5, pp. 6964–6975, 2024.

[3] U. G. K. M. E. A. Gul, "EERBCR: energy-efficient regional based cooperative routing protocol for underwater sensor networks with sink mobility," *Journal of Ambient Intelligence and Humanized Computing*, vol. 14, pp. 6063–6075, 2023.

[4] Z. Liu, X. Jin, Y. Yang, K. Ma, and X. Guan, "Energy-efficient guiding-network-based routing for underwater wireless sensor networks," *IEEE Internet of Things Journal*, vol. 9, no. 21, pp. 21, 702–721, 711, 2022.

[5] Z. Liu, X. Meng, Y. Yang, K. Ma, and X. Guan, "Energy-efficient UAV-aided ocean monitoring networks: Joint resource allocation and trajectory design," *IEEE Internet of Things Journal*, vol. 9, no. 18, pp. 17871–17884, 2022.

[6] C. Xu, S. Song, X. Wu. *et al.*, "A high reliable routing protocol based on spatial–temporal graph model for multiple unmanned underwater vehicles network," *IEEE Transactions on Mobile Computing*, vol. 24, pp. 4434–4450, 2025.

[7] Z. Huang and S. Wang, "Multilink and AUV-assisted energy-efficient underwater emergency communications," *IEEE Internet of Things Journal*, vol. 10, no. 9, pp. 8068–8082, 2023.

Chapter 9
Impact of the provision of constrained energy on UAV-enabled 6G underwater networks

Haseeb Umer[1], Muhammad Umair[1] and Mohaira Ahmad[1]

Unmanned aerial vehicles (UAVs) are increasingly employed in underwater acoustic networks (UANs) to enhance data collection efficiency and overcome the limitations of acoustic communication, such as high latency, limited bandwidth, and elevated energy consumption. However, users cannot rely on UAVs for long missions or broad coverage since their onboard batteries run out quickly. To deal with battery issues in UAVs used in underwater networks, this chapter joins reconfigurable intelligent surfaces, intelligent edge computing and energy-aware planning. By studying the current literature, it is clear that areas such as sustainability, fast responsiveness, and saving energy are not covered enough. Using solar panels, flexible task assignments and teams of UAVs together is beneficial for making the system more efficient and robust. Moreover, having dynamic wireless charging and machine learning assistance allows UAVs to fly independently and for longer in unusual marine settings. The discussion in the literature shows that the recommended ideas can help extended the UAV's flight time, lower energy usage and ensure better communication between ocean vehicles. This research will support future studies on intelligent, sustainable and adaptable UAV monitoring for sensors in the ocean.

9.1 Introduction

Unmanned aerial vehicles (UAVs) in underwater networks operate at low altitude, while being highly maneuverable to collect data from surface buoys through their radio-frequency (RF) communication system. The UAV platform provides efficient services and direct visibility links, and is easily positioned while being economical compared to satellite or high-altitude platform (HAP) systems. UAVs solve the critical problems of underwater networks because the acoustic communication of these networks exhibits high latency while having limited bandwidth and a large number of transmission errors. UAVs serve as bridge instruments, efficiently transmitting data from surface buoys to base stations on land because underwater and surface

[1]School of Electrical Engineering and Computer Science (SEECS), National University of Sciences and Technology (NUST), Pakistan

nodes lack adequate energy storage and struggle with long-distance data transmissions. However, data reach to surface buoys or surface nodes can only be possible through efficient communication of underwater sensor nodes or autonomous underwater vehicles (AUVs). UAVs overcome a number of underwater network issues with their dual-tier communication network, particularly the suboptimal acoustic channel performance, the limited satellite service coverage, and the power supply limitations.

By controlling communication responsibilities and achieving optimal performance outcomes through adaptive trajectory planning, transmission schedule optimization, and power resource distribution, the UAVs reduce the load on the surface buoys. UAVs, despite these advantages, face operational issues in communication networks due to their limited cargo capacity, navigation accuracy, weather sensitivity, and the requirement for real-time adaptive systems. When flying through terrain with restricted line-of-sight pathways, it is more difficult to establish flight stability due to environmental factors and interference signals. For a connection between UAVs and ground control stations to function properly, precise algorithms and high computational computer systems are required. For UAVs involved in underwater network operations, the transmission of slow, error-prone acoustic signals from underwater sensors serves as the primary data source. Accurate trajectory control is necessary when hovering close to surface buoys to collect data, but buoy motions and ocean currents make the process more challenging. Battery restrictions are the most significant operational challenge among all of them because UAV systems only employ on-board power, which limits the number of serviceable buoys per mission, as well as the flight duration and operational distance. Achieving sustainable and reliable underwater data collection requires energy-efficient path planning techniques as well as resource scheduling and allocation plans.

The reason for using the acoustic band in underwater networks is that due to rapid radio signal attenuation rates in underwater networks, acoustic communication remains the main communication method. The channel of acoustic signals faces limitations within reduced speeds along with narrow bandwidths alongside high error rates that result in excessive energy usage. The situation becomes worse because deep-ocean environments make battery replacement for surface buoys and underwater sensor nodes impractical and costly. Therefore, maximizing energy efficiency is not only desirable for UAVs and underwater sensor nodes, but it is necessary for sustainable underwater communications. The aim of this chapter is to present methods for resolving battery-related issues in UAVs. We provide solutions to underwater challenges by using UAVs over the designated area of the ocean so that they can enhance coverage to all underwater sensor nodes. The only way to achieve this is to focus on providing possible solutions for battery issues in UAVs. The rest of the chapter is arranged in the following manner. Section 9.2 discusses related work, Section 9.3 provides solutions to battery issues, and Section 9.4 concludes the chapter.

9.2 Related work

Numerous research investigations within the literature address UANs and aerial transmission individually. The solutions in the literature include hop-based routing and

AUV data collection and satellite communication systems. These methods either fail to address the energy requirements of surface buoys or experience substantial latency and cost issues in satellite-based solutions. Despite their designs, most existing solutions overlook a pivotal trade-off that occurs when sensor nodes extend their transmission duration to increase their own energy conservation thereby causing increased energy consumption for surface buoys negatively impacting the network-wide energy efficiency. [1] investigated the problem of energy-efficient UAV-aided ocean monitoring networks through combined resource planning and flight path creation. They focused on Internet-of-Underwater-Things (IoUT) that enables connection between underwater sensor nodes and surface buoys to achieve essential tasks such as climate monitoring and disaster warning, as well as marine resource management. They answered the question of how a system can be designed that improves the energy efficiency of ocean monitoring networks by integrating UAVs with underwater and aerial communications.

They introduced an approach in which UAVs collect surface buoy data dynamically through optimized flight patterns and enhanced the power management of surface buoys and underwater sensor nodes. UAV flights next to surface buoys enable reduction of power usage within the buoys. Their model implemented a trade-off mechanism to manage sensor node and surface buoy energy usage without separate optimization methods. They improved network efficiency through a combined perspective of the entire system. Through extensive simulations, they demonstrated that their proposed method significantly outperforms traditional fixed-hover or fly-and-hover schemes, achieving faster data collection and better battery conservation. Their contributions are substantial, as they revealed and modeled the energy trade-off between sensor nodes and surface buoys. However, their method leaves some aspects unexplored. For example, it assumes that UAVs can always return to their base or be recharged easily, which might not be realistic in very remote ocean settings. Also, the model does not consider the UAV's own energy consumption in detail nor does it factor in harsh weather conditions that could affect UAV flight stability and communication reliability.

In our opinion, to solve these open issues, future systems could use energy harvesting technologies (such as solar-powered surface buoys, solar-powered UAVs, or reconfigurable intelligent surface (RIS)-based UAVs) to further extend operational life. Cooperative UAV swarms could be used to share the energy burden and ensure continuous operation even if individual UAVs fail or need to be recharged. In addition, adaptive learning algorithms could be used to predict the best UAV trajectories based on real-time environmental conditions, further optimizing energy use. Furthermore, flying ad hoc networks (FANETs) structured from UAVs create a substantial solution to improve connectivity along with surveillance functions across vast ocean regions. Researchers have previously studied remote communication through their work in developing satellite systems for unreachable areas, as well as UAVs for land-based operations and linear sensor networks to monitor pipelines. There is a wide gap in the UAV ocean literature on managing energy along with real-time operations coupled with persistence of network availability. The authors in [2] developed a UAV system with two operating levels in which the lower UAVs create FANETs to connect directly with the assets on the ocean surface.

Real-time UAV deployment optimization was performed based on changing communication requirements through a centralized system for positioning and access management. Their method benefited data delivery efficiency together with network responsiveness through simulation analysis while surpassing traditional static approaches. A primary operational obstacle in UAVs is their battery storage system, which functions as their key restriction. UAVs experience short flight times because of small onboard energy supplies, but they use intensive power whenever conducting extended missions or maintaining position stabilization above the ocean. Network disruption and reduced surveillance effectiveness become risks when UAVs cannot recharge batteries or replace them in isolated marine areas and delivery processes slow down. Their study understood the energy problem but failed to examine the available solution approaches in depth. UAVs powered by solar energy could utilize the bright sunshine over water areas, thus obtaining the ability to recharge while flying. Artificial intelligence (AI)-based predictive battery management systems combined with AI-based dynamic operational control systems reduce the probability of running out of power while UAVs perform their missions.

Their architectural design presented shows better connectivity, but their study did not show a comprehensive battery endurance model appropriate for marine environments. In their study, they also introduced a two-tier FANET design method for aerial ocean surveillance and connectivity, in which they showed effective modeling approaches. The operational sustainability and battery and energy supply solutions that need to be resolved to sustain such systems. Their architecture did not provide sufficient data on UAV recharge cycles or deployment procedures to return and redistribute UAV operations. Their study lacks information on network sustainability and scalability, which prevents answers on their practical deployment in real-life situations. A rotation-based swarm plan enables UAV subsets to function simultaneously while other units perform recharging activities and rest. This approach helps to save collective UAV energy. Floating maritime platforms which integrate solar recharging services enable UAVs to find quick charging facilities. Reducing power usage becomes possible for UAVs through both lighter component construction of UAVs and intelligent flight path optimization.

The construction of a durable connected ocean infrastructure through UAV networks becomes possible through proper integration of energy-effective hardware together with autonomous recharging procedures and intelligent system-level management techniques. The development of future UAV platforms can combine solar-electric propulsion with fuel-cell technology, as well as autonomous floating solar charging stations and UAV systems with battery exchange capabilities. Similarly, a mesh-based communication structure between UAVs can decrease the power needs normally associated with transmitting data over long distances to ground control stations. In addition, the integration of intelligent reflecting surfaces (IRS) technology into UAVs presents an effective approach to enhance the capabilities of 6G network communication systems that work poorly in obstructed areas. According to the authors in [3], regular UAV systems experience three major problems, including signal blockages along with unstable links and restricted power capacity. The implementation of IRS combined with smart signal reflection capabilities enables networks to achieve improved wireless services for remote sensing applications, disaster management, and smart city operations, and large-scale internet-of-things implementation.

9.3 6G-based possible solutions to battery-related issues in UAVs

The primary focus of this section is on discussing the possible solutions in the literature for battery-related issues of UAVs, with a focus on highlighting their challenges and providing insights on how to handle them in the future.

9.3.1 How IRS can solve battery-related issues in UAVs?

IRS operates without transmission activity, so UAVs experience reduced workload and gain better control over wireless channel operations. New challenges emerge in UAV-IRS systems because UAV flights are dynamic, and the systems need quick adaptions with increased energy requirements from UAV motion combined with IRS communication coordination. The dual-layer design of IRS-enabled UAV networks can create reliable communication frameworks that integrate flexibility and energy conservation for the implementation of the 6G standard. Optimization of IRS-assisted UAV networks in 6G requires solutions for various complex problems that include UAV trajectory management and IRS phase shift adjustment, as well as communications quality and energy efficiency trade-offs. The exact hovering position of UAVs represents a challenge because it requires accurate IRS signal reflection, but leads to power efficiency issues and demands powerful computing systems for reactive operations. Flight stability in conjunction with IRS signal alignment becomes more difficult due to UAV battery limitations and environmental interference. Further integration of IRS components makes UAVs more operationally complex and consumes more energy, directly affecting UAV system sustainability along with shorter mission durations.

Mission effectiveness in extensive areas or time spans requires improved energy efficiency and intelligent path planning together with power-aware control systems, since deploying IRS on UAVs comes with critical power constraints. There is a need to create optimization methods to minimize energy consumption in flight while employing AI-based adaptive controls through reinforcement learning, as well as collaborative UAV operational methods that manage energy distribution. The authors in [3] emphasized the potential of energy-efficient IRS design to minimize the power burden on UAVs. However, their study did not explore technologies such as solar powered UAVs, in-flight recharging mechanisms, or energy harvesting solutions that could substantially improve UAV endurance. In addition, practical simulations or experimental validations and real-world deployment scenarios are needed, especially in power-constrained rural or oceanic environments, which currently remain unexplored. Their research contributed by classifying IRS-UAV deployment scenarios and surveying optimization methods such as joint beamforming, trajectory control, and AI-based scheduling.

They identified research gaps in IRS placement, energy trade-offs, security vulnerabilities, and real-time adaptability, laying a foundation for future studies. However, the absence of robust battery models, hardware implementation strategies, or scalable testbeds limits the practical implications of their findings. To address these shortcomings, future research should investigate low-power IRS hardware, solar or hybrid UAVs, federated learning for decentralized control, and dynamic IRS deployment algorithms aligned with UAV flight paths and communication demands. Their study offers a comprehensive theoretical overview of IRS-UAV integration in 6G, but there is a need to test these theoretical models using practical deployment. Energy efficiency, which is crucial for the reliability of the system and mission continuity, remains a key area that requires innovation. To fully realize the potential of IRS-UAV networks, interdisciplinary approaches that combine machine learning, renewable energy systems, and advanced hardware designs are necessary to become sustainable, resilient, and intelligent 6G communication architectures.

Likewise, [4] presented a detailed survey on IRS-assisted UAV communication systems, which combined two of the most promising technologies for advancing future wireless networks. The integration of IRS and UAVs addresses critical issues such as signal blockage, severe path loss, and limited coverage areas that traditional wireless networks often face. IRS technology, which can intelligently control the propagation of wireless signals by adjusting the phase and amplitude of reflected waves, complements UAVs that offer flexible, rapid, and dynamic deployment. Together, they form a highly adaptive communication framework capable of improving coverage, energy efficiency, and service quality in both routine and emergency scenarios. They carefully described the system architecture, where IRS panels could be mounted on UAVs or placed on the ground to assist UAV communication links. They explained key hardware components, control loops, and the interaction between the UAV flight controller and the IRS controller. Upon examination of the existing literature, the authors highlighted the advantages offered by this integrated system, such as improved signal quality, better energy efficiency, and flexible deployment.

Their survey also highlighted critical research challenges, particularly related to accurate channel estimation in fast-changing UAV-IRS environments, real-time joint optimization of UAV trajectories and IRS phase shifts, hardware impairments, and the need to manage communication delays and computational complexity. To address these challenges, state-of-the-art approaches were reviewed that include machine learning and deep reinforcement learning techniques for dynamic optimization, hybrid algorithms that jointly manage UAV path planning and IRS beamforming, and innovations in hardware design to develop lightweight, low-power IRS panels. They further identified future research directions, such as developing robust and realistic channel models, improving energy efficiency, advancing adaptive control mechanisms using AI, and exploring multi-UAV, multi-IRS systems coordinated through swarm intelligence. They made an important contribution by consolidating the current research landscape on IRS-assisted UAV systems, highlighting the unique benefits of this synergy, and offering insights into the challenges that remain open.

Their work sets a solid foundation for further research. In our opinion, possible solutions to these gaps include developing practical testbeds, designing scalable control algorithms, and working on standards that ensure smooth integration into

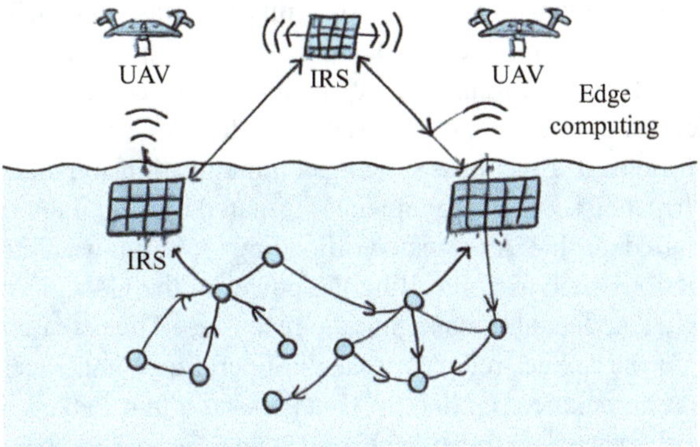

Figure 9.1 UAV-enabled underwater acoustic communications

existing communication infrastructures. Furthermore, incorporating federated learning and edge computing can help reduce latency and improve real-time decision-making. IRS-assisted UAV systems represent a transformative shift in wireless communication, offering the potential to expand connectivity into under-served or disaster-stricken regions, strengthen network resilience in high-demand events, and enable ultra-reliability, low-latency communications envisioned for future 6G networks (see Figure 9.1). However, there are still technical barriers that must be overcome, such as creating robust control algorithms that can operate in dynamic and uncertain environments, improving the design of IRS hardware for lightness and energy efficiency, and ensuring the scalability and interoperability of these systems with current networks. In general, IRS-based UAVs are a powerful and promising research direction that, with continued technological advances and collaborative efforts across disciplines, can revolutionize the landscape of wireless communications in the years to come.

9.3.2 How edge computing can solve battery-related issues in UAVs?

Edge computing, particularly multi-access edge computing (MEC), is a pivotal innovation that enables low-latency, energy-efficient data processing by relocating computation closer to IoT devices. Instead of processing data on centralized servers, this technology does it at nodes near the data source, making it faster and more energy efficient. When it comes to underwater acoustic communications, MEC-equipped UAVs make it possible to access edge servers flexibly and at any time. The study in [5] discussed combining MEC with energy-harvesting IoT devices and UAVs to improve the efficiency of systems in remote areas with limited infrastructure. Some approaches did not address energy fluctuations or require a fixed setup, making them not suitable for underwater environments with many changes. By using UAVs to handle data and computations near the source, the energy required on underwater sensors and surface buoys is much less. As a result, missions last longer, there are fewer delays, and power usage is reduced in UAVs that help with ocean surveillance.

Their idea involves combining an UAV with an MEC to reduce both energy consumption and delays. Drones help transport data for IoT devices to edge servers where it is processed and stored. To address syncing communication and maintaining power, they designed virtual task and energy queues regulated by Lyapunov optimization. A new algorithm called mutli-UAV assisted offloading (MUAO) was introduced to deal with real-time actions such as controlling the power of the UAV, dividing the workload among the IoT nodes, and managing the resources. The research indicated that MUAO speeds up the queue, reduces losses in operations, and saves energy while holding the queue in balance. Unlike previous research that looked only at optimal settings, their work focused on issues such as unstable energy supply and limited computing resources. Still, their model did not take into account how UAVs can continue to fly and whether they can have enough energy or depots to sustain their operations in remote or oceanic areas. Their framework formed a solid base, but further work can be done to make it fully autonomous in marine monitoring tasks.

Incorporating even more energy-focused features in future UAV-assisted edge computing systems can help them perform even better in long-ocean missions. An effective way is to add solar panels to UAVs so that they can collect power while flying or just waiting. A further approach could be to locate recharge stations on ships or floating buoys, so that UAVs can be autonomously refueled or charged during missions. Enabling UAVs to predict and time their operations using AI can optimize how tasks are handled, the energy used, and the routes flown based on the flight environment, the rush of tasks, and the conditions of the network. When using methods such as swarming or scheduling based on rotation, the energy used by UAVs can be distributed more fairly among them, allowing the swarm to continue to perform. Using UAVs alongside edge devices, the suggested architecture is able to provide flexibility and efficient energy usage for to track marine wildlife, offshore maintenance, and disaster management. By combining these technologies, it becomes possible to collect data safely and efficiently under conditions where power is limited in underwater environments.

Intelligent edge computing (IEC) improves the energy efficiency of UAVs by providing local processing of data, minimizing the energy loss in long-distance data transmission. It facilitates UAVs to take decisions in real time such as changing routes or avoiding obstacles. IEC permits machine learning algorithms to predict energy requirements and decides on the suitable charging time. Through dynamic adaptations of the mission plan, UAVs can be routed or delayed depending on the level of battery and the state of the environment. IEC supports intelligent coordination with charging bases or other UAVs that offer dynamic wireless charging (DWC). Maximize resource allocation by prioritizing requirements and thus utilizing resources effectively. In fleet missions, IEC supports collaborative decision making, where UAVs can offload processing to each other and reduce redundant communications. This decentralized intelligence results in more effective task execution. This IEC integration not only increases the robustness of UAVs, but also favors sustainable and energy-efficient operations in dynamic and complex settings. It enables instant decision-making, energy flexibility, and resource efficiency. By reducing dependence on centralized infrastructure and enabling localized intelligence, IEC is a driving force for the evolution of autonomous vehicle technologies.

Early or late UAV arrivals at charging terminals arises several challenges that reduce the efficiency of DWC systems. Delays in charging can disrupt UAV operations and lead to wasted resources and higher system costs. The balance of resources between on-time, ahead-of-schedule, and behind-schedule UAVs is challenging. Dynamic reallocation according to availability and priority is needed. In addition, delays in charging have a negative effect on overall productivity, particularly in time-critical missions. UAVs with critical energy needs might require priority access. Environmental sustainability is also compromised when UAVs hover or idle to wait, wasting energy inefficiently. Random arrival times complicate scheduling algorithms and place additional stress on infrastructure, as simultaneous early arrivals would over-the-limit system capacity. Further, battery capacity reduces due to uncontrolled or multiple charging cycles, and greater airspace utilization around the nodes for recharging is a safety risk. There is a need to mitigate these problems to enable responsive, efficient, and environmentally friendly UAV operations in smart transport systems.

The combination of IEC, DWC, and 6G-supported vehicles increases operational effectiveness, energy efficiency [6]. The infrastructure employs two types of UAVs, laser-beaming UAVs (LBUAVs) for in-flight charging and edge intelligence UAVs (EIUAVs) for data processing. IEC facilitates decision in real-time, minimizes delay in transmission, and allows machine learning at the edge for route calculation and energy forecasting. Layered architecture, with end devices, edge nodes, and central cloud, is used to manage functions such as billing, energy distribution, and authentication. Predictive analytics is applied for charge planning, laser beaming for energy transfer and communication through Li-Fi-based technology, and a secure handshake protocol for session management. Route planning is supported through heuristic algorithms and 3D A* navigation system, while realistic charging behavior is guaranteed through a two-stage battery model. The architecture enables dynamic UAV coordination, task sharing, and efficient resource allocation. Dynamic scheduling and real-time reallocation are employed in managing any challenges introduced by UAVs being too early or too late. Overall, the system facilitates scalable, energy-efficient, and reliable UAVs in smart and intelligent systems.

9.4 Conclusion

This chapter discusses 6G-based solutions to one of the main difficulties in using UAVs to support UANs, the limited battery life. Using a mix of energy sources in the solution, operators can now use UAVs for more time and also rely on them to reliably cover and support sensor nodes. It uses solar energy through harvesting and makes the use of energy efficient algorithms, allowing the UAVs to handle power efficiently through their entire flying duration. The discussion supports the use of energy-aware systems in applications involving marine and remote-sensing. It is particularly noteworthy that the 6G-enabled design helps reduce energy consumption, increases the time the system can work and cuts down on problems with missing data collected from underwater devices. Furthermore, this system can be easily changed and scaled to meet various demands in different situations. The research addresses an important issue by focusing on the overlap between autonomous flight, green energy, and underwater communication coverage. The solution is additionally invaluable because it uses energy-saving routing and adjusts the schedule dynamically, making it useful for long-term monitoring at sea. In the future, these 6G-based frameworks can be tested in practical situations and expanded to include wind energy. Moreover, incorporating better battery and lightweight material technologies can help improve UAVs. In summary, the chapter presents a path that can be entirely practical and expanded to cover all underwater sensor networks with UAVs.

References

[1] Z. Liu, X. Meng, Y. Yang, K. Ma, and X. Guan, "Energy-efficient UAV-aided ocean monitoring networks: joint resource allocation and trajectory design," *IEEE Internet of Things Journal*, vol. 9, no. 18, pp. 17871–17884, 2022.

[2] A. Patil, P. Muthuchidambaranathan, D. N. Jayakody, "Performance of UAV networks over the ocean: a perspective towards connected ocean," *Wireless Personal Communications*, vol. 135, pp. 347–366, 2024.

[3] H. Hamid, G. R. Begh, "IRS assisted UAV communications for 6G networks: a systematic literature review," *Wireless Network*, vol. 31, pp. 779–807, 2025.

[4] H. Shakhatreh, A. Sawalmeh, A. H. Alenezi, S. Abdel-Razeq, and A. Al-Fuqaha, "Mobile-IRS assisted next generation UAV communication networks," *Computer Communications*, vol. 215, pp. 51–61, 2024. https://www.sciencedirect.com/science/article/pii/S0140366423004619

[5] L. Qu, J. Wang, and C. Assi, "Resource scheduling and delay optimization of IoT devices in UAV-assisted multi-access edge computing," *IEEE Internet of Things Journal*, pp. 16998–17011, 2025.

[6] P. W. Shaikh and H. T. Mouftah, "Edge computing-aided dynamic wireless charging and trip planning of UAVS," *Journal of Sensor and Actuator Networks*, vol. 14, no. 1, pp. 1–33, 2025. https://www.mdpi.com/2224-2708/14/1/8

Chapter 10

Optimizing UAV battery usage for supporting underwater acoustic networks

Muhammad Atif[1], *Ahmed Hussain*[1] *and Seung Yeob Nam*[2]

The need for marine surveillance, as well as underwater communication, has brought attention to the use of unmanned aerial vehicles (UAVs) and developed the concept of underwater acoustic sensor networks (UASNs) enhanced by UAVs. The batteries on UAVs often hinder their performance, mainly when they are required to travel far into remote regions or by sea. This chapter discusses the issues connected with batteries and suggests that defining optimal paths is a good way to solve them. Several methods such as smart routing, smart positioning of docking stations, and energy-saving communication methods are studied to reduce power consumption. Clustering algorithms, gaining-sharing knowledge (GSK) and the capacitated vehicle routing problem (CVRP) techniques are used to make planning efficient. The method conserves power, but fulfills requirements such as UAV memory and its allotted airtime. Using adaptation algorithms and wind information made in real time helps reduce energy use. The chapter also considers other options, such as solar drones, platforms in the sky known as high-altitude platforms (HAPs), and charging stations on the water. The emphasis is on using AI-based systems for forecasting, guiding energy management, and correcting the path of a ship on its own. Although the chapter has a good technical model, it does not discuss underwater humidity, team efforts between UAVs and autonomous underwater vehicles (AUVs), what happens if an AUV fails, and the unreliable quality of GPS at sea. The following areas could be improved in the future with swarm intelligence, harvesting of renewable energy, and the use of non-rusty materials. In general, the research in this chapter covers how to make UAVs more sustainable for ocean missions, helping shape the future of collecting marine data and surveilling the environment with intelligent aircraft.

10.1 Introduction

The number of maritime activities grows rapidly in modern times, especially with shipping operations and smart ports together with ocean monitoring systems. Ships,

[1]School of Electrical Engineering and Computer Science (SEECS), National University of Sciences and Technology (NUST), Pakistan
[2]Department of Information and Communication Engineering, Yeungnam University, South Korea

autonomous surface vehicles (ASVs), and surface buoys operate at limited communication standards despite the moderate capabilities of land-based electronic communication systems. The reason is that maritime communication networks operate mainly through shore-towers which have restricted coverage areas along with satellite connectivity that introduces high latency and maintains minimal data rates. In response to these limitations, unmanned aerial vehicles (UAVs) create a new capability for maritime communications. UAVs enable wireless coverage delivery that is adaptable and provides smooth efficient services throughout extensive maritime areas. UAVs provide the capability to connect devices between satellites and ships and underwater equipment and coastal command facilities through the broadband internet while supporting internet-of-things (IoT) devices and search-and-rescue (SAR) operations.

As we move deeper into the ocean, the environment becomes more complex, and more challenges are required to overcome it. The signal attenuation that occurs in seawater causes severe communication problems that restrict the use of radio frequency (RF) signals and slow down data transmission. Standard acoustic methods remain widely used, yet they present the combined problems of slow operation speed together with long transmission delays and environmental interference. The operation of UAVs functions as a bandwidth extending mechanism because they gather information from underwater computing devices before forwarding it through faster RF or optical links to coastal stations or satellites. Both the mobility feature and rapid data aggregation capabilities, as well as backup communication options, are enabled through their flexible structure. UAVs improve network structures by changing their operational patterns in response to immediate operational requirements. UAVs serve as essential elements that address underwater communication problems and enhance network coverage while reducing latency.

With this advancement, researchers need to understand methods for integrating UAVs into underwater networks to solve both coverage and reliability issues, as well as latency problems and especially battery power limitations. Scientific research presents significant positive findings [1] as UAVs serve multiple communication purposes, including data transmission services, edge computing capabilities, and content distribution close to end users, as well as emergency disaster communication support. However, the principal issue that appears in every study pertains to battery constraints. UAVs operate with limited flight duration because they utilize small batteries. Recharging UAVs at sea is difficult and risky. The batteries run out quickly when performing demanding communication operations. Additionally, ocean surveillance operations combined with underwater communication node duties exceed the normal flight duration of UAVs. To address these problems, researchers identified multiple approaches to deal with battery and energy problems [2].

10.1.1 Existing solutions and their challenges

UAVs could receive wireless power supply through the operation of electromagnetic field-based or laser-based wireless power transfer (WPT). The WPT system operates through wireless energy transmission to UAVs for battery recharging, while eliminating the need for physical landings [3]. WPT operations utilize two main techniques, where electromagnetic systems apply the resonant inductive coupling technique and laser-based WPT achieves power generation by using photovoltaic cells to receive light beams. Maintaining UAV alignment with power sources represents the main research challenge researchers face when UAVs need to operate under moving water conditions and harsh weather conditions. Safety controls for high-energy beam systems must be implemented along with efficiency management from beam transmission distance extension. The development of improved energy-saving communication methods through smart handover systems with better routing and power management solutions conserves battery resources.

Smart handover systems choose the best communication links and routes according to the immediate network conditions, so they minimize unwanted power usage while reducing transmission traffic [4]. These systems optimize both network flow and weak links to preserve battery resources of UAV systems. The complex nature of real-time network awareness tracking along with its high computational requirements becomes the primary concern because it consumes excessive onboard energy. Optimization of UAV flight paths focuses on finding minimum-energy routes while maintaining wireless connectivity. Optimization of UAV flight paths depends on algorithms that determine efficient energy-saving routes based on wind conditions, coverage requirements, and distance factors for longer operational durations. The UAV connectivity system consumes minimal battery resources due to its ability to stay connected without excessive movement. The main issue regarding dynamic maritime environments lies in the fact that real-time path adjustments become difficult due to persistent recalculation processes that put processing capabilities under strain and result in slower response times.

Swarm intelligence operates by grouping numerous UAVs to split their assignments into smaller tasks, therefore minimizing the power usage on each drone [5]. Large missions become more manageable by dividing them into smaller activities, which UAVs share while performing basic operations and maintaining power supplies. The swarm manages to execute tasks effectively by coordinating its operations to preserve mission objectives. Constant inter-drone communication and synchronization requirements cause the main challenge because they lead to extra energy consumption while making systems more complicated. UAVs utilizing intelligent reflecting surfaces (IRS) redirect signals without active signal transmission, which conserves power. IRS on UAVs directs incoming signals toward specified directions while demanding no transmission energy from the drones. This technique establishes communication through passive power-saving methods. The main obstacle stems from the need for precise positioning in conjunction with environmental awareness because these requirements are difficult to resolve in dynamic maritime conditions.

High-altitude platforms (HAPs) consist of larger UAVs and floating stations operating at stratospheric heights to function as extended flight platforms that supply continuous energy to smaller UAVs [6]. HAPs function at stratospheric heights, as they provide continuous wireless charging services for UAVs, which add to their operational duration. These HAPs function as energy relay stations that decrease the number of ground returns required for aircraft. The biggest obstacle to stable extended HAP service in harsh atmospheric environments involves both maintenance technology and financial stability. In addition, floating recharge stations consist of mini stations installed on surface buoys, which can also be used by UAVs for quick battery recharge sessions [7]. Building upon these solutions, a combination of existing solutions needs to be applied according to our perspective on system improvement. Special UAVs enable battery swapping operations during flight by carrying spare batteries to exchange with depleted ones. Maritime UAV operations in sunny conditions should use lightweight solar arrays that allow longer mission periods.

Furthermore, AI capabilities enable UAVs to predict their own battery consumption, which allows them to dynamically modify their communication traffic. The section discusses a detailed review of technical developments by organizing current solutions across network layers together with management aspects and computational needs, while evaluating their energy performance and latency capabilities. It provides detailed information about current challenges together with ongoing experimental work. The focus of this chapter is to discuss possible solutions to battery-related issues of UAVs. In the next section, we will provide possible solutions that include optimized path planning for UAVs, which also resolves the battery issue as well.

10.2 Optimized UAV trajectory: a solution to battery-related problems

Rahman *et al.* [8] discussed the issue of limited UAV battery life head-on with a co-optimized plan for trajectory planning and the placement of charging stations. This is of great importance for the general research question of optimizing UAV paths to counteract energy constraints. The focus of the research is to maximize the freshness of the data, which is defined by the age of information (AoI), while ensuring that the UAV can recharge itself effectively through optimally placing several charging stations. In this way, the study offered a clear and mathematically justified procedure to mitigate the main constraint of battery duration in UAV operations. Their most significant achievement was the articulation of a complex but solvable optimization problem that simultaneously integrated the flight trajectory of the UAV and the spatial layout of the energy replenishment infrastructure. They proposed a multi-stage algorithm that gradually refined both the path and placement of charging stations. This double optimization was important because most previous work had been done separately, either concentrating only on path planning or static infrastructure design. By merging the two, they discussed a more practical and scalable approach that fits closer to the practical operational limits.

However, there are some important points that remained out of focus. They assumed a predefined number of charging stations and a given charging station capacity without investigating dynamic or adaptive charging models. Furthermore, the UAV was modeled to have a constant speed and fuel consumption rate that could not represent a change due to wind, on-board loads, or maneuverability. The environment was also assumed to be frictionless, without considering that terrain, buildings, and air traffic could all play a major role in trajectory and the energy consumed. Given these limitations, a number of possible solutions can be suggested. The use of adaptive charging infrastructure, for example, a mobile or drone-based station, can increase flexibility and minimize downtime. An improved optimization can be achieved with a detailed energy consumption model that includes real-time flight conditions and environmental conditions. Moreover, the integration of machine learning methods for predicting future energy demands and for dynamically changing both trajectory and charging schemes can provide a more intelligent automated system. Extending the model to accommodate urban or obstacle-rich environments can also make the findings applicable to practical deployment cases.

All of these extensions may conspire to bring research closer to fully optimized battery-aware UAV mission planning. Amer *et al.* [9] discussed ways to overcome battery-related issues encountered in UAVs used to gather information on wireless sensor networks, especially in remote areas. It is challenging for UAVs to efficiently use energy to check and monitor all dispersed sensors, which causes vehicles to run out of battery faster. The main problem is that UAVs must use a lot of energy to fly to and from the sensors, often leading to shorter operating times. According to the authors, a complete approach was created that helped reduce energy use without going beyond the limitations involving UAV battery life, time in the air, and the amount of memory the UAV holds. This was important in situations such as those found in underwater networks, as long-distance data exchange was not achievable, so UAVs became mobile carriers for data. In order to solve this issue, the authors implemented two steps: the first used novel triangulation-based cluster analysis to reduce the aggregators, and the second mixed gaining-sharing knowledge (GSK) with software for the capacitated vehicle routing problem (CVRP) to design the paths for the drones and strategist dock station placements.

With this approach, energy-efficient route planning and data gathering allow battery-constrained UAVs to last much longer. They mainly focused on designing a two-stage process that organized how aggregators were positioned, where dock stations were placed, how UAVs moved, and how much power was distributed for transmission. They offered a K-means clustering algorithm with a triangulation approach to lower deployment costs by reducing the number of aggregators as long as all sensors were within the maximum range. At this point, the GSK metaheuristic was used to discover the best location for dock stations and to find the highest power-level transmitters that can use them. After creating a candidate solution, GSK uses a CVRP solver to determine the best UAV routes to meet both time and memory constraints during the mission. Reducing the process to two separate steps for the variables improves the computation and produces more desirable results. They studied their approach with varying numbers of sensors and area sizes to ensure that it can be used in any situation. They saved up to 13.25% of energy and reduced both the need for UAVs and the number of aggregators compared to existing ideas.

To further check the results, a Poisson point process was used to generate the specifications and arrangements of the UAVs and sensors. To develop the energy-efficient data collecting system, they had incorporated clustering, metaheuristics, and routing algorithms. However, despite its comprehensive methodology, they did not include coverage of several important issues for implementing the system, especially underwater. The performance of the system was not explored in cases where latency must be managed since it was not designed for real-time operations. The concept did not address methods for reducing UAV collisions or ways for many UAVs to cooperate. Other factors in nature, such as bad weather, obstacles in the environment, or poor signal, were not taken into account, which may reduce the usefulness of these simulations. Furthermore, although system energy use was minimized, no plans for possible UAV or aggregator failures or recovery strategies were included. Since missions differ, the identical abilities of UAVs may not always apply. Dock stations were not designed to change with the surroundings or sensors that are constantly on the move. Reduced the influence of surrounding devices and ignored security issues. This work did not discuss models that combine both UAVs and AUVs for data collection in the sea.

Amer *et al.* [9] mentioned the use of real-time systems or HAPs in the future, they did not describe how this will be accomplished. Moreover, although these areas are not currently suitable for the model, the basics of the theory were solid. In the future, researchers may address these issues by sending communication data uninterrupted through HAPs or meshed networks. Collisions can be avoided in the CVRP by using reinforcement learning or swarm-based models. To make the network more stable, we can set it to reroute traffic when a failure occurs, and we can also depend on the failure features in the tab. If solar or kinetic energy is harvested, the robot can function for longer, yet it will weigh more. If we had various connection interfaces on the docking stations, it would be easier to set up the network. Bringing together UAVs and AUVs in unmanned swarms could support activities in the ocean depths. However, putting AI into use is a difficult task because the costs are very high, extensive training is needed, sensors must be coordinated efficiently, and it should react quickly in all situations. The efficiency of a system should remain the same in any situation. These systems must not fail in the event that something unexpected occurs. In general, the results have improved the energy efficiency of UAVs collecting data in sensor networks on land and in water.

Du *et al.* [10] attempted to solve the major and continuous issue involved in UAVs, especially the lack of energy efficiency and battery power. Their study, aimed at logistics UAVs flying in cities, can be used to solve similar battery problems in other critical UAV roles, such as aiding underwater acoustic sensor systems. Most of the time, these networks are set up in areas too remote and rough for humans to manually manage UAV batteries. They solved the problem by designing a technique that decreased energy use without compromising the aims of the mission, such as delivering goods or gathering data promptly. Real-time wind information was used to design and manage UAV routes, resulting in optimal use of wind flow for flying. This can help us prolong the operation of our battery-powered UAVs. They centered on smart city logistics, useful for underwater settings, especially by improving flight time and range for UAVs. This is key when UAVs help in relaying or charging underwater devices, which send data only during rare moments when they are in contact with the water surface. If the system is fine-tuned, it could help with reconnaissance and sending data back, even when using a small amount of battery energy. As a result, the problems explored in their work can directly assist in the better use of UAVs in water-rich environments.

They demonstrated that making the most of wind can definitely help save battery power in UAVs. Based on this principle, engineers can work on marine energy-resilient systems that continue to support underwater sensor networks as long as current UAV batteries allow. They created a valuable AI model that can do part of what a typical GPS system tool does. As the winds change during flight, they make necessary changes in the drone's path, considering the drop in energy, the added time, and the ability to complete the task. As a result, they complete deliveries on time, adhere to weight restrictions, and ensure efficient use of electricity and low energy waste. The simulations conducted in an intelligent model showed that the approach of their work could lower the distribution of network cost by 9%. They also proved that wind energy planning leads to greater energy savings and higher customer satisfaction. It is clear from these updates that the limited availability of batteries is affecting what system engineers design. In the future, this approach might be expanded using adaptive learning technology. They proved that operating with several variables as input in UAVs leads to more sustainable operations. Also, UAVs that come and go to the surface level for batteries and packages help to maintain the reliability of the model.

With adapted UAVs that move between water stations, the support provided to underwater acoustic sensors is greatly boosted. To sum up, this is possible by using AI, optimization, and real-life constraints in a model that improves the use of UAVs and assists in fixing energy-related problems experienced during marine sensing. Even so, their study tackles several positive points but ignores the important subject of how UAVs operate in water. Another point to note is that the discussion has yet to address how UAVs could operate over the long term without being able to stop and recharge on the ground. Experts mention that the UAV service should have a reliable battery and be able to fly a long distance, but they did not consider using different power sources or refueling mid-flight. In addition, the wind model in their work studied just the urban wind, not the turbulent winds found in coastal and sea areas. Again, it was assumed that only a few key nodes, kept in the same spots, were part of the system. Since nodes in the ocean can move or be rearranged, the target needs to be able to track them on the move. The gathering of data from underwater sensors is also an overlooked matter; since they do not transmit information as fast as land nodes, more frequent UAV visits are necessary to provide current information.

They did not discuss how UAVs deal with humid environments, salt spray, or in emergencies situations where they came into contact with water. The algorithms they proposed must consider these points while being used in oceanographic or underwater work. Since the algorithm had only been applied in simulations, there was no way to judge how it performed in real situations. Knowing the gap, there is uncertainty as to whether such UAVs are suitable to handle the varying winds and occasional inaccuracy of GPS in the sea. These gaps can be closed using a variety of technological and planning approaches. The integration of solar panels on UAVs can help them collect energy as they fly, allowing them to stay on mission for a longer time without having to land often. Another option is to use floating stations or surface vehicles that allow UAVs to power up or change batteries while they are flying. They can be used as mobile transmitters and receivers for supplying information and electricity to underwater sensors. Additionally, the navigation algorithm could be adjusted to take into account disruptions from sea spray, humidity, or magnetic disturbances during navigation. The ocean planning process can be helped by teaching AI historical marine weather information, allowing it to respond to changing weather over the sea.

A predictive scheduling algorithm can be used with the genetic algorithm with large-scale neighborhood search (GA-LNS) system, allowing it to recommend when and how to collect data according to the behavior of the sensor nodes. UAVs designed for water operations should be made from materials that are both corrosion resistant and waterproof. For communication purposes, having both acoustic and RF signals is ideal for forming stable links with sensors located underwater and in the air. Researchers may also be able to set up feedback loops in the UAV that enable it to automatically respond to new environmental changes by changing its course. Yet, there are problems that need to be solved, such as dealing with more complex systems and the cost of adding multi-modal sensors, as well as making sure resource-limited UAVs respond quickly. The setting of guidelines for maritime airspace and advances in wireless energy transfer are currently major problems. Even so, the plan offered by their work provides a good start for these new ideas. Considering AI-planned paths along with renewable and adaptive systems may provide a solution for UAV battery issues in underwater and marine explorations and sensing in the not-so-distant future. Sornek *et al.* [11], solar-operated drones are used in a variety of areas; most notably, they play a part in monitoring and surveillance to upgrade communications.

Their panels provide energy to charge the UAV, and the remaining electricity is stored in batteries so that the drone can function day and night. They sort UAVs by size, range, modes of travel, and the jobs they perform. Fixed-wing solar UAVs are mentioned in their work as being able to fly longer than their multi-rotor counterparts. Besides the solar panel problems, the presence of panels on a wing can create issues for airflow and the aircraft's durability. Both carbon fibers and composites are applied to help make the airplane more stable and fast, as they also focus on altering the solar panel surface to increase or reduce lift and drag. Some believe that these UAVs can make it possible for civil aviation to achieve new flight records. This research outlines the history, beginning with sunrise and ending with the systems in use today. It is important to note that choosing the best wings' size and shape helps to design solar drones more efficiently. The researchers believe that modeling and practical testing prove that the design was done correctly. For managing power supplies and their use in energy management, both maximum power point tracking (MPPT) and extremum seeking control (ESC) are used. It also involves creating ways to strengthen a bridge without compromising its safety due to weight.

Having made these wings with both 3D printing and Z-infinite technology, there were no joints. Engineers design solar cells so that they contribute to the greatest lift and least drag. Their research outlines a guide to addressing common issues that affect most USB-powered UAVs. It will involve placing a solar power unit in the airplane, making the design more suitable for solar panels, and selecting energy-efficient propellers. Their work presents a system for aircraft, with wings that are long and narrow and are built using lightweight composites, so the aircraft has more solar cells, yet flight stability is not affected. The materials, gallium arsenide, perovskite, and copper indium gallium selenide, are talked about because they manage to work in various situations and are efficient. Furthermore, they explored approaches such as attaching solar panels to the sides of buildings or using flexible solar material wrapped around objects. In checking the wings, the researchers looked at the amount of stress they can withstand and their aero and photovoltaic efficiency. Usually, the performance of an airplane is tested by using computational fluid dynamics and particle image velocimetry (PIV). Their work applying AI and genetic algorithms was examined in their survey to learn about new developments in boat design, driving, and providing power on boats. The knowledge of field work and high flight cases demonstrates that predictions for the AtlantikSolar UAV are correct.

Drones are tested close to the ground or at a vertical angle because these may better suit some tight spaces. To compare, batteries are tested according to their efficiency, the weight they carry, and how quickly they can be charged. They offered details on how drones are designed to work from solar or battery power, giving them greater reliability. Flight control and responsiveness are achieved in the control system using proportional integral derivative (PID) and linear quadratic regulator (LQR) algorithms. Observe that the design of a UAV must consider its mission, weight, speed, and the environment in which it is working. Even though they detailed notable achievements and UAVs, they found no solutions to the main problems in the area. While detailed aerodynamics was proven via testing, putting solar cells on UAV wings still did not resolve issues with microcracks caused by vibrations. There was little information about how solar UAVs could be maintained and operated in various difficult settings. The problems related to data security when sending data from UAVs to their stations were not discussed. Since UAVs were not allowed to fly alongside other aircraft in controlled air traffic, researchers were concerned about their safety on civilian flights. The feasibility analysis and budgets for the use of solar energy in large projects were not included in their work.

There was also no explanation for how changes in climate may affect UAVs in places with very different amounts of sunlight. Their system did not address how to manage many UAVs at the same time. The procedure did not ensure that a backup copy of important applications was in place. Their work did not include details on AI, hot topic predictive maintenance, or how to navigate in varied circumstances. Not enough attention was paid to checking the damage caused by frequent flying. Furthermore, their work did not focus on retrieving crashed UAVs, designing their landing systems, or managing emergencies. No one has yet addressed the issues that arise during long-distance solar-powered UAV transborder flights. For this reason, further experiments and studies are needed for different military operations. Due to the identified gaps, the findings of this research should encourage multidisciplinary cooperation between material scientists, experts in power electronics, and those who control UAVs. For example, using graphene solar panels, AI programs to oversee the solar energy process, and automatic systems to find any drone component problems are some of the alternatives proposed.

Using machine learning in flight planning helps us improve plane efficiency by adjusting both its altitude and route based on weather, light from the sun, and current wind. Building planes with lithium-sulfur or solid-state batteries increases storage capacity, but the batteries are not heavier. They allowed several UAVs to band together to gather and send vital information. With blockchain, UAVs and control centers have a secure way of communicating with each other. A single standard should be used for communication among all UAVs to deal with UAV rules. Due to modular construction, improving and repairing the building is simple. Using AI on autopilots could result in more efficient flying for tomorrow's planes. Working with UAVs and underwater relays leads to the development of networks for collecting information in sea environments. With fewer resources, green tech-equipped drones may be set up to carry out ongoing monitoring. To address contemporary issues, solar UAVs have to go through a variety of trials, get money from sources, and ask for help from organizations around the world.

10.3 Conclusion

This chapter identifies that the use of batteries in UAVs to enhance underwater communications limits missions and can reduce confidence in underwater UASN. By reviewing wireless power transfer, intelligent reflecting surfaces, the concept of swarm intelligence, and working from high altitudes, the researchers demonstrate the pros and cons of each strategy. Planning flights for UAVs and locating charging stations is done to help save energy and reduce waiting times. AI models installed in UAVs enable the aircraft to adjust their journey, predict power, foresee changes in the sea, and act accordingly. Moreover, clustering, metaheuristics, and routing schemes for CVRP help to find the best possible route and ways to organize the data. Lighter UAV parts, flexible solar panels, and the use of solar energy extend the amount of time UAVs can operate after charging. At the same time, drones only play a small role in real everyday work, and most researchers ignore how issues with winds, GPS, and oceanic humidity could affect them. Besides, research recommends adding capabilities such as modular design, no corrosion, acoustic communication, and reinforcement learning to UAV automated flight systems. Reliable and extended missions could be carried out with the help of floating docks, the use of energy from the environment, drones, and autonomous equipment. Nonetheless, solar systems are still looking to sort out standardization, how long each part lasts, and if large-scale solar can become more reasonable. Researchers suggest that specialists in different fields should work together to enhance the strength and intelligence of UAVs. As a consequence, it provides clear instructions for using AI, improves energy use, and helps plan ahead for upcoming UAV systems. As soon as they are completed, they might support research and exploration in the oceans, while also tracking environmental shifts and assisting in the exchange of information.

References

[1] N. Nomikos, P. K. Gkonis, P. S. Bithas, and P. Trakadas, "A survey on UAV-aided maritime communications: deployment considerations, applications, and future challenges," *IEEE Open Journal of the Communications Society*, vol. 4, pp. 56–78, 2023.

[2] M. Rahman, N. I. Sarkar, and R. Lutui, "A survey on multi-UAV path planning: classification, algorithms, open research problems, and future directions," *Drones*, vol. 9, no. 4, pp. 1–32, 2025. https://www.mdpi.com/2504-446X/9/4/263.

[3] P. P. Ray, "A perspective on 6G: requirement, technology, enablers, challenges and future road map," *Journal of Systems Architecture*, vol. 118, p. 102180, 2021. https://www.sciencedirect.com/science/article/pii/S1383762121001302.

[4] P. Rahimi, C. Chrysostomou, I. Kyriakides, and V. Vassiliou, "An energy-efficient machine-type communication for maritime internet of things," in *2020 11th IEEE Annual Information Technology, Electronics and Mobile Communication Conference (IEMCON)*, 2020, pp. 0668–0676.

[5] W. Wu, F. Zhou, B. Wang, Q. Wu, C. Dong, and R. Q. Hu, "Unmanned aerial vehicle swarm-enabled edge computing: potentials, promising technologies, and challenges," *IEEE Wireless Communications*, vol. 29, no. 4, pp. 78–85, 2022.

[6] Z. Jia, M. Sheng, J. Li, D. Zhou, and Z. Han, "Joint hap access and LEO satellite backhaul in 6G: matching game-based approaches," *IEEE Journal on Selected Areas in Communications*, vol. 39, no. 4, pp. 1147–1159, 2021.

[7] S. Guan, J. Wang, C. Jiang, R. Duan, Y. Ren, and T. Q. S. Quek, "Magicnet: the maritime giant cellular network," *IEEE Communications Magazine*, vol. 59, no. 3, pp. 117–123, 2021.

[8] J. Liu, F. Yang, X. Wang, L. Qu, M. Jin, and H. Dai, "Joint optimization of charging station placement and UAV trajectory for fresh data collection," *IEEE Internet of Things Journal*, vol. 11, no. 14, pp. 25057–25073, 2024.

[9] A. A. Amer, R. Ahmed, I. S. Fahim, and T. Ismail, "Energy optimization and trajectory planning for constrained multi-UAV data collection in WSNs," *IEEE Access*, vol. 12, pp. 9047–9061, 2024.

[10] P. Du, Y. Shi, H. Cao, S. Garg, M. Alrashoud, and P. K. Shukla, "AI-enabled trajectory optimization of logistics UAVs with wind impacts in smart cities," *IEEE Transactions on Consumer Electronics*, vol. 70, no. 1, pp. 3885–3897, 2024.

[11] K. Sornek, J. Augustyn-Nadzieja, I. Rosikoń, R. Łopusiewicz, and M. Łopusiewicz, "Status and development prospects of solar-powered unmanned aerial vehicles – a literature review," *Energies*, vol. 18, no. 8, pp. 1–31, 2025. https://www.mdpi.com/1996-1073/18/8/1924.

Chapter 11

Crypto-secure UASN: secure communication for the network layer of underwater acoustic sensor networks

Muhammad Ali[1], *Ali Taimoor Khalid*[1] *and Menaa Nawaz*[2]

Underwater acoustic sensor networks (UASNs) are vital in oceanography, environmental surveillance, and military operations, but they also present considerable security and operational challenges. This chapter is dedicated to the study of methods for protecting UASNs from Sybil assaults, paying attention to the sustainability of energy consumption and communication aptitude. Highlights the vulnerability of typical encryption solutions for underwater networks and discusses recent ideas such as lattice-based cryptography, certificateless aggregate signature schemes, and edge computing approaches. Moreover, the chapter explores machine learning-based intrusion detection systems, including isolation forest, in order to allow efficient and effective anomaly identification. The research highlights the importance of coming up with hybrid solutions that combine cryptographic methods and adaptive and context-aware solutions. The study provides an in-depth mapping of resource-sociable and secure approaches customized to ensure that underwater nodes in restricted and fluid marine environments are secure from ever-sophisticating threats. The chapter also discusses recent approaches and provides a holistic set of rules for Sybil attack protection in systems of unmanned aerial vehicles. A detailed analysis of efficient cryptographic procedures is presented, and an analysis of their practicality in dynamic underwater environments is provided. In addition, the chapter describes the way in which the combination of machine learning algorithms and lightweight security protocols can enhance the speed and accuracy of threat detection. Furthermore, the role that edge-based and decentralized systems play in reducing communication overhead is also discussed.

11.1 Introduction

Due to its courtesy, what would previously have been impossible communication tasks below water surfaces, underwater acoustic communication (UAC) has become

[1]School of Electrical Engineering and Computer Science (SEECS), National University of Sciences and Technology (NUST), Pakistan
[2]Department of Biomedical Engineering, Air University, Pakistan

a vital technology. Since it has been proven useful in multiple sectors thus far, including defense operations, earthquake detection, and oil and gas exploration, UAC is essential. However, there are technical obstacles that prevent the field of research in underwater networks from being slow. Currently, it has never been a more opportune time to speed things up, as the spread of machine learning and neural networks has never been so fast. The chapter describes technology and encourages emerging scientists and engineers to close the gaps in current research. The real concept of this study is underwater communication that remains hidden from hostile forces and interferes with marine life as little as possible. Liu *et al.* [1], different solutions for ocean communications have been formulated, which are advantageous in one way or another. In this regard, getting examples of acoustic communication offers a good chance of extending the transmission of distance since it is not detected by third parties hundreds of thousands of meters.

Acoustic communication methods have remained a fundamental component of underwater communication systems that depend heavily on them. Although high, the speed capabilities of optical communication are still hindered by transmission length and the need for direct visual sight, preventing its application for standard and distance use. At a depth of about 20 m, radio frequency (RF) and very low frequency (VLF) signals can work fairly well in shallow water, but transmission is clearly one-way because of strong losses. The operation of this communication technology has high speed, but the communication distance requires adjacent distances. For very low-detection and extended underwater coverage operations, acoustic waves are significantly better than any other submarine communication method. UASNs are a special subclass of communication networks developed for the underwater environment with specific characteristics. Introducing these networks helps in a number of applications, ranging from monitoring aquatic ecosystems, tracking water quality, detecting oil spills, and underwater navigation.

In contrast to terrestrial wireless networks, underwater communication is characterized by acoustic signals that transmit rather than electromagnetic waves, which are heavily attenuated in water and practically unusable for reliable transmission. Although limited by low bandwidth and high latency, the acoustic medium is the most feasible way to have meaningful underwater communication over long distances. However, the underwater world is vast, hostile, and rather inaccessible, and therefore it is convenient to employ autonomous underwater devices with sensing and communication capability. Such environments are costly, risky, and sometimes impossible to intervene in humanely over long periods of time. Therefore, these networks must be able to work independently, and an important aspect that needs to be improved is reliability, energy efficiency, and, chiefly, fault security. However, these systems are threatened by several threats until robust security mechanisms are implemented, one of them being the Sybil attack [2]. In these attacks, a malicious node creates multiple spoofed identities (see Figure 11.1), and this may cause compromised data integrity, routing failure, or even the collapse of the network communication.

Thus, securing underwater networks against Sybil attacks is much more than desirable, it is essential to secure underwater operations. This research then poses the key question of how under the severe restrictions imposed by underwater conditions, UASNs and UAC can attempt to prevent, detect, and mitigate Sybil attacks. Because

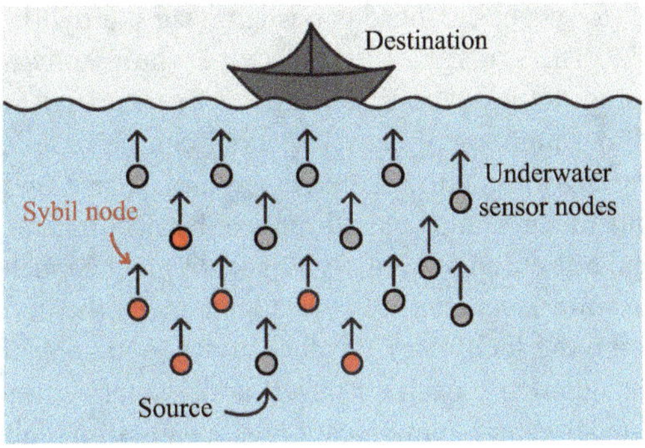

Figure 11.1 The Sybil attack on underwater acoustic communications

underwater devices tend to be energy limited, computationally weak, and in highly dynamic environments, traditionally deployed security solutions from terrestrial networks are not applicable. The focus of this chapter is on protecting the network layer, with the Sybil attack being our primary focus and providing all possible methods to prevent such attacks in underwater networks. The rest of the chapter is arranged in the following manner. Section 11.2 discusses various schemes to protect underwater networks from the Sybil attack, Section 11.3 discusses recent advancements to improve network security and provides solutions based on the cryptography algorithms to handle challenges in the underwater environment, and Section 11.4 concludes the chapter.

11.2 How can underwater networks be secured against Sybil attack?

There are several schemes discussed in the literature to handle Sybil attacks. Some are the same as those used for terrestrial networks, but due to the challenges of underwater environments, special schemes need to be proposed to make underwater networks more secure. There are methods in which nodes validate each other using pre-shared keys or certificates. Others use location-based techniques, where they exploit the fact that it is practically impossible to enforce that a single device can assert it is located in more than one place at any one time. Moreover, there are behavior-based methods that monitor the transmission patterns of nodes in order to detect anomalies that signal Sybil behavior. However, most of these efforts fail when applied underwater. Sybil attack protection has been thoroughly studied in terrestrial networks, such as ad hoc, Internet-of-Things (IoT), and vehicular networks; however, little effort has been made in the underwater environment. As underwater localization systems introduce inaccuracies, location-based approaches can tolerate the jittery nature, but not the accuracy of the output. However, behavior-based solutions usually do not deal with environmentally induced variable communication patterns driven by current and obstacle variations [3].

We believe that to overcome these persistent issues, a hybrid approach may be the right path to take. The synergy of lightweight cryptography with physical layer security, such as using the acoustic fingerprint of the transmit nodes, can improve Sybil detection without high-energy cost. Sparsity can be introduced in the design of machine learning algorithms to take advantage of sparse underwater data, and dynamic adjustments of thresholds would reduce the occurrences of false positives due to environmental noise. In addition, Sybil attacks can be much harder to execute successfully if nodes are leveraged to work together for the collaborative defense mechanism, validating the legitimacy of other nodes, and spreading computational loads. There is no standard approach for covert underwater communication. Various strategies exist for underwater communication, since mission requirements combine with environmental limitations. The core domains in covert underwater communication consist of low probability of intercept (LPI) and low probability of recognition (LPR) with the addition of low probability of detection (LPD) [4]. The main objective of LPI concerns the encryption of messages.

Encrypted signals make it impossible for interceptors to decipher the contents even if they detect the communication. The communication interception process remains possible through this method because encryption controls what recipients understand but does not stop interception. The security enhancement process relies on digital signal encryption through various cryptographic transformation methods. LPR has its origins in natural processes that scientists have used to develop bionic communication systems. In this technique, researchers duplicate the sound rhythms that various sea creatures emit, such as dolphins producing whistling sounds. The research used equally combined receiver techniques and rake receivers with dolphin-like sonic patterns to hide and retrieve binary information. The benefit of this technique is the absence of any defined transmission power threshold, so communication distances can be expanded without being detected. The processing procedure excludes all marine sounds that help hide covert operations while protecting marine ecosystems from disturbance. LPD techniques work by decreasing the levels of the signal-to-noise ratio (SNR) until the transmitted signal merges with background noise [5].

The communication system merges with natural acoustic elements, making it virtually imperceptible. This technique is successful solely because the intruder can detect signals through background noise, but such detection becomes harder because of frequency hopping and spread-spectrum techniques. Several steps must be devised to ensure that such things do not occur for true LPD communication. As the distance between the transmitter and the intruder is greater than between the transmitter and the receiver, the risk of exposure increases. To make the signal spread more, techniques such as PN sequences and frequency hopping spread spectrum would spread the signal over frequencies where the detectability of the signal would be less. Furthermore, orthogonal frequency division multiplexing (OFDM) is used to spread the signal energy even more, nearly making it impossible to detect the signal at a single point. The receiver sensitivity and intruder detection capability have to be analyzed with probability. As the SNR drops, the receiver's problem is to correctly detect and decode the message, while the intruder problem increases as well to detect that some communication even exists.

However, these concepts date back to the 20th century, specifically with regard to World War II and techniques like binary phase shift keying (BPSK) as well as early spread spectrum communication to avoid enemy detection. The first type of receivers are only simple matching receivers, which try to match a template to a known signal; the second type of receivers are dual sensor receivers, which use alternative methods to raise the signal detection. It also allows for other innovative possibilities, such as energy harvesters based on high-speed currents or heat differences, to name a few, that would enable the lifetime extension of the system without reducing stealth. The research in Zukarnain *et al.* [2] seeks to fill this gap by collecting tactics specifically tailored for UASNs and UAC. Many have been proposed in the literature as solutions, including authentication-based methods such as lightweight fuzzy logic systems, multifactor authentication methods based on MAC addresses, angle of arrival (AoA), and hop counts, and cryptographic protocols, such as Advanced Encryption Standard (AES) and Digital Signature Standard (DSS). Cluster-based secure synchronization communication (CLUSS) [6] was designed to improve time synchronization and detect malicious activity in the cluster formation process.

Physical layer defenses have also been investigated in which species of legitimate nodes can be identified based on the features like channel impulse response (CIR) and AoA. Recent systems based on dynamic threat detection have been built with confidence-based systems using adaptive neuro fuzzy inference systems (ANFIS) [7] and support vector machine (SVM) and structures using blockchain. However, although these efforts have been made, many problems remain. In fact, several of the very critical issues that remained unsolved. Since the dynamics in the ambient medium in the dynamic is high, we find that the conventional approaches are not well suited for this case of application. To date, the real-world validation of the many assumptions concerning intruder capabilities is incomplete and theoretical. Second, covert systems usually require more time to operate with a short energy supply. We reckon there are some solutions to these challenges. Because predicting environmental changes or adapting communication parameters on the fly based on adaptive machine learning signaling, this can be used to improve communication efficiency. Covert communication systems should be rigorously tested using enhanced intruder simulation models.

Numerous proposed methods are very energy-intensive and are not energy-constrained friendly; hence, they are unsuitable for underwater nodes with energy constraints. Packet loss and multipath propagation render communication reliability low, resulting in reduced detection accuracy. In addition, localization and trust estimation become even more complicated for the mobility of nodes and sparse network deployments. Furthermore, it is also shown that the vast majority of these solutions are only tested in a simulated environment and have not been validated in a real oceanic environment. To address these challenges, hybrid models, that is, lightweight authentication intermixed with machine learning-based trust evaluation, enable the secure system while at the same time being efficient. An additional feature could enhance anomaly detection, namely, context-aware detection that includes environmental information such as tides and noise patterns. Energy-adaptive protocols are protocols that should dynamically adjust the sensing intensity as a function of the energy state of the node, and at base stations, edge intelligence can reduce the computational burden on underwater devices.

Finally, modular, application-specific security frameworks for implementing system-level security and decentralized, real-time consensus that penalize prevention rather than recovering from costly post-incident are a suitable way to improve system performance overall. Acoustic covert communication underwater is not purely a technological need, but an increasing imperative. Traditionally, spread spectrum and frequency hopping give us the ability to achieve truly hidden underwater messaging by combining the simplicity of the first generation and the strength of modern encryption. However, the story is not yet finished. The next sections will focus on the innovations that will take place in artificial intelligence, environmental adaptability, and energy management. Undoubtedly, the prospect of having underwater communication that is invisible and reliable is something that the new generation of engineers and scientists must embrace to meet this challenge and usher us into this future.

11.3 Recent advancements in UASN network layer security

The deep oceans are characterized by challenges that protect underwater sensor and device communication, allowing maritime sovereignty alongside resource development operations. Ultrasonic communication networks used for environmental and military applications encounter various security risks. Sensitive data become vulnerable to attackers because it passes through edge servers as well as base stations without proper encryption. The threat is addressed by developing new security measures that protect underwater networks by preserving privacy without changing the original information while providing dual security functions. A new aggregation privacy scheme is being used as an innovative solution which implements advanced cryptographic methods to defend underwater communication. Wang *et al.* [8] developed a privacy-preserving aggregate signature scheme that brought optimizations to computational resources in addition to decreasing bandwidth usage along with [9] who analyzed an elliptic curve-based cryptosystem to achieve reduced computational expenses. Research pursuits, together with multiple studies, prioritize the protection of aggregated data privacy, which remains secure from the beginning to the end of the transmission and encryption procedures.

Traditional encryption methods face a major threat because quantum computing is emerging as an upcoming advancement. The development of quantum algorithms [10], specifically the Shor algorithm, puts classical encryption schemes, including Rivest–Shamir–Adleman (RSA), at great risk due to their vulnerability. The industry is currently making a transition to quantum-resistant cryptographic systems as protection against these emerging threats. The NTRU encryption system [11] shows potential as a lattice-based method because it demonstrates strong protection against quantum attack tactics. These lattice-based schemes operate within post-quantum cryptography (PQC) to achieve a key position in defending communication systems throughout the quantum age. The underwater domain requires increased efficient, secure solutions, which edge computing seems to provide an effective prospect for advancement. NTRU lattice encryption is a component of an edge aggregation scheme that works for underwater acoustic communication applications. NTRU lattice encryption system connects edge differential privacy and certificateless aggregated signature verification for the creation of a powerful, secure data transmission framework.

The aggregation system applies Laplace noise to maintain privacy while protecting sensitive information, which remains protected even when edge servers analyze and transmit data. The NTRU-based scheme achieves its strength through its adoption of the short-integer solution (SIS) problem in NTRU lattices because it remains secure against quantum adversaries. To protect data during transmission, the process first adds noise to each entry, making intercepted information impossible to decrypt. Multiple edge servers use batch validation approaches to verify the accuracy of aggregated data, which protects system privacy during the validation process. The implementation of key management solutions together with Oracle queries is essential to achieve successful outcomes through this scheme. The system maintains its security regardless of whether the sensor nodes have public keys and constructs correct aggregated signatures that block unauthorized system entry. Using NTRU encryption together with differential privacy techniques, the system protects itself against multiple attack vectors that try to perform forgery, replay attacks, or quantum-based attacks. The aggregation scheme in Li and Xu [11] shows superior performance through its efficient operation.

This method is superior to current methods because it requires less network usage and performs faster cryptographic operations at reduced computational costs. Under such constrained circumstances, the method keeps underwater communication networks both effective and responsive. The NTRU-based key encapsulation method stands as the core component of this solution because it addresses underwater communication requirements. The protection of the trapdoor of the scheme in He and Xu [12] achieves efficiency through ring sampling mechanisms that decrease the size of the trapdoor and reduce the amount of necessary communication. The characteristics of the underwater acoustic channel become a crucial part of the encapsulation process through the NTRU-based key encapsulation scheme due to their use as identity information, which improves confidentiality and reliability during key generation. Underwater communication systems face three major operational problems that make traditional encryption methods, including AES and RSA less effective. This key encapsulation scheme employs lattice-based cryptography to provide quantum-resistant security along with improved efficiency, because of which it suits well for resource-limited underwater applications.

The entire procedure consists of four stages that combine key generation with encapsulation and digital signature creation, along with key decapsulation to maintain both the integrity and authenticity of transmitted data. Security investigators analyze this scheme through rigorous mathematical proof, which demonstrates its resistance against adaptive chosen-message attacks for various sophisticated threats. Under the random oracle model, this key encapsulation scheme demonstrates robust security that protects it against quantum attacks. [12] developed a sophisticated model for the underwater acoustic channel, which addresses multipath distortion and Doppler shift effects for improved application of their scheme. The compensation matrix they proposed helps users overcome communication distortions through a system that maintains reliable data transmission in challenging environments. Their final outcome produces a crypto scheme that combines high-security capabilities with practical features needed for ocean communication systems. They show superiority over existing methods by citing their quantum resistance, lower storage needs, and faster computing capabilities.

They demonstrate excellent suitability as an underwater communication solution, yet management of extreme environmental conditions promotes specific research directions to enhance system efficiency and quantum threat resilience. A revolutionary certificateless aggregate signature (CLAS) scheme [13] specifically designed for underwater communication is also presented in their study. Lattice-based cryptography, together with polynomial fitting, creates secure pseudonyms by extracting identity characteristics from the acoustic signal properties. This new technique makes it possible to aggregate signatures while reducing operational complexities, signature size, and computational requirements. The advances described in their work deliver essential security improvements for underwater communication systems by providing a flexible solution that maintains operational efficiency and computational tractability in modern complex cryptographic environments. Mandal and Chatterjee [14] addressed the detection of Sybil attacks in UASNs using an intrusion detection system (IDS) based on the isolation forest algorithm. Sybil attacks pose a major threat to UWSNs by allowing a malicious node to impersonate multiple identities, leading to resource exhaustion, routing disruption, and compromised network integrity.

Traditional methods have largely focused on general energy efficiency or attack taxonomy but lacked targeted solutions for Sybil detection. The proposed IDS leverages isolation forest's unsupervised anomaly detection capabilities to identify abnormal patterns in node behavior, such as irregular temperature, pressure, or oxygen level readings. Their experimental validation shows that the system can effectively differentiate between normal and malicious activity, making it a promising, lightweight, and scalable solution for securing UASNs. Considering the limits of energy, bandwidth, and computational power in subaqueous contexts, we promote a security strategy based on isolation forest intrusion detection, lattice-based encryption, and context-aware trust models. The isolation forest stands out by enabling the use of unsupervised learning, its compactness, and its capacity to successfully detect anomalous sensor data (e.g., temperature or pressure) injected by Sybil nodes. Unlike signature-based methods, which require knowledge of past attacks in order to identify an attack, it operates autonomously independently of attack information, which can be suitable for the dynamic and sparse UASNs.

Simultaneously, NTRU-based lattice cryptography resists quantum attacks and supports efficient encryption, outperforming RSA and AES in constrained underwater environments when applied under the key encapsulation and aggregate signature protocols. In addition, the use of certificateless aggregate signatures in conjunction with acoustic signal-based pseudonymization reduces the weight of signatures while maintaining authenticity, which is highly important in real-time situations in marine settings. Edge computing further extends this strategy through the decentralization of computational tasks and tolerance of energy-based protocols for continued trust verification with minimal scripting on batteries. Furthermore, by incorporating the mechanism of anomaly detection based on assessing the environmental signals such as noise and tide variations, the reliability of Sybil detection can be enhanced. The most promising approach to future-proof UAC systems is grounded in the synthesis of machine learning, quantum-safe window of cryptographic technique, and energy-conscious algorithms.

11.4 Conclusion

The chapter discusses the challenges and solutions to securing UASN to secure it from Sybil attack and make it power efficient as well as reliable in communication. Our focus is on the primary concerns in underwater situations, for the most part, low power provision, low rates of data, and signal transfer in underwater environments, due to multipath distortion and Doppler shifts. Well-documented security practices such as AES and RSA are burdened by high processing demands and are vulnerable to new quantum challenges in cryptography. A recent trend has been that the research has moved in the direction of replacing basic cryptographic methods with lightweight variants such as NTRU-based lattice encryption and certificateless aggregate signature schemes for low-cost quantum-resistant security. The focus is on edge computing and energy-adaptive protocols with regard to their contribution to reducing the computational burden of nodes while not negatively affecting overall performance. Furthermore, algorithms such as isolation forest offer effective, label-free, real-time Sybil attack across underwater networks. The use of environmental sensing, with information on tide and noise, in detection algorithms improves the accuracy of Sybil attack detection. The chapter finally suggests the deployment of a hybrid framework that combines energy-saving cryptographic techniques, intelligent trust frameworks, and flexible detection systems to provide a strong security scheme for the future underwater networks.

References

[1] S. Liu, M. A. Khan, M. Bilal, and H. H. Zuberi, "Low probability detection constrained underwater acoustic communication: A comprehensive review," *IEEE Communications Magazine*, vol. 63, no. 2, pp. 21–30, 2025.

[2] Z. A. Zukarnain, O. A. Amodu, C. Wenting, and U. A. Bukar, "A survey of sybil attack countermeasures in underwater sensor and acoustic networks," *IEEE Access*, vol. 11, pp. 64, 518–564, 543, 2023.

[3] K. Saleem, L. Wang, A. Almogren, *et al.*, "Cognitive intelligence routing protocol for disaster management and underwater communication system in underwater acoustic network," *Scientific Reports*, vol. 15, no. 10004, pp. 1–24, 2025.

[4] H. Wang, X. Wu, Z. Wang, *et al.*, "Low-resource generation method for few-shot dolphin whistle signal based on generative adversarial network,"

Journal of Marine Science and Engineering, vol. 11, no. 5, pp. 1–18, 2023. https://www.mdpi.com/2077-1312/11/5/1086.

[5] S. Yan, X. Zhou, J. Hu, and S. V. Hanly, "Low probability of detection communication: opportunities and challenges," *IEEE Wireless Communications*, vol. 26, no. 5, pp. 19–25, 2019.

[6] M. Xu, G. Liu, D. Zhu, and H. Wu, "A cluster-based secure synchronization protocol for underwater wireless sensor networks," *International Journal Distributed Sensor Networks*, vol. 10, no. 4, pp. 1–13, 2014.

[7] M. M. Arifeen, A. A. Islam, M. M. Rahman, K. A. Taher, M. Islam, and M. S. Kaiser, "ANFIS based trust management model to enhance location privacy in underwater wireless sensor networks," in *2019 International Conference on Electrical, Computer and Communication Engineering (ECCE)*, 2019, pp. 1–6.

[8] H. Wang, L. Wang, K. Zhang, J. Li, and Y. Luo, "A conditional privacy-preserving certificateless aggregate signature scheme in the standard model for VANETs," *IEEE Access*, vol. 10, pp. 15, 605–615, 618, 2022.

[9] Y. Chen and J. Chen, "CPP-CLAS: Efficient and conditional privacy-preserving certificateless aggregate signature scheme for VANETS," *IEEE Internet of Things Journal*, vol. 9, no. 12, pp. 10354–10365, 2022.

[10] K.-A. Shim, "A survey on post-quantum public-key signature schemes for secure vehicular communications," *IEEE Transactions on Intelligent Transportation Systems*, vol. 23, no. 9, pp. 14, 14025–14042, 2022.

[11] C. Li and M. Xu, "A NTRU lattice edge aggregation scheme based on underwater acoustic communication," *IEEE Internet of Things Journal*, vol. 12, no. 12, pp. 22325–22336, 2025.

[12] P. He and M. Xu, "An NTRU-based key encapsulation scheme for underwater acoustic communication," *Electronics*, vol. 14, no. 3, pp. 1–17, 2025. https://www.mdpi.com/2079-9292/14/3/405.

[13] M. Xu and C. Li, "An NTRU-based certificateless aggregate signature scheme for underwater acoustic communication," *IEEE Internet of Things Journal*, vol. 11, no. 6, pp. 10, 10031–10039, 2024.

[14] A. Mandal and P. Chatterjee, "Intrusion detection system to counter sybil attacks in underwater wireless sensor networks using isolation forest," in *2025 International Conference on Emerging Systems and Intelligent Computing (ESIC)*, IEEE: India, 2025, pp. 146–151. https://doi.org/10.1109/ESIC64052.2025.10962762

Chapter 12

Secure underwater acoustic communication using 6G technologies

Muqaddas Masood[1], Saad Farooq[1] and Pham Duy Thanh[2]

Underwater communication networks (UCNs) have significant applications such as oceanographic research, military operations, and environmental monitoring. However, they suffer from a few issues, such as signal loss, low bandwidth, high delay, and strong security threats in the form of jamming, spoofing, and eavesdropping. Traditional cryptographic methods cannot guarantee sufficient security in dynamic and resource-constrained environments. This chapter investigates a multi-layer security paradigm with the latest technologies, such as federated learning (FL), intelligent reflecting surfaces (IRS), and digital twin (DT) platforms to counter these challenges. FL enables a decentralized, privacy-preserving intrusion response and anomaly detection, while IRS technologies offer greater signal control and stealth in turbulent underwater environments. Moreover, the DT architecture mimics the performance of live systems and forecasts attacks using long short-term memory (LSTM)-based models so that adaptive actions can be produced without disrupting live network services. The combination of such approaches creates an intelligent, scalable, and resilient security platform. The study in this chapter verifies that combining AI-based analytics, power-efficient hardware, and virtual simulation platforms can actually make underwater networks more robust and reliable for civilian and defense applications. The chapter ends with an emphasis on the need for next-generation systems to be able to support secure underwater communications systems in increasingly hostile environments.

12.1 Introduction

Modern underwater communication technologies increase the security needs of the system, whether it is for military operations, marine environmental monitoring, underwater exploration, or the emerging Internet-of-Underwater-Things (IoUT) applications. Such networks are based on four main technologies: radio-frequency (RF), acoustic band, optical band, and magnetic induction (MI). RF-based UCNs

[1]School of Electrical Engineering and Computer Science (SEECS), National University of Sciences and Technology (NUST), Pakistan
[2]Big Data Integration Research Center, National Institute of Information and Communications Technology (NICT), Japan

have advantages such as tolerability to underwater turbidity and turbulence, but at the cost of high attenuation, limited communication ranges, and the requirement for large antennas. Consequently, they are not ideal for long-distance communication. Acoustic-based systems, which are used in sound navigation and ranging (SONAR) applications, have the capability of long-range transmission, but with drawbacks such as low speed, noise, path loss, and multipath effects, restricting them from use in real-time high-volume data transfer. Optical-based UCNs support high-speed, high-bandwidth communication, and hence are suitable for applications such as underwater video transmission, but their performance degrades due to misalignment and changing environmental conditions. MI-based UCNs, on the other hand, support low transmission delay and stealth features without impacting marine life, but are prone to orientation sensitivity and range restrictions. This system also has air-water communication, which has two types: direct and indirect.

Direct communication sends signals from the water to the air interface, whereas indirect communication employs both water and air mediums to pass signals between underwater devices and surface devices, so that communication is more flexible and longer in range. Given the broadcast nature of these wireless systems and the security environments and interfaces, the key question posed by many researchers, including us, is the following. What are the main security threats, and how can we ensure secure communication in such complex and dynamic environments? The underwater and air–water communication networks are critically vulnerable to security threats, such as eavesdropping (where attackers intercept signals, particularly in spreading RF and acoustic channels), jamming (where communication is blocked by flooding the network with noise, which is effective in RF systems), spoofing and impersonation (where attacking nodes masquerade as authentic ones to steal information or disperse false data), and denial-of-service (DoS) attacks such as wormhole and black-hole attacks, which render networks immobile by tunneling or dropping information. To combat these threats, the literature has used multi-layered security solution that begins with conventional encryption techniques such as advanced encryption standard (AES) and Rivest–Shamir–Adleman (RSA) [1]. These techniques offer basic security, but they are challenged by key distribution in underwater environments.

There are more advanced methods, such as physical layer security (PLS), that benefit from the unique characteristics of underwater wireless channels, such as channel impulse response (CIR) and received signal strength (RSS), to detect, generate secure keys, and inject fake noise to mislead eavesdroppers. For added security, the quantum key distribution (QKD) employs quantum mechanics to generate essentially unhackable encryption, although it remains experimental for underwater applications. Secondly, artificial intelligence (AI) assists in enhancing security through the application of machine learning (ML) to identify strange patterns in network traffic and detect probable eavesdroppers or jammers in real time. Neural networks identify signal distortions and device properties to prevent spoofing, while reinforcement learning dynamically adapts defenses, creating power and frequency adjustments to evade attackers. These AI solutions render security faster, more intelligent, and more energy efficient compared to traditional approaches. In this chapter, we discuss the opportunities that enhance underwater network security. In our opinion, the integration of different technologies discussed above can meet security requirements and improve the performance of UCNs. The chapter outlines how the integration of various technologies can open new doors for researchers in this field to enhance underwater network performance. The rest of the chapter is arranged in the following manner. Section 12.2 discusses related work, Sections 12.3, 12.4, and 12.5 discuss the provision of security for UCNs with the aid of IRS, FL, and DT, respectively, and provide solutions to challenges with the integration of these technologies, and Section 12.6 concludes the chapter.

12.2 Related work

AI and ML help to make real-time adjustments to improve security and speed. Building on earlier encryption techniques, QKD offers an additional layer of security, particularly effective against quantum-level threats. Also, since devices have limited power, using energy-efficient protocols makes them last longer. Finally, the creation of sets of devices that work in tandem can expand the range while reducing interference to provide better communication efficacy and reliability. In large IoUT deployments, security must extend from the hardware to the application software. Encryption techniques such as AES and RSA are ineffective if hardware-level vulnerabilities remain unaddressed, such as insecure firmware updates or physical device tampering. Tamper-resistant encasement and hardware security modules (HSMs) onboard prevent physical tampering and safeguard cryptographic operations [2]. Similarly, hashing algorithms (e.g., SHA-256) ensure that any unauthorized modification of stored or in-transit data becomes readily identifiable. Since the underwater communication channel is broadcast in nature, acoustic, eavesdropping, or disruption of the transmission by attackers is easy.

To counter these threats at the physical layer, spread spectrum, and frequency hopping are commonly employed. The spread spectrum spreads the signal across a broad band of frequencies, making it more difficult to intercept and jam [3]. Frequency hopping, a form of spread spectrum, rapidly switches the transmission frequency on a pseudo-random sequence known only to the transmitter and receiver [4]. This makes the signal more immune to interception and interference, which is ideal for underwater use where security and signal integrity are the most important. Following this, further sophisticated methods are in the process of development. For example, smart surfaces [5], arrays of small, tunable reflectors, can dynamically direct acoustic beams. This guarantees that authorized receivers receive a strong signal (with a high signal-to-noise ratio (SNR)), whereas would-be eavesdroppers receive only noise. These developments introduce yet another layer of physical security into underwater communications. In addition to strengthening fundamental protections of the physical layer, such as spread spectrum, frequency hopping, and smart acoustic surfaces, advanced beamforming technologies, such as parametric acoustic arrays, enable highly directional transmission without the need for large antennas.

This focused signaling not only improves energy efficiency but also renders it much harder for attackers to jam or intercept communications. However, static defense mechanisms fail in the dynamic and variable contexts of underwater settings. ML offers the versatility here [6]. For example, light-weight neural networks that are trained on physical-layer channel parameters, tap delays, power allocation, and Doppler shifts can authenticate nodes in real-time by detecting real nodes from attackers even if nodes drift or change conditions. In addition to authentication, ML supports trust management by employing evidence fusion and support vector machine (SVM) models. The models examine various behavioral metrics such as energy consumption, data consistency, and packet delivery rates to detect malicious activity and dynamically adapt the route or label of suspect nodes for isolation. To implement these local defenses, decentralized intelligence is emerging. He *et al.* [7] allow each node to train security models locally from its own data and only transmit encrypted model updates to a central server. This preserves data privacy while reducing communication overhead.

At the same time, virtual copies of DTs based on cloud-based underwater systems allow real-time simulation and testing [8]. By creating models of what-if scenarios, teams can actively explore reactions to cyber-attacks or system failures without impacting live operations. Despite its potential, FL has unique challenges to overcome in the sea. The periodically inherent nature of acoustic links, added to the Doppler shifts, multipath fading, and low bandwidth, makes synchronous parameter aggregation impractical. To address this, future models will have to use asynchronous update protocols, in which edge nodes can push model deltas whenever they want and the server can opportunistically aggregate. Prioritized scheduling for straggler nodes and gradient update compression light weighting can also help mitigate channel bottlenecks, and even low-capacity sensors meaningfully contribute to the global model. Our focus in the following sections will be on how IRS, FL, and DT technology can help secure UCN and the challenges that remain unresolved. Our discussion of these challenges will provide young researchers with a way to conduct research in these areas.

12.3 Security of UCNs based on intelligent reflecting surfaces (IRS)

Operations rely on secure wireless communication between an autonomous underwater vehicle (AUV) and a float sonobuoy using buoyed-to-AUV (BA) connections. UAC is used the most to create BA connections because it provides for long distances, but it has issues like being intercepted by an enemy, attenuation, limited bandwidth, noise, and interference. Security measures such as cryptography, steganography, and authentication employed in traditional security suffer from computational inefficiency and complexity in real-time underwater networks. PLS, however, exploits the dominance of the underwater channel to enhance security without intensive processing using methods, such as frequency hopping, spectral scrambling, and covert communication to reduce eavesdroppers. Also, novel devices such as IRS are capable of modifying the direction and form of sound waves as they travel in the water. IRS are initially designed for radio communication but can now be researched for underwater communications. Scientists have researched various methods of using IRSs underwater to enhance the way sound or light signals pass through, even when there is delay, obstruction, or distortion in the water.

One such system is known as the buoyed-IRS-AUV (BIA) link [9], in which IRSs are situated between floating objects and underwater vehicles. These IRSs dynamically adapt to track water motion induced by wind and waves, ensuring signals remain strong and confidential. The acoustic IRS utilized in BIA links consisted of two components: a controlling desk and a reflecting desk. The controlling desk modifies the beamwidth in real time according to changes in wind and tide. The reflecting desk directs the beam toward the receiver (Rx). It consists of components such as head mass, piezoelectric rings, and tail mass. These components function through electrical and mechanical means. The primary controller alters the phase and amplitude of the signal. This forms the beam through interference. Because of waves and currents, the beam deviates from its direction. The IRS only modifies beamwidth and does not amplify the signal, preserving drone battery. In the transmitter (Tx) model, like Tx-IRS-Tx BIA link model, the Tx emits an acoustic Gaussian beam to the IRS. It modifies beamwidth through a digital chip. The Rx gets the reflected signal by the IRS.

Their model considered the position and motion of every component: Tx, IRS, and Rx. For the analysis of the BIA-based UAC system, important factors like DC gain (channel strength at zero frequency) and probability played significant roles in performance. The DC gain is the capability of the system to increase low-frequency signals, which in turn directly influences the clarity and strength of the communication. The probabilistic path loss model captures the signal strength variation caused by environmental conditions such as water conditions, distance, and reflections. The performance analysis of BIA links aims to maximize the channel secrecy rate ratio (CSRR) through a max-min optimization method, tuning the beamwidths and IRS depth to maximize legitimate AUV communication and reduce eavesdropping threats. The CSR is calculated for legal and eavesdropping AUVs, and optimal beamwidths are determined using the exhaustive search approach. The real-time running is divided into offline and online stages, whereby optimal settings are precomputed and dynamically tuned on the basis of wind and sea current velocities.

Their numerical outcomes paid attention to outage probability and CSRR in comparison between BIA links and BA links under the dynamic environmental variations of sea currents, wind velocity, and tidal velocity. According to their results, greater IRS components and ideal IRS depth enhanced link reliability and confidentiality, especially in adverse conditions such as high tide or wind speeds. The findings showed that BIA links were superior to BA links, especially under adverse conditions with slow tide currents and high wind-induced currents. The most important factors were IRS size, optimized incident and hardware surface (OIHS) values, wind and tide speeds, and link length. The BIA link structure utilized acoustic IRS to provide secure and dependable connections among AUVs and sonobuoy nodes. It adjusted beam widths and IRS depth according to changes in the wind and tidal currents to gain a 300% improvement over BA links in secrecy. The method performed better for long-range applications, particularly in conditions where the wind rate was higher than the tidal rate. The reinforcement learning, multipath fading, cascaded IRS systems, and cooperation with eavesdroppers can be considered as future work.

12.4 Security of UCNs based on federated learning (FL)

Underwater drone networks are subject to severe limitations by the extreme communication conditions of underwater environments, that is, high latency, low bandwidth, attenuation, and sporadic connectivity. Standardized centralized data processing approaches do not apply to these situations since they are based on the continuous transmission of raw data to the central server, and this introduces not only performance inefficacious but also increased security threats. In order to tackle these restrictions, a FL method was proposed in Singh Popli *et al.* [10] as a promising substitute. FL enables underwater drones to learn on their own using data onboard and send model updates to a central server without revealing their local data, thereby conserving privacy and saving communication overhead. This decentralized process of FL best matches the distributed and dynamic nature of underwater networks. The FL model is well-suited to intrusion detection as it can identify known and unknown (zero-day) attacks without the need for direct exposure to sensitive information. By allowing each drone to train its model on local data and contribute to a global model, FL improves detection accuracy while still being contextually aware of anomalies.

In addition, more advanced FL methods like FedProx [11] deal with the non-independent and identically distributed (non-IID) aspect of underwater datasets, wherein each node can have highly dissimilar kinds of data [12]. Experiments based on the SOLIDS laboratory underwater dataset illustrate that this framework can efficiently identify benign versus malicious behavior even when training is only done with benign data, indicating robust potential for the detection of zero-day attacks based on anomalies from learned normal behavior. Moreover, the FL-based intrusion detection system is not only efficient but also resource-friendly. Whereas methods such as FedAdam, which adapts the widely used Adam optimization algorithm to federated settings, may gain marginally improved accuracy through its momentum-weighted updates and adaptive learning rates, they do so at the cost of substantially increased computational and memory requirements. This makes them less than optimal for memory-constrained and power-constrained systems such as underwater drones, where light operation is critical. On the other hand, FedProx, which is an adaptation of the vanilla federated averaging (FedAvg) algorithm, has a proximal term in the loss function employed for regularization during training when the data of the clients are non-IID.

In fact, it reduces client drift and results in more convergent behavior between heterogeneous nodes. The ensuing FL structure, combined with light-weight network models, can be highly accurate while considerably reducing memory footprint and run-time. It becomes much more viable and scalable to use in real-world IoUT applications, where resources are always constrained. Moreover, it also scales with larger numbers of clients, exhibiting consistent performance even when the data set is split across 10 nodes. These findings highlight FL's promise as a secure, scalable, and efficient method for protecting UCNs from future cyber threats. In our opinion, the combination of IRS, FL, and DT technologies is a necessary step that the design of secure underwater network's needs (see Figure 12.1). The security mechanisms by themselves are no longer adequate and are vulnerable, considering the increasing

Figure 12.1 *6G-based secure underwater acoustic communications*

complexity and the size of IoUT systems. Adopting decentralized intelligence and virtual duplication will not only future-proof these systems against new threats but also provides new opportunities for adaptive, self-healing, and mission-resilient underwater operations.

12.5 Security of UCNs based on digital twin (DT)

The use of DT technology in UAC networks introduces a paradigm shift in underwater network security through enabling real-time local and global system state emulation. By continuous data harvesting and simulation, DTs yield high-fidelity approximations that enable threat detection and evasion in advance [13]. At the local level, nodes that host lightweight DTs can identify performance anomalies from past patterns and ML algorithms like LSTM, which is specially designed to prevent the neural network output for a given input from either decaying or exploding as it cycles through the feedback loops. Such deviations are likely to indicate security aberrations like jamming, spoofing, or intrusion. Through the simulation of a variety of operating and environmental conditions, local DTs can infer whether performance degradation is due to internal faults or external intrusions, enabling instant and autonomous security reactions without drowning communications links. For maximum resilience to cyber-attack, the DT resource allocation paradigm (RAPD) serves both purposes of performance tuning and anomaly sensing. RAPD uses the measures of forecasted versus actual performance as indicators for anomalous conditions suggestive of malicious entry or compromised nodes.

The loop enables preemption reconfiguration of communication parameters in order to preserve continuity despite an attack. Since the optimization is executed in the virtual duplicate prior to affecting the real network, RAPD is also a sandbox environment where nodes can test and deploy secure configurations without risking real-world failure or data disclosure. This ability comprehensively minimizes the attack surface and limits the spread of harm across the network. Building on this capability, RAPD not only provides greater operational safety but also accelerates the rolling out of adaptive defense responses. By simulating multiple attack scenarios and evaluating responses in advance, the system allows for faster decision-making under real-time scenarios. Proactively taking the initiative, this significantly improves the network's overall resilience such that during coordinated attacks or sudden anomalies, the system remains operational and recovers speedily with minimal disruption. In our view, the use of DT technology to protect underwater acoustic communication networks is not only revolutionary but also necessary. Being able to simulate the physical world within a high-fidelity virtual environment offers a strong basis for preventive security, as opposed to the reactive schemes we have thus far utilized.

Another of the main strengths of the DT strategy is that it can facilitate cooperative security through learning centralization. By consolidating data from local DTs into a global DT framework, the system can find larger patterns of coordinated anomalies that does not be apparent at the node level. This collective knowledge is particularly valuable in defense against dispersed threats such as synchronized spoofing or distributed DoS attacks. This is further encouraged by the collaborative multi-agent framework in DTs (CMFD) that combines training data from real operations and DT simulation, thus enhancing model generalizability and resilience. Thus, not only is every individual node better equipped to deal with local attacks, but the network as a whole enjoys a collective, current sense of possible vulnerabilities. This generalized knowledge enables flexible re-configuration of policy and resources, safeguarding the network from sophisticated and multi-layer threats without affecting the efficiency of communications and response time. What appeals to us the most is the two-way utility of the DT paradigm, both as an online analyzer tool and as a risk less testbed using the RAPD mechanism. This too we believe will define next-generation secure, resilient IoUT infrastructures.

12.6 Conclusion

Underwater communication systems are essential for oceanographic research, naval operations, and ecological surveillance; however, they face unique challenges, including signal degradation, multipath interference, and security vulnerabilities, such as jamming, spoofing, and eavesdropping. Conventional cryptographic methods prove inadequate against these dynamic underwater threats, creating demand for AI-enhanced solutions that provide real-time anomaly detection, adaptive threat response, and intelligent signal analysis. ML algorithms excel at identifying sophisticated intrusions such as manipulated data injections or signal interference patterns, while FL architectures preserve data privacy through collaborative model training without raw data exchange. The hardware ecosystem incorporates innovative solutions, such as dynamically configurable IRS with buoyancy-assisted stabilization, to maintain reliable signal paths in turbulent marine environments. Persistent obstacles, including severe energy constraints, propagation latency, and corrosive seawater effects, necessitate an integrated approach combining AI-driven security protocols, robust pressure-resistant hardware, and ultra-efficient power management systems. Emerging DT technology revolutionizes threat mitigation by creating virtual network replicas that employ neural networks to detect and classify anomalies in real-time, enabling autonomous countermeasures. The digital twin resource allocation paradigm further enhances system resilience through continuous performance monitoring and virtual protocol validation. This synergistic integration of AI-powered analytics, reconfigurable surface technologies, and predictive digital twin frameworks establishes a comprehensive security architecture for underwater networks. Future advancements must focus on developing self-optimizing systems to address the evolving challenges of subaquatic communication environments, ensuring reliable operation for critical military, scientific, and commercial applications in increasingly contested undersea domains.

References

[1] W. Aman, S. Al-Kuwari, M. Muzzammil, M. M. U. Rahman, and A. Kumar, "Security of underwater and air—water wireless communication: state-of-the-art, challenges and outlook," *Ad Hoc Networks*, vol. 142, p. 103114, 2023. https://www.sciencedirect.com/science/article/pii/S1570870523000343.

[2] N. Adam, M. Ali, F. Naeem, A. S. Ghazy, and G. Kaddoum, "State-of-the-art security schemes for the internet of underwater things: a holistic survey," *IEEE Open Journal of the Communications Society*, vol. 5, pp. 6561–6592, 2024.

[3] L. Milstein, "Interference rejection techniques in spread spectrum communications," *Proceedings of the IEEE*, vol. 76, no. 6, pp. 657–671, 1988.

[4] M. K. Hanawal, M. J. Abdel-Rahman, and M. Krunz, "Joint adaptation of frequency hopping and transmission rate for anti-jamming wireless systems," *IEEE Transactions on Mobile Computing*, vol. 15, no. 9, pp. 2247–2259, 2016.

[5] S. Kisseleff, S. Chatzinotas, and B. Ottersten, "Reconfigurable intelligent surfaces in challenging environments: underwater, underground, industrial and disaster," *IEEE Access*, vol. 9, pp. 150214–150233, 2021.

[6] L. Bragagnolo, F. Ardizzon, N. Laurenti, P. Casari, R. Diamant, and S. Tomasin, "Authentication of underwater acoustic transmissions via machine learning techniques," in *2021 IEEE International Conference on Microwaves, Antennas, Communications and Electronic Systems (COMCAS)*, 2021, pp. 255–260.

[7] Y. He, G. Han, A. Li, T. Taleb, C. Wang, and H. Yu, "A federated deep reinforcement learning-based trust model in underwater acoustic sensor networks," *IEEE Transactions on Mobile Computing*, vol. 23, no. 5, pp. 5150–5161, 2024.

[8] J. Liu, C. Li, J. Bai, Y. Luo, H. Lv, and Z. Lv, "Security in IoT-enabled digital twins of maritime transportation systems," *IEEE Transactions on Intelligent Transportation Systems*, vol. 24, no. 2, pp. 2359–2367, 2023.

[9] A. S. Ghazy, G. Kaddoum, and S. Satinder, "IRS-aided secure reliable underwater acoustic communications," *IEEE Transactions on Vehicular Technology*, vol. 73, no. 11, pp. 16861–16875, 2024.

[10] M. Singh Popli, R. P. Singh, N. Kaur Popli, and M. Mamun, "A federated learning framework for enhanced data security and cyber intrusion detection in distributed network of underwater drones," *IEEE Access*, vol. 13, pp. 12634–12646, 2025.

[11] A. Das and D. Saha, "Fedprox-based federated transfer learning for efficient model personalization in healthcare," in *2025 International Conference on Ambient Intelligence in Health Care (ICAIHC)*, 2025, pp. 1–6.

[12] Z. Lu, H. Pan, Y. Dai, X. Si, and Y. Zhang, "Federated learning with non-IID data: a survey," *IEEE Internet of Things Journal*, vol. 11, no. 11, pp. 19188–19209, 2024.

[13] S. Song, B. Huangfu, J. Guo, J. Liu, J. Cui, and X. Shen, "A digital twin-based intelligent network architecture for underwater acoustic sensor networks," *IEEE Transactions on Mobile Computing*, no. 01, pp. 1–18, 5555. https://doi.ieeecomputersociety.org/10.1109/TMC.2025.3555640.

Chapter 13

AI for securing data integrity in acoustic communication links

Amna Siddiqui[1], *Aima Ghaffar*[1] *and Hoang Thi Huong Giang*[2]

Talking about ocean monitoring using acoustic sensor nodes is difficult due to tough conditions, unexpected signal problems, and security risks. Underwater acoustic networks (UANs) have trouble with delays, the use of power wells, and managing trust, unlike networks on land. This chapter looks at new ways to protect underwater acoustic sensor nodes. The focus includes trust systems that use artificial intelligence (AI), adaptive learning methods, such as Q-learning, and verification of identity at the physical level. These methods address jamming attacks, fake nodes, and changes in the surrounding environment. The work in the literature has been done to make underwater networks more adaptable and safer. The chapter ends with a look ahead. It points out the need for systems that can adapt and save power to keep underwater communication secure for things such as protecting the environment and national defense.

13.1 Introduction

Imagine trying to send secret messages across the deep, unpredictable ocean. Unlike what we do on land, where we have stable networks and solid cryptography (coded information) to keep our data safe, underwater communication presents numerous new challenges. The devices and networks beneath the waves have to deal with harsh conditions, poor signal quality, and constant interference, and all of these factors make it cumbersome to guarantee that even encrypted messages stay unharmed. That is why we need a trustworthy and secure network, and that is where trust management comes into play. It's quite similar to building a trustworthy and cooperative team, where every device (node) must prove its reliability. Researchers primarily focused on securing stable land-based communications, such as in smart cities and homes (the

[1]School of Electrical Engineering and Computer Science (SEECS), National University of Sciences and Technology (NUST), Pakistan
[2]Graduate School of Information Science and Technology, Tokyo City University, Japan

Internet-of-Things (IoT) concept). However, when it comes to the underwater world, progress is not up to the mark.

Trust models that have always worked well on land do not easily adapt to the complicated ocean environment. Poor sound signals, delays in encryption (data hiding), and harsh conditions in the deep sea make things much tougher, increasing latency. To handle these challenges, experts are turning to creative solutions. AI, especially learning by trial and error, can help nodes "learn" whom to trust over time. Repeaters, devices that strengthen and relay signals, can extend communication over larger distances without losing signal quality. These smarter, logic-based models can make the decision-making process quicker and more effective, helping to prevent eavesdropping and hacking attempts. The future of underwater communication will rely not just on technology, but on trust itself, designed carefully beneath the waves. As more and more information travels through underwater networks, ranging from scientific research to national security, it is clear that building truly secure and trustworthy systems is highly important.

UANs are an important step in research and are of paramount importance for everything from ocean exploration to environmental monitoring. The underwater things are unique. Instead of swift, invisible Wi-Fi or Bluetooth signals, information is sent using sound waves such as dolphins communicating with each other or submarines. But the sea is a messy and unpredictable place. Waves, currents, and even sea creatures can mess with these sound signals, making it difficult to tell whether a message was lost because of nature or because someone was secretly listening in. That is the mystery that scientists are trying to solve. The objective of this chapter is to offer possible ways to save UANs and their acoustic communications. The rest of the chapter is arranged in the following manner. Section 13.2 discusses-related work, Section 13.3 provides solutions to secure UANs, and Section 13.4 concludes the chapter.

13.2 Related work

Underwater communication is an entirely new ball game compared to what we are used to on land. Forget Wi-Fi or Bluetooth: beneath the ocean's surface, data is transmitted using sound waves. The sound does not travel smoothly underwater. It gets disturbed by currents, temperature changes, and even marine life, making communication latent and unreliable. This creates a major problem when it comes to keeping underwater data safe. Imagine trying to figure out whether a signal was mixed up because of a hacker or just a pesky school of fish. That is the kind of problem researchers face with in a recent study [1]. The chapter focused on securing underwater data through a trust model. Their main objective was to determine whether poor communication is due to natural interference or something more sinister, such as eavesdropping. One enticing solution involved a technique known as the hidden Markov model (HMM). This model continues to check signals from different underwater nodes (as in seafloor Wi-Fi hotspots) and uses subjective logic to decide which ones are reliable and which ones are not.

In order to verify the system, researchers simulated a variety of underwater cases with both calm and chaotic conditions. The results were promising; their trust-based model maintained security ranging from crystal clear to completely mixed communication channels. Additionally, with future improvements, these trust models could even handle mobile underwater nodes, allowing secure, flexible communication as they move through the deep sea. This could change the future of UANs that help submarines talk, monitor ocean pollution, or even explore shipwrecks. UANs help monitor underwater activity, but are exposed to eavesdropping and data tampering due to the open nature of the sea and the unpredictable transmission of sound waves. To combat these shortcomings, another security protocol was proposed in Kumar *et al.* [2]. They introduced SafeCom, a protocol specifically designed to protect underwater communication. Its ultimatum was to ensure that only trusted users could send data across nodes, significantly reducing the risk of unauthorized access and data breaches. At its core, SafeCom used mutual authentication and secure key data protection.

Before data were exchanged, both users went through a two-way verification process. Once identities were confirmed, a temporary secret key was generated and shared. This time-limited key ensured that the data remains confidential and protected; even in the most hostile underwater conditions. Secure encryption then occurred between the nodes. SafeCom ensured that only authentic users register using a key and password and used random numbers to share secret information. Thus, the secret key and the secure encryption among users made it nearly impossible for an outsider to hack data. They used the Scyther simulator to test the accuracy of their protocol. This tool simulated various harsh underwater conditions that might interfere with data transmission. By exposing SafeCom to these scenarios, the researchers were able to polish its functionality, so the result is a faster, more secure, and highly resilient protocol for underwater data transmission. The work on SafeCom can be extended for future use to make this protocol a usable option for cloud-based and mobile nodes.

Sensor nodes are used underwater to track and monitor sea life and activities such as marine transmitting hidden data, etc. Instead of traditional WiFi or other wireless networks, sound waves are used for these sensors, building a setup for UAN. Now, here the sensors face the problem due to the unbounded and obnoxious sea world. The continuous loss of data packets and the delayed acknowledgment of the data between the users make this network setup practically inefficient. Signori *et al.* [3] proposed one of the solutions, which was to build the trust model. The model was trained to distinguish between the types of nodes in a way that separated the honest and the dishonest nodes. But how did this make it different from previously used models? Each node was manipulated to have awareness of its neighboring nodes, eventually creating a channel-based model. The structure of the proposed trust models was smart enough to work in both behavioral and channel-based scopes. But what if it occurred in the undesired situations? A simulator was used to test the undesired situations.

It trained the model to adapt accordingly to the sea environment, thus flawlessly spotting the good and bad nodes even in the amendments of behavior. One of the potential defects of this model was the use of sound sensors, which made it harder to keep track of the signals. Moreover, if the conditions were dynamic instead of static, these channel models did not work as expected. A lot of training is required for these models, for doing experiments in the real sea, for a better identification of the authentic nodes, and to make the encryption more secure. Sometimes a node might behave strangely, not because it is bad, but just because the water becomes too noisy or rough, and for that the trust system also needs to be aware of misleading nodes. That is why their model was built to weigh each condition individually, relying on the dependence of the quality of node connection, whether the connection was good or bad. However, if a smart attacker changed their behavior slowly, their system might take longer to detect this. To make the system even resilient, future versions will let the nodes communicate with each other about what they see, but that brings new risks such as the spread of false trust. To summarize, while these trust models make huge strides forward, they also show that underwater networks are full of tricky trade-offs.

13.3 Underwater challenges and their possible solutions to secure UANs

In the realm of maritime surveillance and security threats, the center of attention is certainly communication networks. Autonomous underwater vehicles (AUVs) and static sensor nodes are essential parts of these networks, and they help monitor everything from marine wildlife to national borders. Underwater communication is not a cakewalk. The most reliable signals in depth of the sea are acoustic signals and face deliberate disruptions called jamming attacks. Jammers can emit noise or fake traffic to interrupt communication, drain batteries, and cause sensor nodes to fail. These networks are very important for applications such as intrusion detection, so any error in these networks is fatal. Mertens *et al.* [4] introduces a smart fix to this: using machine learning, particularly Q-learning, to make the network adapt and fight back. When the network faces jamming, Q-learning comes to the rescue and helps nodes learn from experience which paths work best in real-time. This approach is not dependent on universal knowledge or any fancy predictions. The sensor nodes decide which neighboring node to forward a packet by weighing factors such as residual energy and prior success rates. This ability to make decisions helps balance energy usage across networks, keeping more nodes alive longer. It also avoids jammed paths, finding alternative routes even if they are longer in path.

By modeling underwater communications using a 2-state Markov model (good or bad signal states), researchers simulate real-world challenges such as noise, long delays, and signal interference from the sea or vessels. Jammers are versatile in nature; they can be constant, sweep, reactive, and deceptive, and they are all chaotic. Constant jammers blast signals continuously, rapidly draining their own energy, but are highly efficient. Reactive jammers are smarter than constant jammers, detecting only when a signal is detected. Deceptive jammers mimic legal traffic to confuse nodes, while random jammers act unpredictably. To defend against this variety, the network routing protocol must be both proactive and responsive in nature. Their system systematically updates routes, senses neighbor status, and identifies possible failures or jamming activity. A void-handling method is also used when a node hits a dead end, redirecting packets around blocked areas. This combination of proactive path setup and reactive path switching makes the network significantly stronger. Reinforcement learning-based opportunistic routing protocol (RLOR), a well-known protocol, was compared with the solutions tested by researchers. They implanted underwater networks of 10, 20, and 30 nodes and placed jammers in different positions. Their findings were crystal clear: the proposed Q-learning-based approach reduced energy waste, delivered more packets successfully, and kept latency lower than RLOR.

Even under severe jamming conditions, the system found ways to adapt. Larger networks performed better, with more options to route around jammed areas. The fairness of energy use across nodes also improved with network size, avoiding the pitfall where some nodes die early from overuse. Metrics such as energy efficiency, delivery rate, and average latency favored their method. Finally, their study shows how a touch of AI can make underwater networks smarter and tougher. By teaching nodes to "learn" the best paths based on current conditions, they can achieve stable, energy-efficient communication even when bad actors try to jam the system. This has for sure consequences for secure border monitoring and environmental sensing, especially in remote or hostile marine zones. Their solution does not just work; it learns and improves over time. Whether it is spotting a sneaky submarine or tracking ocean pollution, such smart networks could become the backbone of underwater surveillance in the years ahead. In the world of wireless communication, 6G networks play an important role in building communication between a large number of networks by incorporating multiple technological applications such as IoT, Internet of Underwater Things (IoUT), and Internet of Vehicles (IoV) [5]. These can establish a network phenomenon between sensor nodes or vehicles for both communications, intra- and inter-communications.

The vehicles are all mobile, making it challenging for the nodes to constantly track secure data transfer, leading to high risks of attacks on data security. Sedjelmaci *et al.* [5], the authors proposed a solution for the dynamic and jamming stress among the nodes, applying the use of AI and deep learning to make encryption adaptable. With mobility in the underwater environment, real-world strategies should be used based on adaptive learning. Traditionally, security systems are designed just enough to make static and limited-scope networking secure and efficient. For the integration of IoV or IoUT with a 6G heterogeneous network, data transmissions face critical cyber issues. Their solution is based on the use of adaptive learning and AI, as presented in Mertens *et al.* [4]. The 6G-IoV network uses hierarchical federated learning and the Stackelberg game to combat internal and external cyber-attacks, as well as to prevent AI-specific security issues. The entire working framework of their solution includes devices used in IoV, specific servers and routers, various MAC devices, monitoring devices, etc. For the purpose of detecting core data privacy, a system called security information and event management (SIEM) is used. The use of machine learning algorithms, as discussed in Mertens *et al.* [4] for underwater mobile nodes and jamming, they propose FL, which is supposed to be trained with various prompts, to make it work efficiently in the threatening and dynamic conductivities.

This whole strategy starts with node detection, followed by secure encryption between them, using advanced encryption standard (AES) for integrity. The leaders, who are the trustworthy agents, track any eavesdropping done by an outsider. 6G-based IoV key performance indicators (KPIs) are stressed in the discussion. In underwater jamming stress, Q-learning ensured a scalable and balanced energy. The KPI in IoV integrates with machine learning to reduce AI-based hacking, thus making the routing protocols secure over a vast range. Although underwater jamming focuses mainly on fixed conditions, while the 6G-based IoV integration deals mainly with mobile and more ambulant situations, one of the main goals for both is to make sure that the system has balanced energy consumption. They use a unique training algorithm and KPIs, but their desired purpose is to enhance the physical equilibrium for a loss-proof energy system. For real-world applications, this trained framework efficiently manages traffic networks and easily detects malicious agents among the dense number of nodes. When encryption processes are faced with security attacks, there is a chance that latency and scalability are reduced. The 6G KPIs work on autonomous devices and sensors to make the system stable. The training is not public. These adaptable learning systems are mainly based on private data prompts, securing real-time data confidentiality. Today, smart networks are using IoT or IoUT, with the integration of FL, to make reliable, secure systems without cyber-attacks.

The core of underwater jamming research was to make encryption flawless with improved performance. The focus was mainly on the survival and efficiency of the trained machine learning-based system for transmitted data, rather than the security. For 6G-based IoV networks, their focus was on managing trust-based architecture for more resourceful innovations, leading to a threat-free routing analogy. Without exposing the data, raw, real-time models are to be trained to enhance their performance for the future. In the advent of AI, a lot of work is to be done on these frameworks to make them more adaptable, thus making them useful beyond just the vehicles. Various machine learning models have to be established for an optimized connectivity, even encountered with the high bandwidths or more data rates. Aman *et al.* [6] focuses mainly on verifying whether a device is truly which says it is in underwater communication networks. The main goal is underwater acoustic communication, which is used on a large scale because of its long-range capabilities. Despite all this, underwater channels are exposed to spoofing attacks, where a harmful node tries to pretend to be a trusted one. Traditional ways of locking and securing data work well today, but might not be safe in the future, especially as powerful new technologies such as quantum computers are being developed. To deal with that they propose an alternative: using physical-layer features for authentication.

Focusing on the location of a device as a fingerprint to confirm its identity, their approach is tested in static and mobile 3D underwater environments using smart signal timing and filtering techniques. They have developed two main cases. First, a static environment where all nodes, both good and bad, stay in fixed positions. Second, a mobile setup where the nodes move underwater. In both, they used time-of-arrival (ToA) signals to estimate the position of the sending device. By comparing this position with known legal coordinates, they can decide if the message is trustworthy and safe. For the moving case, they also estimate speed using changes in position over time. A Kalman filter is used to track the movement of legal nodes and improve effectiveness. This is vital as underwater movement and noise can cause significant location errors if not handled with care and caution. Binary hypothesis testing is a central part of the method. The system checks whether the estimated location matches a familiar legal position. If it is in proximity, the message is accepted; if not, it is rejected. This test involves two key metrics: false alarm rate, which is when a good node is falsely tagged as bad, and missed detection rate, which is when a bad node gets through the system without being recognized. Surprisingly, their method worked better than other techniques that rely on things such as signal strength, the direction the signal comes from, or how the signal bounces around.

Researchers in the literature have shown that location-based authentication is extremely effective, backed by simulations. Even when the fake device is very close to the real one, the system can still tell them apart accurately and rarely makes mistakes. As the distance between them increases, detection becomes even easier. They also highlight that features such as distance or channel responses can be faked by an attacker, especially in the underwater setting, where there are fewer scattering objects. This makes the location a more error-free fingerprint. Their work also shows that adding more reference nodes improves accuracy and that the method works well even with moderate noise levels. Honestly, we found their work to be a smart and timely take on a growing issue in underwater communication. The idea of using the physical location of a device as a kind of "digital fingerprint" is both clever and refreshingly practical. What really stood out to us is how it avoids the usual weaknesses of traditional encryption, especially with all the talk about quantum computing possibly breaking current methods. In addition, the fact that it works for both still- and moving-devices makes it much more flexible and closer to real-world use. We also second that it does not break even when a fake node is close and near. In future, we will see how this holds up in an actual underwater test, beyond the simulations. If it performs as well in the ocean as it does in simulations, this will be a game changer for secure underwater networks.

In order to establish a communication network for several underwater activities like marine life monitoring, confidential data transmission for cyber purposes, transportation and shipment tracking, etc., UANs play a big role, as mentioned above. These sensors are used to ensure flawless sensing and connectivity between nodes in the sea world. But here the issue arises in terms of security. Cyber-attacks are quite common in oceans due to interrupted acoustic signals and continuous interference among nodes. This open boundary network may result in packet loss and failed acknowledgments. One of the main security challenges might be the impersonation of illegal nodes. An unauthorized fake node interferes with the communication, causing hacking. The generally used cryptography and IoT security methods are not fast enough to detect faulty and fake nodes. Moreover, these methods are not energy efficient and are less resourceful. Hence, the study in [7] suggests a proposal to combat the underwater jamming and node authentication issue in sea networking. They also discuss channel enhancement for dynamic systems and use a hybrid approach for this problem. According to their study, for better trust management, the mechanism should be based on the behavioral characteristics of the nodes as well as the physical ones. Their trust system, using the evidence, is established, creating an adjustment for the trust values in the ocean's unpredictable and noisy environment.

The mechanism begins with node detection, further differentiating between the false nodes and the authentic ones. Stable and resourceful energy consumption is ensured along with node authentication. This is done by cooperative jamming that adjusts the node trust values and monitors the behavior. The network architecture is designed using clusters, nodes (members and helpers). These are made to train by prompting in the conditions where an eavesdropper or an attacker tries to invade the data confidentiality. The cluster heads build a trust model by establishing authentication, followed by the assistance of the member nodes. Their framework uses a training-based algorithm that uses real-time as well as past data storage. The previously used prompts are made to store the data. As attackers have an enormous amount of energy, the management forces the outsiders to waste their packets, which in turn minimizes their energy to a great extent. The reliability of trust managers is determined by the signal-to-noise ratio (SNR). The false and detection rates of the performances identify an illegal and legal node, in turn making it easy to build a set of behavioral trust values. The jamming frameworks misguide the routing for unauthorized nodes, thus disrupting their pathways. This whole process is done outside the domain of the correct/authentic routing, leading to resourceful energy consumption and minimizing the interference.

This is done to make sure that the hacker's assumed substantial amount of energy is lost as much as possible, thus making the encryption save from severe cyber-attacks. The probability of the data transferring rate and security, using the SNR is set according to the metrics. The performance of the nodes for data storage capacity and secure communication is measured. Using Bellhop design, this proposed solution can be simulated. This can be done by the comparison among the methods for the real-time as well as for the historical data. The performance metrics previously used keeps on monitoring the UANs devices for security point of view. The underwater trust management sensors make reliable to the changing capacities for real-world data storage. The jamming attackers constantly disrupts the linkage among the nodes, leading to a significant energy loss. The energy of the connectivity network is kept under check using probability during the simulation. Thus, the Bellhop framework assists in monitoring and further detection of the nodes, in the alarming and noisy environment. This trusted and evidence-based simulated model was trained to increase the secrecy, that has much higher capacity as compared to other strategies. The principle of the correlation and the hybrid trust models integrates with the smart jamming, that recognizes intentionally only the neighboring nodes, that is supposed to be a threat.

The whole network, other than, remains untouched. In case of a malicious agent, the eavesdropper is forced to lose its packet using secrecy outage probability (SOP). The security for the node is inversely proportional to the SOP; the larger the SOP, the more suspicious the homer is. The traditionally designed cyber schemes are not dynamically adaptable to noisy surroundings and signals, while this targeted jamming and hybrid model adjusts the SNR appropriately. Although false node detection relies heavily on the probability of linkages, it is still not flawless. When we see the threshold, the nodes are only good at misguiding the nearest node, which might be efficient for a fixed threshold. The model has to be trained using a better adaptive learning algorithm. In that case, to track the previous prompts and history, AI algorithms might be implemented. Still, their work is quite efficient in dealing with fake data packets and keeping in check the power control of the targeted nodes only. Their model works flawlessly and smartly for a predefined threshold. Secure encryption in the sea world is a challenging task. The solution proposed in their work is altogether a good approach when dealing with the sea world's cyber issues. However, future work is needed to make this model more adaptable in the case of a dynamic threshold. In the era of AI and trained models, a suitable adaptive learning algorithm, such as reinforcement learning, needs to be integrated with this cooperative jamming. Moreover, the systems need to be trained to reduce the errors from the legal nodes. This will ensure a more resourceful and energy-efficient network for the future.

13.4 Conclusion

Beneath the waves lies a world rich in mystery and information. Monitoring this world is of paramount importance for marine biology, underwater exploration, and even military defense. Furthermore, ensuring data safety in underwater environments is not easy. Unpredictable weather conditions and the unique behavior of sound signals beneath the surface make communication problematic. Underwater communications present unique challenges that require specialized solutions beyond traditional land-based approaches. Harsh ocean conditions, signal interference, and latency can affect the security and reliability of UANs. Trust management proves to be a critical component in ensuring reliable collaboration between nodes in unpredictable environments. AI, especially Q-learning, provide adaptive strategies to combat interference and optimize routing. Physical layer authentication, such as location-based authentication, provides strong protection against spoofing attacks. Hybrid models that combine behavioral- and channel-based trust metrics improve node reliability. Energy efficiency remains a primary goal, and solutions such as cooperative jamming can reduce unauthorized interference with the nodes. Simulation tools such as the Bellhop ray tracing channel model and the Scyther validate these approaches under realistic conditions.

Dynamic thresholds and adaptive learning algorithms are needed to address evolving threats. The integration of reinforcement learning can further refine trust models and energy management. Future UANs must balance security and scalability to support diverse applications. Real-world testing is essential to validate advances in theory and simulation. The role of UANs in scientific research and national security highlights their importance. With the advent of quantum computing, traditional encryption methods may become obsolete. Location-based authentication offers a promising alternative for future-proof security. Collaboration between nodes can mitigate risks such as eavesdropping and data manipulation. The solution discussed in this chapter highlights the potential for smarter self-learning underwater networks. Challenges such as mobile nodes and dynamic environments require continuous innovation. Energy-efficient protocols ensure longer node lifetimes and sustainable network operations. This chapter highlights the need for interdisciplinary research to address the complexity of UANs. As underwater communications increase, protecting these networks will remain a top priority. The future of UANs lies in adaptable and resilient systems that can thrive in unpredictable oceans.

References

[1] R. Zhu, A. Boukerche, L. Long, and Q. Yang, "Design guidelines on trust management for underwater wireless sensor networks," *IEEE Communications Surveys & Tutorials*, vol. 26, no. 4, pp. 2547–2576, 2024.

[2] C. M. Kumar, R. Amin, and M. Brindha, "Safecom: Robust mutual authentication and session key sharing protocol for underwater wireless sensor networks," *Journal of Systems Architecture*, vol. 130, p. 102650, 2022. https://www.sciencedirect.com/science/article/pii/S1383762122001679.

[3] A. Signori, F. Campagnaro, I. Nissen, and M. Zorzi, "Channel-based trust model for security in underwater acoustic networks," *IEEE Internet of Things Journal*, vol. 9, no. 20, pp. 20479–20491, 2022.

[4] J. S. Mertens, A. Panebianco, A. Surudhi, N. Prabagarane, and L. Galluccio, "Network intelligence vs. jamming in underwater networks: how learning can cope with misbehavior," *Frontiers in Communications and Networks*, Vol. 4, pp. 1–16, 2023. https://www.frontiersin.org/journals/communications-and-networks/articles/10.3389/frcmn.2023.1179626.

[5] H. Sedjelmaci, N. Kaaniche, A. Boudguiga, and N. Ansari, "Secure attack detection framework for hierarchical 6g-enabled internet of vehicles," *IEEE Transactions on Vehicular Technology*, vol. 73, no. 2, pp. 2633–2642, 2024.

[6] W. Aman, S. Al-Kuwari, and M. Qaraqe, "A novel physical layer authentication mechanism for static and mobile 3D underwater acoustic communication networks," *Physical Communication*, vol. 66, p. 102430, 2024. https://www.sciencedirect.com/science/article/pii/S1874490724001484.

[7] Su, Y., Lu, Y., Fan, R., *et al.* "A cooperative jamming scheme based on node authentication for underwater acoustic sensor networks," *Journal of Marine Science and Application*, vol. 21, pp. 197–209, 2022.

Chapter 14

Underwater acoustic communications powered by machine learning

Mehwish Awan[1]

Underwater acoustic communication (UAC) is the cornerstone of data transmission in submerged environments, where traditional radio and optical waves fail due to high attenuation. However, UAC systems face unique challenges, including limited bandwidth, high propagation delays, severe multipath propagation, Doppler shifts, and pervasive ambient noise. This chapter explores how machine learning (ML), particularly deep learning and reinforcement learning (RL), addresses these challenges to enhance UAC reliability and performance. It provides a comprehensive overview of ML techniques, such as supervised, unsupervised, and RL, and their application in denoising, channel prediction, Doppler mitigation, and delay reduction. The chapter highlights recent advancements, including adaptive modulation, multi-agent reinforcement strategies, and deep neural architectures, such as convolutional neural networks (CNNs), recurrent neural networks (RNNs), bidirectional long short-term memory networks (Bi-LSTMs), and generative adversarial networks (GANs). By integrating intelligent and data-driven models, the chapter demonstrates that ML enables robust, adaptive, and efficient underwater communication systems, paving the way for scalable and resilient marine networking technologies.

14.1 Introduction

UAC is concerned with the transmission of information underwater using acoustic (sound) waves. This method is widely used because radio and optical waves are strongly absorbed in water, making them ineffective for long-distance communication. Additionally, due to the high attenuation of radio and optical waves in water, acoustic waves are typically used. In today's world, underwater communication has emerged as an imperative subject of research and development, driven by its crucial applications in scientific discovery, exploration, security, and environmental conservation. Acoustic communication (AC) is fundamental to underwater communications, supporting diverse applications ranging from navigation and exploration to surveillance and environmental research. By utilizing sound waves, it enables

[1]School of Electrical Engineering and Computer Science (SEECS), National University of Sciences and Technology (NUST), Pakistan

effective data transmission through water, making it well-suited to overcome the limitations that water imposes on other communication methods, such as electromagnetic waves [1].

The major challenge of underwater communication is environmental variability. The underwater acoustic (UA) channel presents a major challenge for reliable communication because of its extensive multipath propagation and rapid time variations caused by Doppler effects. Despite this fact, AC is more reliable than radio frequency (RF) communication underwater. This method leverages sound waves for data transmission through water, capitalizing on their ability to propagate efficiently over long distances in underwater environments. Unlike electromagnetic or optical waves, which suffer from rapid attenuation, scattering, and absorption in water, acoustic waves maintain signal integrity over kilometers. Their relatively low propagation speed (~1500 m/s) is offset by their robustness against the physical and chemical properties of water, making AC the most viable and reliable solution for underwater applications, such as sonar, environmental monitoring, and coordination of autonomous underwater vehicles (AUV).

UAC systems face a unique set of challenges that differ significantly from those of terrestrial or aerial wireless networks, including the following.

- Severely limited bandwidth: the available frequency spectrum for UA signals is extremely narrow, typically limited to a few kilohertz, especially for long-range communication. This is orders of magnitude smaller than what is available in RF or fiber-optic systems, which can span hundreds of megahertz to terahertz.
- High propagation delay (latency): acoustic signals travel through water at approximately 1500 m/s, which is significantly slower than electromagnetic signals in air or fiber (which propagate at nearly 3×10^8 m/s). This results in high latency, especially over long distances, posing a serious challenge for real-time or synchronized applications.
- Severe multipath propagation and signal fading: due to reflections off the sea surface, seabed, and submerged objects, multiple delayed versions of the same signal arrive at the receiver, causing inter-symbol interference (ISI) and unpredictable signal fading, which complicates decoding and degrades performance.
- Doppler effects and time variability: the relative movement between communication nodes (for example, submarines, underwater drones, or surface buoys) introduces Doppler shifts and time-varying channels. These effects distort the frequency and timing of the received signals, making synchronization and channel estimation more complex.
- High ambient and impulsive noise: underwater environments are inherently noisy. Natural sources, such as wind, waves, rain, and marine animals (e.g., whales or shrimp), contribute continuous background noise, while man-made sources, such as ship engines, propellers, sonar pings, and underwater drilling, introduce strong impulsive and nonstationary noise components. These degrade the signal-to-noise ratio (SNR) and increase bit error rates (BER).

As a result, there is a growing need for advanced technologies and effective techniques to mitigate these challenges. In the last two decades, ML has emerged as an effective tool for tackling complex tasks such as pattern recognition, including object detection and recognition, speech recognition, and object tracking. In particular, deep learning has caused significant performance demands in areas such as computer vision [2] and natural language processing. This chapter focuses on integrating ML techniques in UAC to overcome the challenges mentioned above. These techniques enable task-specific learning from the data, allowing the system to progressively improve its performance. Research in AC focuses on improving system performance by developing sophisticated coding and transmission methods, along with implementing techniques that minimize noise and enhance signal clarity within underwater environments.

14.2 Machine learning techniques: an overview of recent insights

ML encompasses a variety of techniques that enable systems to learn from data and make decisions. These techniques are primarily categorized into supervised, unsupervised, and RL, each with distinct methodologies and applications. ML constitutes a transformative paradigm to address the inherent problems of underwater communications through intelligent signal processing, adaptive decision-making, and data-driven optimization.

14.2.1 Supervised learning

Supervised learning algorithms are trained on labeled datasets and are primarily used for classification and regression tasks. In UAC, models such as support vector machines (SVMs) and CNNs have been effectively used for classifying different types of underwater noise, detecting signal presence, and estimating signal parameters. These methods rely heavily on feature extraction techniques such as power spectral density and Mel-frequency cepstral coefficients.

14.2.2 Unsupervised learning

Unsupervised learning deals with unlabeled data and focuses on discovering hidden patterns and structures in the underlying dynamic environment and data. Techniques such as clustering (e.g., *k*-means, density-based special clustering of applications with noise (DBSCAN)) have been explored in anomaly detection and acoustic event classification. However, their use in UA is still emerging due to the complex nature of underwater environments.

14.2.3 Deep learning

Deep learning models, particularly CNNs, RNNs, and LSTM networks, have gained significant traction in underwater signal processing. These models excel at learning temporal and spectral features from raw data, enabling robust denoising, source separation, and feature extraction. Hybrid architectures, such as Dual-path RNNs

(DPRNN) and Bi-LSTM with time-frequency masking, further enhance performance by capturing both short-term and long-term dependencies in the signal.

14.2.4 Generative models

GANs and other generative models are being explored to synthesize clean acoustic signals from noisy data. These models are particularly useful for data augmentation and improving signal quality in the presence of background noise or signal loss.

14.2.5 Transfer learning

Transfer learning involves reusing a model trained in one domain (e.g., simulated underwater noise) and fine-tuning it in a different, often data-scarce, target domain (e.g., real underwater recordings). Techniques such as 1D CNN with multi-attention mechanism (1DCNN-MAM) leverage transfer learning to improve impulsive noise suppression in orthogonal frequency division multiplexing (OFDM) systems.

14.2.6 Reinforcement learning

Reinforcement learning (RL), drawing inspiration from behavioral psychology, focuses on learning tasks where immediate reward feedback is available. It explores how agents should act within a given environment, using this feedback to make decisions that maximize their total accumulated rewards over time. While still relatively underexplored in UAC, RL holds potential for adaptive communication strategies, dynamic channel estimation, and real-time decision-making in autonomous underwater systems.

14.3 Enhancing underwater acoustic communication: challenges and ML-driven solutions

ML, particularly deep learning, has developed with evolutionary advancements for addressing challenges intrinsic to UAC. ML models can learn complex patterns from data and adapt to varying noise conditions, making them well-suited for tasks such as noise reduction, signal enhancement, source separation, and signal classification. Techniques like CNNs, RNNs, GA models, and transfer learning frameworks have been successfully applied to improve signal clarity and robustness. These models can suppress background noise, separate overlapping signals, and generalize well across different underwater environments. By leveraging ML-driven solutions, underwater communication systems are becoming more resilient and reliable, enabling more accurate and consistent data transmission even in harsh acoustic conditions. As research progresses, integrating ML with domain-specific knowledge in acoustics and oceanography is expected to further advance the capabilities of underwater communication networks.

14.3.1 Machine learning solutions for challenges of UAC

ML methods can help address limited bandwidth issues in networks, including UAC systems where bandwidth is intrinsically constrained. By using ML methods, existing bandwidth can be optimized, resulting in more efficient and intelligent 6G communication underwater networks that are valuable in a highly dynamic environment. In addition, UAC systems inherently suffer from high propagation delay due to the slow speed of sound in water, which is significantly lower than the speed of electromagnetic waves in air or vacuum. This delay is exacerbated in deep-sea or long-range communications and poses major challenges to real-time data transmission, medium access control, routing protocols, and feedback-based error correction. Traditional solutions often rely on conservative scheduling, long guard intervals, or delay-tolerant network (DTN) architectures, but these approaches can be inefficient. To overcome these challenges, recent research explores ML-based strategies that offer proactive, adaptive, and delay-resilient communication in UAC environments.

One key ML approach involves predictive learning for proactive communication, where models are trained to anticipate future channel conditions or data requirements, reducing reliance on round-trip feedback. For example, Jing *et al.* [3] developed an LSTM model to predict future link quality and latency based on past transmission records in UA channels. Their LSTM-based predictor enabled autonomous adaptation of modulation schemes and transmission power levels before the onset of delay, thereby enhancing throughput and reducing packet loss without real-time feedback. Another ML-driven strategy is delay-aware routing using RL [4]. In UAC networks, route selection must consider both link quality and latency, which are often dynamic and difficult to measure directly. To address this, Su *et al.* [5] proposed a deep Q-network (DQN) approach that learns optimal routing policies by interacting with the underwater environment. Such approaches ensure timely update of the model while reducing the delay and overhead associated with centralized communication [6].

In addition to the obstacles mentioned above, underwater networks are faced with Doppler spread and strong multipath effects caused by reflection from the oceans. These factors lead to frequency-selective and time-varying fading, which severely degrades signal integrity, particularly over long distances or in dynamic underwater conditions. Traditional signal processing methods, such as adaptive equalization, time-reversal processing, and diversity combining, often become computationally intensive or ineffective under such complex and nonstationary conditions. To overcome these limitations, ML techniques are increasingly being employed to model, predict, and mitigate multipath and fading effects in UAC systems. One of the most prominent ML applications in UAC is in adaptive channel equalization. Traditional equalizers struggle with the highly time-varying and nonlinear characteristics of underwater channels. To address this, Gao and Liu [7] proposed a deep neural network (DNN)-based equalizer trained on simulated and real underwater multipath channels.

Moreover, as previously mentioned, the Doppler effect has a significant effect on the performance of underwater networks. This phenomenon can significantly impact the performance and accuracy of UAC networks. The Doppler effect, in particular, occurs when a sound source and a receiver are moving relative to each other: (1) if the source and receiver move toward each other, the frequency increases (compressed wave) and (2) if they move away from each other, the frequency decreases (stretched wave). In sonar communication, this shift alters the characteristics of the received signal, which can cause problems in the detection, tracking, and decoding of the information. In UAC systems, alleviating the Doppler effect is a critical challenge due to the frequency shifts caused by relative motion between the transmitter and receiver. Recent advances in ML have introduced data-driven approaches that offer promising solutions. ML offers data-driven adaptability and the ability to model nonlinear and nonstationary dynamics.

One prominent line of research uses supervised learning to estimate and compensate for Doppler shifts. In these approaches, ML models such as CNNs or RNNs are trained on labeled datasets where input features (e.g., channel estimates, signal spectrograms) are mapped to Doppler shift values. A CNN-based Doppler estimation framework is proposed for OFDM systems that learns to infer Doppler spread directly from time-frequency signal representations, demonstrating improved robustness in high-speed scenarios [8]. Another promising direction leverages deep learning for end-to-end learning of communication systems, where Doppler mitigation is implicitly learned as part of the modulation and demodulation process. In this paradigm, autoencoders are trained to jointly optimize the transmitter and receiver functions over channels affected by Doppler. O'Shea and Hoydis [9] laid the foundation for this approach, while subsequent works extended it to fading and Doppler-impacted channels using deep neural networks trained with stochastic channel models.

RL has also been applied, particularly in scenarios where the channel conditions evolve over time and require continuous adaptation. This is particularly important in 5G and 6G systems, where devices operate in diverse and dynamic mobility conditions. In summary, ML-based Doppler mitigation represents a significant advancement over classical signal processing by offering adaptability, data-driven modeling, and the ability to handle nonlinearities in high-mobility channels. However, challenges remain, including generalization to unseen channel conditions, interpretability of learned models, and the need for large datasets. As underwater wireless systems become more complex and mobile, the integration of ML techniques for Doppler compensation is expected to play a central role in next-generation underwater communications. Furthermore, by integrating effective and robust ML techniques, UAC systems can significantly enhance their resilience against ambient and impulsive noise, leading to improved data transmission reliability and efficiency.

ML methods offer effective solutions to mitigate these challenges in UAC systems through advanced noise suppression, classification, and signal enhancement techniques. These noise sources, stemming from natural phenomena such as marine life activities and environmental conditions, as well as anthropogenic factors such as ship engines and underwater drilling, significantly degrade the SNR and increase BER. Advancements in ML have significantly improved the resilience of UAC systems against ambient and impulsive noise. Deep learning-based denoising techniques have been pivotal in this regard. For instance, dual-path RNNs utilize multiscale convolutional networks to extract higher-order features and employ bidirectional and unidirectional RNNs for effective noise reduction, demonstrating improvements in SNR by up to 12 dB in challenging underwater environments. Additionally, CNNs have been implemented in denoising autoencoders to process time-frequency representations of acoustic signals, effectively suppressing impulsive noise and enhancing signal clarity.

Furthermore, GANs have been applied in ambient noise-free frameworks to generate clean underwater ship engine audios from hydrophone data, significantly improving the SNR by removing background ambient noise. Likewise, transfer learning approaches have also been employed to address the scarcity of labeled real-world data. A notable example is the use of a 1DCNN-MAM, which combines simulation-based training with real-world data to enhance impulsive noise suppression in OFDM systems [10]. This method reduces the need for extensive labeled data and improves generalization in practical UAC scenarios. In addition to denoising, noise classification and source separation are crucial for robust communication. SVMs and CNNs have been utilized to classify various underwater noise types based on features such as power spectral density and Mel-frequency cepstral coefficients, achieving high classification accuracy even under low SNR conditions. Moreover, deep learning models leveraging time-frequency masking with Bi-STM networks have demonstrated effective separation of overlapping signals, isolating desired signals from background noise in multivariate and high-noise environments.

Zhou *et al.* [11], a deep learning model is proposed aimed at enhancing the quality of UA signals by effectively mitigating noise interference. They propose a noise-aware deep learning model termed NAFSA-Net, which integrates a full-band–sub-band attention (FSA) network. Another study, the DPRNN, integrates multi-scale CNNs with RNNs to achieve superior noise reduction performance. The model employs a multi-scale CNN to extract higher-order nonlinear features from the noisy input signal. Furthermore, recent studies have introduced novel architectures to enhance UA signal processing. For example, a dual-stream deep learning-based acoustic denoising model with frequency characteristics transformation (DS-FCTNet) Gao *et al.* [12] have been proposed to enhance the perception of underwater information. This model predicts amplitude and phase masks separately and employs an information exchange module to swap learned features between the amplitude and phase spectra, aiding in better phase information reconstruction and signal denoising. Collectively, these ML-driven methodologies offer promising solutions to the challenges posed by noise in UAC networks, paving the way for more reliable and efficient underwater communication technologies.

14.4 Conclusion

In summary, this chapter underscores the transformative role of machine learning (ML) in overcoming the inherent limitations of underwater acoustic communication (UAC). By leveraging advanced ML models, ranging from traditional supervised approaches to cutting-edge deep and RL techniques, researchers and engineers can significantly enhance signal clarity, reduce latency, and adapt to dynamic underwater environments. These intelligent systems not only improve current communication capabilities but also lay the groundwork for the development of future autonomous and large-scale marine networks. As ML continues to evolve, its integration with UAC networks promises a new era of reliable, efficient, and resilient underwater communication.

References

[1] Theocharidis T, and Kavallieratou E. Underwater communication technologies: a review. *Telecommunication Systems*. 2025;88(2):54.

[2] Awan M, Whangbo TK, and Shin J. Deep learning methods for autonomous driving scene understanding tasks: a review. *Expert Systems with Applications*. 2025; p. 128098.

[3] Jing L, Dong C, He C, *et al.* Adaptive modulation and coding for underwater acoustic communications based on data-driven learning algorithm. *Remote Sensing*. 2022;14(23):5959.

[4] Geng X and Zheng YR. Exploiting propagation delay in underwater acoustic communication networks via deep reinforcement learning. *IEEE Transactions on Neural Networks and Learning Systems*, 2022;34(12):10626–10637.

[5] Su Y, Fan R, Fu X, *et al.* DQELR: an adaptive deep Q-network-based energy and latency-aware routing protocol design for underwater acoustic sensor networks. *IEEE Access*, 2019;7:9091–9104.

[6] Yurdem B, Kuzlu M, Gullu MK, *et al.* Federated learning: overview, strategies, applications, tools and future directions. *Heliyon*, 2024;10(19):e38137. https://www.sciencedirect.com/science/article/pii/S2405844024141680.

[7] Gao L, and Liu S. Underwater acoustic channel estimation based on sparsity-aware deep neural networks. In: 2021 *OES China Ocean Acoustics (COA). IEEE*; 2021. pp. 544–549.

[8] Yoon E, Kwon S, Yun U, *et al.* Doppler spread estimation based on machine learning for an OFDM system. *Wireless Communications and Mobile Computing*, 2021;2021(1):5586029.

[9] O'Shea TJ, and Hoydis J. An introduction to deep learning for the physical layer. *IEEE Transactions on Cognitive Communications and Networking*, 2017;3:563–575.

[10] Wan L, Xu S, Diao Y, *et al.* Impulsive noise mitigation for underwater acoustic OFDM systems based on 1DCNN With multiattention mechanism and transfer learning. *IEEE Internet of Things Journal*, 2025;12(8):9917–9926.

[11] Zhou A, Zhang W, Li X, *et al.* A novel noise-aware deep learning model for underwater acoustic denoising. *IEEE Transactions on Geoscience and Remote Sensing*, 2023;61:1–13.

[12] Gao W, Liu Y, and Chen D. A dual stream deep learning-based acoustic denoising model to enhance underwater information perception. *Remote Sensing*, 2024;16(17):3325.

Chapter 15

Machine learning techniques for optimizing underwater acoustic signal transmission

Muhammad Hassan[1], Syed Shaheer Raza[1] and Hoang Thi Huong Giang[2]

Underwater acoustic communications (UAC) are required for a range of mission-critical applications, including oceanographic monitoring, disaster detection, off-shore exploration, and naval operations. However, the underwater environment is extremely challenging: the acoustic channel offers extremely limited bandwidth and high latency due to the slow speed of sound, and signals are subject to attenuation, multipath interference, and ambient noise. These factors, combined with dynamic ocean environments and the finite battery life of underwater devices, significantly hinder UAC performance. Machine learning (ML) has been an extremely promising tool to counter these challenges. Unlike fixed models, ML models can learn from complex and time-varying underwater data and adapt to dynamic channel conditions in real time. This adaptive capability improves communication reliability, optimizes routing, and improves signal processing efficiency. In this chapter, supervised learning algorithms, such as support vector machines (SVM), k-nearest neighbors (KNN), and decision trees, are utilized for operations such as channel estimation, adaptive routing, and node localization. Likewise, unsupervised techniques such as K-means clustering and principal component analysis (PCA) enable network topology control, feature reduction, and energy-efficient operation. All of these ML-based methodologies combined contribute significantly to the robustness and efficiency of UAC networks. The chapter concludes with a vision for the future of ML-empowered UAC. Key research directions are the design of scalable, low-power ML models suitable for underwater deployment and the use of advanced techniques such as deep learning and semi-supervised learning to address limited labeled data. Such advances are of extreme significance for making intelligent and reliable underwater communication systems a part of real-world applications.

[1]School of Electrical Engineering and Computer Science (SEECS), National University of Sciences and Technology (NUST), Pakistan
[2]Graduate School of Information Science and Technology, Tokyo City University, Japan

15.1 Introduction

UACs are the process of conveying information through sound waves in water to facilitate essential applications in the submerged environment. They constitute the core of oceanographic data gathering systems, pollution monitoring systems, disaster warning networks, offshore exploration, navigation aid, and tactical naval surveillance. All of these functions are important in environmental monitoring, disaster early detection, national defense, and deep-sea exploration missions. However, UAC systems have to deal with extreme challenges presented by the aquatic environment. The underwater acoustic channel is probably one of the toughest wireless communication environments in the natural world, resulting in serious performance degradation. This acoustic channel has unique features that differ from the radio channels applied to terrestrial wireless networks. Radio frequency waves are attenuated quickly in seawater and are not practical for long-distance underwater communication; therefore, acoustic waves (and sometimes optical waves for very short ranges) serve as the carrier of choice. Using sound allows communication over kilometers, but at the cost of very limited bandwidth and slow propagation speed. The acoustic frequency band is narrow compared to radio, which severely restricts data rates; in practice, most underwater acoustic links support on the order of tens of kilobits per second at best. These acoustic signals are heavily attenuated over distance and frequency (with severe propagation loss), and reflections at the sea surface and seabed generate multipath interference that can result in abrupt fading of the received signal.

In addition, sound propagation is five orders of magnitude slower than electromagnetic waves, approximately 1500 m/s in water compared to 3×10^8 m/s in air, creating significant latency in signal exchange. This low velocity results in substantial propagation delays, as a signal may take seconds to reach a few kilometers [1]. It also causes multipath echoes to arrive spread out in time: acoustic waves reflect off the surface, seabed, and various obstacles, and these delayed replicas can persist for tens or hundreds of milliseconds, leading to inter-symbol interference and fading. Additionally, higher-frequency acoustic signals attenuate more rapidly, so only low-frequency (hence lower data rate) signals can be used for long-range transmission, further limiting channel capacity. The ocean environment itself is extremely dynamic and unpredictable, with currents, waves, and temperature gradients constantly changing the conditions of the channel. Consequently, conventional model-based communication designs tend to be ineffective, as it is not easy to build strong analytical models for such time-varying channels. The underwater acoustic propagation becomes even more complex because of its sensitivity to environmental conditions. The sound speed varies with water temperature, salinity, and depth, creating stratified sound speed profiles; as these factors fluctuate with location and time (diurnal cycles, currents, weather), the channel characteristics also change dynamically, making the link highly variable and site-dependent. In underwater sensor networks, these challenging channels interconnect diverse node types:

- *Source nodes*: For example, seafloor sensors or autonomous underwater vehicles (AUVs), which generate data and transmit it to collection points.

- *Relay nodes*: Intermediate devices that forward packets in a multi-hop fashion to extend coverage.
- *Destination (sink) nodes*: Typically surface buoys or vessels that collect and process data.

All underwater nodes are powered by batteries and are deployed in environments where replacing or recharging batteries is a great challenge. Indeed, powering submerged equipment is a major obstacle; solar panels are impractical underwater, and battery retrieval requires costly device recovery, so energy consumption is a critical design constraint. Moreover, node mobility (e.g., drifting buoys or moving AUVs) introduces Doppler shifts and dynamic network topologies, while ambient noise from marine life (e.g., snapping shrimp) and human activities (e.g., ship engines, sonar) adds further interference. These factors, limited bandwidth, high latency, multipath varying in time, stringent energy constraints, node mobility, and heavy noise, make the design of efficient and resilient UAC systems a formidable challenge. Sophisticated methods such as machine learning (ML) are therefore being explored to mitigate these effects and enhance UAC performance [2]. ML has thus been seen as a promising solution to counter these UAC challenges. In contrast to being dependent on static theoretical models, ML algorithms can learn directly from empirical underwater data and adjust to new conditions in real time, offering a degree of adaptive intelligence that traditional approaches cannot match.

For instance, an ML-based system could automatically modulate its scheme or coding rate in real-time to compensate for channel fading, or optimize routing and transmission scheduling according to real-time feedback from the network, thus enhancing link efficiency and reliability under dynamic ocean conditions. By doing so, ML methods give UAC networks the capability to repeatedly improve their communication methods (e.g., adaptive modulation, dynamic routing, and careful transmission timing) and continue functioning regardless of the ocean's uncertainties. The focus of this chapter is on various ML models that can be utilized to enhance UAC performance. The chapter then outlines possible solutions to each of these challenges, which will open up new opportunities for researchers in this field. The rest of the chapter is arranged in the following manner. Section 15.2 discusses related work, Section 15.3 provides solutions to challenges with the aid of ML, and Section 15.4 concludes the chapter.

15.2 Related work

Underwater acoustic sensor networks are plagued by low energy, as poor channel conditions result in excessive packet retransmissions that quickly exhaust batteries. Chen *et al.* [3] intended to address this problem by proposing an ML-based underwater channel quality prediction approach and deferring transmissions during poor channel conditions. In their approach, termed ML-based environment-aware communication channel quality prediction (ML-ECQP), each underwater link employed a logistic regression predictor to forecast the quality of the communication channel (in terms of bit error rate) as a function of quantifiable environmental variables such as signal-to-noise ratio (SNR), water temperature, and wind speed. During poor predicted channel conditions, the transmitter defers rather than transmits, and hence avoids transmissions during bad times, as well as the wasted energy due to unsuccessful retries. They implemented this solution using real measurements of underwater acoustic experiments in a campus lake to train and cross-validate the prediction model, and then by extensive simulation to compare the performance of their solution. They showed that transmitting only when scheduling a packet on the channel that is predicted to be good can significantly lower the transmitter's energy cost and packet loss ratios on a system without channel prediction. These findings illustrate how environment-aware prediction can enhance network lifetime by avoiding unnecessary retransmissions. The authors focused on improving the quality of the underwater channel, but there are still limitations and some aspects of their proposed method are an open problem.

The logistic regression predictor for the underwater channel is never optimal; for example, it might mark a good channel as "bad" and miss an available transmission opportunity, resulting in unnecessary delay and additional latency to data transmission. In a similar way, when a bad channel is incorrectly predicted to be good, a packet then transmits too early, resulting in a failed transmission and wasted energy spent transmitting. Hence, although ML-ECQP avoids retransmissions, it results in additional latency compared to state-of-the-art methods due to these errors. Another limitation is that only one ML algorithm and test platform were investigated. The authors used logistic regression because it was inexpensive and computationally simple, but they did not experiment with more complex algorithms that can model more complex underwater channel behavior. Moreover, the model was trained on a single lake experiment dataset, so one has absolutely no idea how general the applicability of the model is to varied underwater situations or varying conditions. To address these limitations, more complex learning algorithms (or other environmental parameters) can be utilized in the future to make more accurate predictions and prevent false actions. An online learning adaptation scheme can allow the predictor to adapt as conditions evolve and further improve both prediction accuracy and energy-efficiency trade-offs.

Another research, Mannepalli *et al.* [4] solved the issue of having stable high-rate orthogonal frequency division multiplexing (OFDM) links in underwater acoustic channels with a high Doppler shift. The authors introduced a low-complexity Doppler compensation scheme where the integer and fractional time scaling of every OFDM block were jointly estimated (through cyclic-prefix correlations and sample-rate offsets) to compensate for carrier frequency offsets and Doppler stretch. The tracking of intra-block Doppler variation was achieved using a one-pilot-tone mechanism and a smoothing filter to deliver improved bit error performance and throughput compared to conventional block-based compensation. The technique required only moderate computation and was thus implementable on energy-limited underwater nodes. The results of simulation and shallow-water experiments showed that the scheme efficiently eliminates errors in typical fading environments. The authors demonstrated that in the absence of prior knowledge of target motion, their receiver recovers orthogonality among subcarriers and maintains data integrity with a single-tone pilot. Overall, their main contributions were practical design and performance demonstration of an OFDM receiver, which efficiently compensates for Doppler and carrier frequency offsets (through integer/fractional estimation) and was demonstrated to work in field tests. However, the algorithm shown is predicated on only modest Doppler variation per packet; if the scale varies strongly between blocks, performance suffers. Their testbed is also constrained (fixed node spacing, shallow water, single marine noise source), so the robustness of the solution in heterogeneous ocean conditions has not been determined.

They mentioned ways to improve more dynamic Doppler tracking as a future work, for example, iteratively refining the scaling estimate or applying adaptive filters to cope with fast-varying channels. However, from the user's point of view, incorporating data-driven or learning-based optimizations could enhance adaptability further, e.g., a simple ML predictor can predict channel scaling from experience, or extra pilots or tones can be adaptively turned on when there is strong Doppler. Overall, they proposed a decent low-complexity UAC-OFDM design, but left open the question of how to scale the approach to highly dynamic Doppler and multi-path conditions beyond those shown. Likewise, Saravanan *et al.* [5] applied ML to predict underwater wireless sensor network performance. Specifically, they compared a novel logistic regression method with standard linear regression to predict the performance of an underwater acoustic sensor network. From a very limited dataset (20 total samples) and rigorous statistical testing, they determined that logistic regression performs significantly better than the linear model. To give an example, their novel logistic method had approximately 87.8% accuracy (12.2% loss) while linear regression had approximately 80.2% accuracy (19.3% loss), a difference that is statistically significant ($p < 0.05$). This indicates that the nonlinear logistic model better models the underlying structure in their acoustic data. They also mentioned a clustered network topology with advanced cluster head nodes to resist attenuation, but this aspect was not quantitatively compared to their ML results.

Their model showed improvement, but the fundamental questions remain unanswered. Their experimental design was very narrow, with only 20 data points, and two algorithms were examined, so the results are unlikely to apply to other situations. Moreover, it is not entirely clear what communication efficiency metric was being predicted or how the training data were generated (e.g., measured field data or simulation) in their scheme, so questions remain about noise, overfitting, and relevance to real-world deployment. In actual networks, underwater environments (nonuniform temperature, salinity, multi-path, etc.) and network characteristics (node mobility, energy limitations) could render prediction impossible. So, their results can be more persuasive if shown on larger, real-world datasets and if more models were tested. As a trivial example, follow-up work could include more features (for example, channel SNR or physical sensor readings), use cross-validation to avoid overfitting, and experiment with other ML algorithms (decision trees, SVMs, or even neural networks) for comparison. Adding these features, larger sample sets, richer feature sets, and more sophisticated algorithms can further make the case that ML can actually improve UAC prediction. In its current form, their main contribution was to show that logistic regression yields better accuracy than linear regression in a narrow case; further work along this line of research could yield more useful tools for underwater network planning and adaptation.

15.3 How ML can improve UAC?

ML has emerged as a highly effective way to address the challenges of UAC. Advances in ML and neural networks quickly enabled the processing of severely degraded, low-SNR underwater signals [6]. By learning sophisticated channel patterns and optimal combinations of modulation and spreading techniques, ML-based receivers are able to demodulate weak signals and improve covert communication reliability. Essentially, ML adds data-driven adaptability, such as neural network–aided signal processing, which can remove multipath and ambient noise, enabling more reliable waveform design and detection. Thus, the integration of ML into UAC systems improves performance and security in the hostile low-bandwidth underwater environment (see Figure 15.1). We divide this section into supervised and unsupervised ML algorithms that can enhance UAC performance.

15.3.1 Supervised ML algorithms

Zhang *et al.* [7] proposed an adaptive SVM-based routing (ASVMR) protocol for underwater acoustic sensor networks (UASN) that directly addresses the issues of excessive packet loss, excessive delay, and wasteful energy expenditure. In this design, the SVM was trained to select the relay nodes based on four features (remaining energy, link quality) of the state of the nodes rather than a single parameter. Classifying a candidate set of neighbors, their SVM model adaptively selects good next hops to minimize end-to-end delay and balance load. Their primary contributions were (1) an SVM-based multi-attribute forwarding decision that yields significantly better routing decisions, (2) a better waiting time mechanism with node distances to reduce delay and packet loss, and (3) an adaptive routing pipe radius based on packet

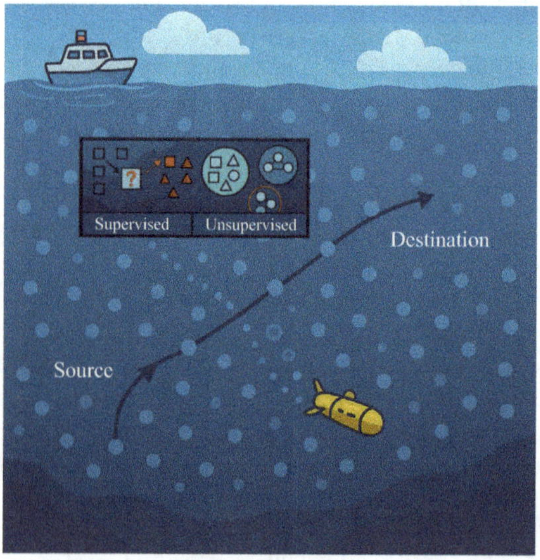

Figure 15.1 The machine learning-based underwater acoustic communications

delivery ratio (PDR) to reduce unnecessary transmissions. Their simulation results showed that ASVMR achieves high PDR with lower delay and energy consumption. However, the method requires training and computation of the SVM model at sensor nodes; indeed, the authors of ASVMR stated that the complex mechanism and computational overhead involved may be challenging for real underwater deployment. However, lightweight or incremental SVM training, integration with reinforcement learning for online adaptation, or hardware-accelerated implementations to enable SVM routing in resource-constrained UASNs can be explored as future work.

Fang *et al.* [8] utilized random forest (RF) to detect modulation modes in non-cooperative UAC signals, bypassing the difficulty of classifying multipath and noise-corrupted signals. A new spectral peak feature extraction that is resistant to underwater channel effects was proposed. The feature, pre-processing, and waveform optimization were fed into an RF classifier, which had been trained to distinguish between OFDM and a number of frequency-shift keying (FSK) schemes. The key contribution was this hybrid of optimized feature extraction and ensemble classification, which achieved very high recognition accuracy. Their simulations and sea trial data showed recognition rates greater than 95–98% for all modes when SNR is reasonably high. However, the approach is not without its limitations; it employs empirical parameter settings for different signal bandwidths, so performance suffers if the FSK bandwidth is very small or very large. In other words, the spectral feature extractor must be calibrated to the scenario. Future research can learn or automate these parameters, generalize to other modulation types, or combine RF with deep neural networks to improve generalization. This RF approach paves the way for adaptive classification of UAC waveforms and can be a building block for blind demodulation or cognitive acoustic systems.

Shah *et al.* [9] proposed trust estimation model based on the decision tree for UASN optimized for underwater environments. The decision tree utilized a set of contextual features; sensor node energy, mobility (water currents), link quality (SNR or bit error rate) and reliability feedback history, to determine trustworthiness. Its strengths were a dynamic trust update system, if a link decision fails, trust values are updated (with a learning rate factor) such that the tree learns and adapts over time. Their simulations and experimental results demonstrated that their approach achieved around 96% trust estimation accuracy with very low false positives, with reduced communication overhead and energy consumption. In short, the tree-based method offers an energy-efficient, feedback-based decision-making solution to UASNs (as routing or data acceptance) by taking each cluster of features as a tree split. The weaknesses in their scheme are its dependence on reliable measures of trust (which may be difficult to acquire) and that it was tested on specific scenarios; real-world networks may have dynamic threat models. The online learning or anomaly detection in the tree and investigation of decentralized or federated tree training such that nodes can independently maintain trust under mobility can be extended as future work.

Shamshad *et al.* [10] used KNN for node localization and orientation prediction in dynamic underwater sensor networks. They introduced a KNN-based algorithm that solves the localization error and the energy cost due to currents. The algorithm was operated using the location of the closest neighbors to predict the position and orientation of each node, efficiently computing the shortest distance between sensors in real time. Their main contribution was a remarkably high localization accuracy; in simulation and tank experiments they achieved a reduction in error from an initial 4.59 m to effectively zero (3.88×10^{-8} m) and a general accuracy of 99.98%. They also significantly minimized the energy cost and the time cost of localization, that is, the energy per localization of 0.0045 J, compared to considerably higher in previous approaches. In practice, the KNN solution optimizes which anchors to use and how to interpolate positions to achieve maximum precision. However, such near-ideal outcomes are theoretical; real ocean deployments may introduce noise, non-line-of-sight, or need dynamic updating of communication between nodes. The online adaptation of KNN (like adapting with node density), inclusion of motion models for mobile nodes, and integration of KNN with learning-based approaches (e.g., deep networks) can be included as future work to deal with more complex environments and noisy measurements.

15.3.2 Unsupervised ML algorithms

An energy-efficient clustering and cooperative routing based on improved K-means and Q-learning (ECRKQ) was used in a hybrid ML scheme for mobile UASN in Zhu *et al.* [11]. The authors adapted the base K-means by choosing cluster heads (CHs) by distance from the centroid and in terms of node residual energy. The adapted K-means made clusters depend upon centrally located nodes that possess sufficient energy. The clustering process was further optimized with Q-learning for cooperative routing, but the main ML contribution was adaptive clustering using K-means for the CH election. Their simulation results (with nodes that drift) indicated that ECRKQ achieves better balance in network energy consumption compared to existing methods. Their key contribution was a combined clustering and routing approach in which K-means were adapted to the node energy. However, there are still open issues that involve the latency associated with iterative learning and the model assumption of prior knowledge of the node energy level. The fully unsupervised or adaptive clustering, like ML choosing the number of clusters, as well as application of the concept to re-clustering for dynamically varying node energy levels, can be considered as future work.

Likewise, a centralized control-based clustering scheme (CCCS) was suggested in Tian *et al.* [12] to balance energy in UASN. Their network was two-level; the sink had a global view, balancing energy between clusters, and each cluster had a cluster controller (CC) to balance energy within the cluster. During CH selection, CCCS considered both the residual energy of a node and the total energy budget of the cluster after role assignment. Clusters were treated as super nodes at the top level and CCs managed nodes at the bottom level. This centralized solution significantly prolonged network lifetime; in simulations, CCCS always produced longer node survival and more packets delivered than the benchmarks. However, depending on centralized control presumes the sink can know everything, which may not be feasible in very large or dynamic networks. Hence, hybridizing this with distributed learning such as CH election based on local artificial neural network or exploring how to extend CCCS to node failure and mobility might be considered future work.

Zhu *et al.* [13] employed principal component analysis (PCA) via principal component regression as a feature selection step for the location of the underwater acoustic source. PCA is a traditional dimensionality reduction technique that maximizes variance and eliminates correlations between raw features. Using PCA, they identified the most informative spectral features with respect to the source location and eliminated redundant information. Then these principal components were input into a deep learning localization pipeline. Their contributions were that this PCA-based feature selection provided both interpretability, indicating which frequencies matter, and significantly speeds up training; the neural network learned much more quickly with only a small reduction in localization accuracy. The method also used a semi-supervised autoencoder or multi-layer perception architecture to leverage unlabeled data. The limitations of their work are PCA's linearity, which may miss nonlinear relationships, and the fact that the approach was tested on only one dataset. The nonlinear variants, kernel PCA or autoencoder-based feature learning, multi-sensor arrays, or adaptive feature selection that updates as more data arrive can be included as future work. These PCA methods open the door to faster, more efficient ML pipelines for UAC tasks where data are high-dimensional, but labeled examples are limited.

15.4 Conclusion

In summary, UAC is central to the internet of underwater things and ocean monitoring, but it faces severe challenges from limited bandwidth, high delay (sound speed of approximately 1500 m/s), powerful fading or noise, and stringent energy constraints. ML has been a powerful antidote to these challenges by enabling environment-aware channel estimation, adaptive routing, modulation recognition, and other intelligent features. This chapter reviewed both supervised and unsupervised ML approaches to UAC. For example, supervised classifiers like SVM, KNN, decision trees, and random forests have been used for routing, localization, and trust evaluation; in fact, an SVM-based routing protocol that adaptively adjusts paths to reduce end-to-end delay and maximize network lifetime was discussed. Unsupervised approaches have been used to control network topology and energy. Energy-efficient clustering uses modified K-means (and Q-learning) to optimize node energy usage in mobile UASNs, and centralized cluster controllers control relay selection to optimize energy usage in the network. Dimensionality reduction using PCA was also discussed. The selection of PCA-based features for source localization achieved about 95% faster training in the literature and improved robustness when labeled data are scarce. In general, these ML-powered techniques have great potential to greatly improve the reliability, efficiency, and adaptability of UAC systems. However, there are several important open challenges that need to be addressed in future work. Underwater datasets have limited labeled data (so purely supervised models might not generalize), and the complexity of the ML algorithms must be traded off against device energy and computational resources. Future research should also explore semi-supervised and deep or reinforcement learning methods to leverage unlabeled data, lightweight online adaptation to dynamic channels, multi-modal sensing integration, and large-scale field trials to evaluate ML-based UAC solutions.

References

[1] O. Onasami, M. Feng, H. Xu, M. Haile, and L. Qian, "Underwater acoustic communication channel modeling using reservoir computing," *IEEE Access*, vol. 10, pp. 56550–56563, 2022.

[2] L. Huang, Y. Wang, Q. Zhang, J. Han, W. Tan, and Z. Tian, "Machine learning for underwater acoustic communications," *IEEE Wireless Communications*, vol. 29, no. 3, pp. 102–108, 2022.

[3] Y. Chen, W. Yu, X. Sun, L. Wan, Y. Tao, and X. Xu, "Environment-aware communication channel quality prediction for underwater acoustic transmissions: A machine learning method," *Applied Acoustics*, vol. 181, p. 108128, 2021. https://www.sciencedirect.com/science/article/pii/S0003682X21002218.

[4] K. Mannepalli, K. B. Raju, J. Sirisha, K. Saikumar, and K. S. Reddy, "Low complex OFDM channel design using underwater-acoustic-communication using machine learning techniques," in *2021 5th International Conference on Electronics, Communication and Aerospace Technology (ICECA)*, 2021, pp. 1505–1513.

[5] M. Saravanan, V. Naveen, S. P. Kumar, V. Dilli Ganesh, S. Sivashankar, and K. V. Bhaskar, "Enhancing the prediction of underwater wireless communication using machine learning techniques and related issues," in *2023 International Conference on Data Science, Agents & Artificial Intelligence (ICDSAAI)*, 2023, pp. 1–5.

[6] S. Liu, M. A. Khan, M. Bilal, and H. H. Zuberi, "Low probability detection constrained underwater acoustic communication: a comprehensive review," *IEEE Communications Magazine*, vol. 63, no. 2, pp. 21–30, 2025.

[7] S. Zhang, H. Chen, and L. Xie, "ASVMR: Adaptive support-vector-machine-based routing protocol in the underwater acoustic sensor network for smart ocean," *Journal of Marine Science and Engineering*, vol. 11, no. 9, pp. 1–22, 2023. https://www.mdpi.com/2077-1312/11/9/1736.

[8] T. Fang, Q. Wang, L. Zhang, and S. Liu, "Modulation mode recognition method of non-cooperative underwater acoustic communication signal based on spectral peak feature extraction and random forest," *Remote Sensing*, vol. 14, no. 7, pp. 1–19, 2022. https://www.mdpi.com/2072-4292/14/7/1603.

[9] S. Shah, A. Munir, A. Salam, *et al.* "A dynamic trust evaluation and update model using advance decision tree for underwater wireless sensor networks," *Scientific Reports*, vol. 14, no. 22393, pp. 1–22, 2024.

[10] N. Shamshad, L. Wang, K. Saleem, *et al.* "Advanced KNN-based cost-efficient algorithm for precision localization and energy optimization in dynamic underwater sensor networks," *Scientific ReportsSci*, vol. 15, no. 2182, pp. 1–17, 2025.

[11] J. Zhu, Y. Chen, X. Sun, J. Wu, Z. Liu, and X. Xu, "ECRKQ: Machine learning-based energy-efficient clustering and cooperative routing for mobile underwater acoustic sensor networks," *IEEE Access*, vol. 9, pp. 70843–70855, 2021.

[12] W. Tian, Y. Zhao, R. Hou, *et al.* "A centralized control-based clustering scheme for energy efficiency in underwater acoustic sensor networks," *IEEE Transactions on Green Communications and Networking*, vol. 7, no. 2, pp. 668–679, 2023.

[13] X. Zhu, H. Dong, P. Salvo Rossi, and M. Landrø, "Feature selection based on principal component regression for underwater source localization by deep learning," *Remote Sensing*, vol. 13, no. 8, pp. 1–16, 2021. https://www.mdpi.com/2072-4292/13/8/1486.

Chapter 16

Exploring the underwater acoustic environment through deep learning

Mohaira Ahmad[1]

There is an entirely different world beneath the water's surface, where sounds play their roles. Sound waves travel in water for surprisingly long distances instead of fading away quickly as light or radio waves do, carrying messages through the darkness where sight is useless. As sound has better propagation in water, it has become the main method for everything from communication underwater, exploring the underwater environment changes, to keeping an eye on what's happening below, and understanding underwater ecosystems. The development of sound navigation and ranging (SONAR) technology has spurred advances in underwater acoustics. SONAR and hydrophones are now rapidly being used to navigate the tricky environment of underwater, study marine life, explore geological structures, and monitor the impact of human activities such as shipping and resource extraction. However, the underwater world is not straightforward; it is filled with unpredictable echoes, swirling currents, and constant noise that make it a challenging place to work with sound.

The SONAR technology came into attention in the mid-20th century, in particular during and after World War II. Over time, scientists and engineers have had to find creative ways to use this technology so that all these acoustic signals start making sense. The latest leap forward has come thanks to deep learning – an area of artificial intelligence (AI) that can spot subtle patterns hidden in huge amounts of data. Classic methods of breaking down underwater sounds work well up to a point, but these methods suffer from challenges like persistent noise, signal overlap, or the shifts caused by moving sources. Deep learning approaches, such as convolutional neural networks (CNNs), have changed the game by being able to tell apart the quiet buzz of a distant ship from the lively chatter of ocean animals. Other structures, like recurrent neural networks (RNNs) and transformers, help in understanding sounds that play out over time or involve complex interactions.

This chapter takes you through these new developments, starting with a basic look at how sound moves through the sea and how it is traditionally analyzed. The main focus of this chapter is to see how the powerful deep learning models are designed specifically for underwater use, along with the tools and datasets, and how

[1]School of Electrical Engineering and Computer Science, National University of Sciences and Technology, Pakistan

they are being used by the researchers to judge their progress. These models support a wide array of applications: from aiding ships in safe navigation, to enabling environmental researchers in monitoring marine wildlife, and assisting the military in detecting and tracking underwater activity.

All technological advancements come with their challenges, especially in their application in a complex environment like underwater. There is a limitation in the ability to deploy and maintain the equipment for data collection in deep water that often results in gaps in the data. Sometimes, the data collected from various sources creates a difficulty in differentiating between natural and anthropogenic noise levels, affecting the accuracy of methods. Furthermore, the underwater environment changes constantly, so the methods that work in one region might not work elsewhere. As researchers are developing these methods, there is an increased interest in making these deep learning methods more efficient and combining them with sensors that can actually "see" in different ways.

Overall, this chapter will be both an introduction to the underwater acoustic environment and a beginning point for anyone keen to understand the possibilities made by deep learning in studying the underwater acoustic environment and work beyond that. It brings together what we know, highlights where we are still searching for answers, and invites fresh ideas for the future.

16.1 Underwater acoustic environment-exploration

16.1.1 Influence of the propagation medium

The underwater kingdom accounts for more than 70% of the Earth's surface. In this ever-changing world of underwater, sound serves as the main messenger because electromagnetic waves, which dominate the telecommunication above water, fail in this part of the world. The water, particularly the salt water, exhibits a strong conductivity and is therefore highly dissipative. Thus, electromagnetic waves attenuate extremely rapidly, due to which they have limited range and usefulness in underwater communications. The only practical way to carry information underwater is through acoustic waves. The journey of every acoustic wave is shaped by a combination of physical, chemical, and even biological forces. They can propagate easily in seawater, allowing creatures and sensors to communicate and sense their surroundings in ways that would otherwise be impossible [1].

Sounds are an important part of our aerial environment. We intuitively master their physical characteristics using our physiological senses and use them constantly, mostly for communication. Things are different when we talk about the underwater environment. The ocean is never quite the same from moment to moment. These acoustic waves propagate with four to five times higher speed in water, reaching higher levels sometimes and with less attenuation, thus covering larger distances. One of the most remarkable features of this underwater world is the existence of sound channels. These are the natural pathways, such as the famous sound fixing and ranging (SOFAR) channel in deeper waters, where sound waves can travel for thousands of kilometers with very little loss. Marine mammals have been using these pathways to stay in touch across long distances, and human beings have tried to mimic this feature using technology to use them for everything from research to underwater surveillance. The underwater world is also alive with a tapestry of sounds: the rhythmic snaps of shrimp, the whistles of dolphins, and the rumble of passing ships. Not all noise is coming from natural sources, activities like commercial shipping, drilling, and military SONAR add more layers to this acoustic tapestry, and their influence changes depending on both place and time. The common sources of anthropogenic noise underwater are shown in Figure 16.1.

The advantages of sound wave propagation in an underwater environment are overshadowed by some constraints. The propagation of sound underwater depends a lot on the constantly changing nature of the sea itself. While the speed of sound in air usually stays the same, things are much more complicated underwater. There are certain factors like temperature, the amount of salt, and the pressure as you go deeper, that combine to change how fast and in what direction sound moves. Sometimes, these changes bend sound waves, creating mysterious "shadow zones" where certain sounds can barely be heard or not at all heard and "convergence zones," where sound seems to gather and become unexpectedly loud at specific spots.

One of the trickiest hurdles for underwater communication is a phenomenon called multi-path propagation. Here, a single sound does not take just one straightforward path to its destination. Instead, it bounces off the seafloor, the surface, and anything else it encounters, arriving at the receiver at slightly different times. This creates overlapping echoes, which can jumble up or mask the original message. This is especially common in shallow waters, where such reflections are much more likely

Figure 16.1 Anthropogenic sources of underwater noise

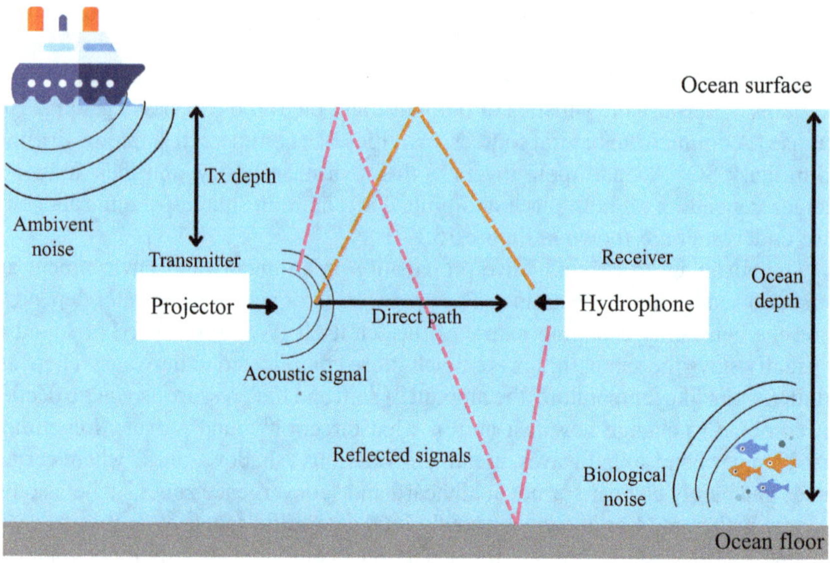

Figure 16.2 Underwater acoustic propagation scenario

to pile up. A typical scenario of the underwater acoustic propagation is shown in Figure 16.2.

Another challenge is that sound gradually loses its strength as it travels through water, a process known as attenuation. High-pitched (high-frequency) sounds, in particular, fade faster than low-pitched ones, meaning they cannot travel as far. On top of that, tiny particles, bubbles, and even schools of fish scatter the sound, especially in areas near coastlines where the water tends to be murkier. To make things even more complicated, the ocean is always in motion. Currents, tides, and underwater waves constantly reshape the way sound travels from one moment to the next. Moving objects, whether they're animals, ships, or the receiver itself, can shift the pitch of sounds which is known as the Doppler effect which has to be carefully corrected in communication and navigation systems.

For those who design SONAR and underwater communication technology, understanding this complex and ever-changing soundscape is crucial. Making sense of all this requires a team effort, drawing on insights from oceanographers, acousticians, and signal processing specialists. Every detail, from how sound bends and bounces to how it fades and shifts, has to be taken into account to ensure messages are delivered clearly beneath the waves. New technologies, especially those powered by deep learning, are now making it possible to sift through this complex soundscape and pick out meaningful signals with greater precision than ever before, opening up new opportunities for exploration, communication, and discovery beneath the waves.

16.1.2 Basics of underwater acoustics

Underwater acoustics is the study of how sound propagates through water. Since light is absorbed and scattered rapidly by water, sound becomes the primary means for communication, navigation, and exploration underwater. This field plays a critical role in numerous oceanographic and naval applications such as SONAR, submarine communication, underwater navigation, and marine biology. Understanding how sound behaves in a water body like an ocean requires insight into the ocean's physical properties, primarily how they affect the speed of sound in water. Unlike air, water causes the speed of sound to vary significantly due to its physical properties, making underwater acoustics an essential domain. The speed at which sound travels underwater is not constant. It depends on temperature (T), salinity (S), and pressure, which is a function of depth (z). These relationships are captured in an empirical formula, which provides the sound speed c in meters per second (m/s):

$$c = (1449.2 + 4.6T - 0.055T^2 + 0.00029T^3 + (1.34 - 0.01T)(S - 35) + 0.016z) \tag{16.1}$$

where T is the temperature in degrees Celsius, S is the salinity in parts per thousand, and z is the depth in meters.

The property that most influences the speed of sound underwater is temperature. As temperature increases, so does sound speed. In deeper regions, where temperature stabilizes, pressure becomes the dominant factor increasing sound speed. Salinity has a moderate effect but is usually less variable compared to temperature.

In tropical regions, the water is generally warmer due to solar heating near the ocean surface. This results in higher sound speeds at the surface. As depth increases, the temperature decreases, which causes a rapid decline in sound speed. This region is called the thermocline. Below the thermocline, the temperature becomes relatively constant, but pressure continues to rise, which in turn increases the sound speed again at greater depths. This creates an almost U-shaped sound-speed profile in temperate waters, where the sound speed is highest at the surface, reaches a minimum at mid-depths, and gradually increases in deeper layers. This minimum speed region is known as the deep sound channel or SOFAR channel. It acts as a natural waveguide for sound, allowing acoustic waves to travel thousands of kilometers with very little attenuation. In polar regions, the sound-speed profile is different. The water is coldest at the surface, and the temperature increases slightly with depth. Therefore, the minimum sound speed occurs near the surface, and the deep sound channel is either absent or behaves differently compared to equatorial waters. These sound–speed profiles can be seen in Figure 16.3:

Shallow water bodies also exhibit different sound-speed profiles in different seasons. In summer, surface heating leads to a higher sound speed at the top, which then decreases with depth. This creates a downward refracting environment for sound waves. During winter, wind, and wave activity mix the water column thoroughly, creating an isothermal and nearly isovelocity profile, meaning the sound speed remains fairly constant throughout the depth.

Sound propagation can also be affected by ocean boundaries. The sea surface, though reflective, is often rough due to wind and waves, leading to sound scattering. The seafloor, with its layered composition and varying geo-acoustic properties, can reflect, absorb, or transmit sound. Together, these boundaries form an acoustic waveguide, where sound behavior is shaped by both the ocean's internal sound speed structure and its external limits [2].

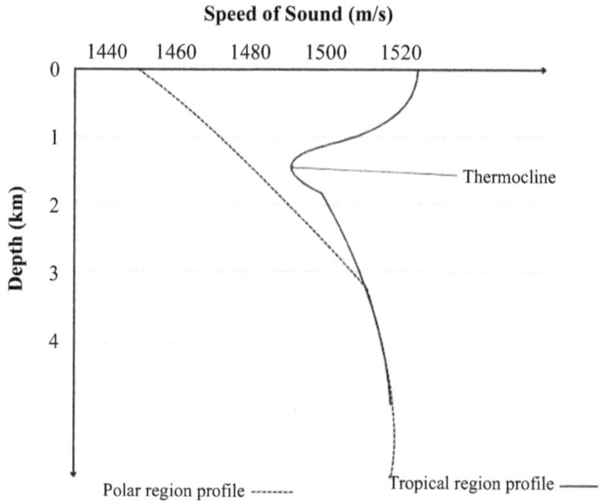

Figure 16.3 Underwater sound speed profile

16.1.3 SONAR systems

Acoustic waves find their application in many avenues underwater [3]. They are used to:

- Detect and locate the targets: SONARs are mostly used for this purpose.
- Determine the characteristics of the marine environment, finding the location and/or velocity of a moving object underwater.
- Transmit signals which may act as data acquired by the underwater scientific instruments, or the messages between submarines and surface vessels, or to command the remotely operated systems.

Sonar systems are the primary tools for underwater detection, navigation, and communication. They operate in two main modes: active SONAR; they emit signals, and these signals will be reflected on a target or transmitted directly to the receiver. Analyzes their echoes and passive SONAR, which is designed to listen for underwater sounds coming from other sources/targets.

Active Sonar Systems are widely used for object detection, seabed mapping, and underwater imaging. These systems transmit carefully designed acoustic pulses such as linear frequency-modulated (LFM) chirps or coded sequences and process the returning echoes to extract information about targets. The echo principle of active SONAR is shown in Figure 16.4.

The key challenges of active SONAR include:

- The target discrimination: Separating echoes from clutter (e.g., rocks, fish schools).
- Doppler compensation: Correcting for frequency shifts caused by moving targets.

(a)

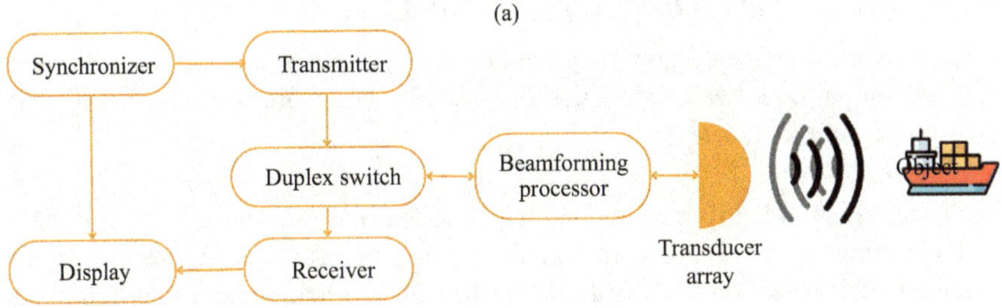

Figure 16.4 Echo principle of active SONAR

(b)

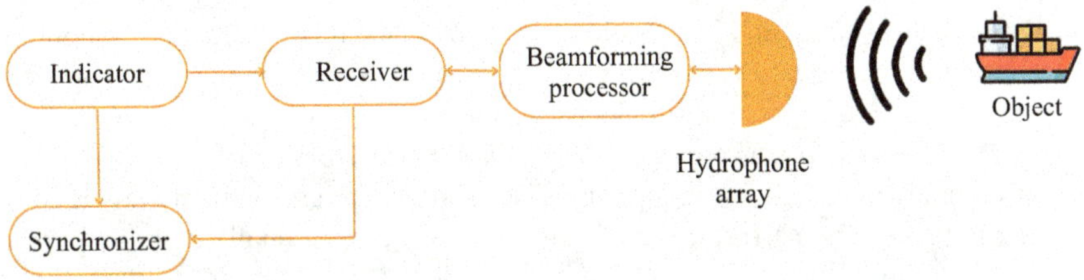

Figure 16.5 Echo principle of passive SONAR

- Beamforming: Using hydrophone arrays to localize sound sources with high precision.

Modern active SONAR systems increasingly rely on machine learning techniques to improve detection performance. Deep neural networks, for example, can learn to recognize subtle patterns in echo returns that traditional algorithms might miss.

Passive Sonar Systems do not emit signals but instead listen for sounds generated by vessels, marine life, or other sources. This mode is crucial for stealth applications, such as submarine tracking, as well as for environmental monitoring. The passive SONARs have the echo principle as shown in Figure 16.5.

Passive systems face unique challenges, including:

- The separation of weak signals from noisy backgrounds.
- The identification of signals, whether a sound comes from a ship, whale, or other emitter, is known as signal classification.
- The determination of the direction of arrival of incoming sounds using hydrophone arrays.

Advanced signal processing techniques, such as time–frequency analysis and adaptive filtering, are essential for passive SONAR. Recent developments in deep learning have enabled breakthroughs in automated target recognition, allowing systems to classify sounds with human-like accuracy.

16.1.4 Traditional SONAR signal processing

The designs of signals and how they are processed directly affect the performance of the traditional underwater acoustics systems [3]. The main functions of SONAR signal-processing include:

- Detecting and identifying a signal that is buried in noise.
- Estimating specific parameters like the propagation time of the acoustic signal, the angular direction of arrival. This is helpful in tracking the target relative to the SONAR.
- Target identification or characterization by estimating some of its intrinsic parameters.

Detection of signals and localization have been made possible by using antennas, which increase the signal-to-noise (SNR) ratio and measure the angles. The conventional signal processing methods to detect signals are based on the Neyman–Pearson criterion [4]. This method is based on the probability distribution of the received signals in comparison to the noise signals. The probability ratio is then compared to the present value. The performance of these methods is continually being improved by the large number of studies over the last few decades, in the various domains of SONAR signal processing.

Acoustic signals are one-dimensional signals that oscillate in amplitude with respect to time. The temporal information of audio signals is very useful and is used in many applications but some of their characteristics can be best explored in the frequency domain [5]. Fourier transforms (FT) are used to obtain the information in the frequency domain from these time domain signals [6]. The FT of an autocorrelation function yields the power spectrum, which quantifies signal energy at each frequency [7,8]. While amplitude and frequency content can be directly read from the power spectrum, it lacks temporal resolution. To capture how frequencies evolve over time, one method known as the short-time Fourier transform (STFT) is used. In this method, the signal is windowed in time, each segment is FT-analyzed, and the results are assembled into a spectrogram plotted with time (x-axis), frequency (y-axis), and magnitude (z-axis) [7,8]. Further insight into harmonic structure comes from the cepstrum, the inverse FT of the log-spectrum, effectively the power spectrum of the log-power spectrum [9,10]. Human hearing is nonlinear across frequencies [11], so the Mel scale, a perceptually motivated, logarithmic frequency axis, often replaces linear frequency in feature extraction. Applying a Mel-filter bank to the spectrum, taking logs, and then a discrete cosine transform produces Mel frequency cepstral coefficients (MFCCs), which compactly represent perceptually salient spectral features [12,13]. Neither the FT nor STFT localizes abrupt time–frequency changes well, since sinusoids extend infinitely in time. Wavelet transforms overcome this by decomposing signals into finite-duration "wavelets" (e.g., Morlet, Mexican hat), parameterized by scale (frequency) and shift (time), yielding a time–scale matrix that captures both high- and low-frequency events [14–16]. Inspired by mammalian visual processing, Gabor filters, band-pass kernels tuned to specific frequencies and orientations, have found use in sound-spectrum preprocessing. Gabor-filter banks enhance texture-like patterns in spectrograms and have improved performance in tasks ranging from environmental–sound classification to music-genre and speech analysis [17,18].

These time frequency and wavelet representations transform raw audio into richly structured data, but they do not themselves classify or interpret signals. Traditionally, humans or basic statistical models performed that task; today, machine learning – and especially deep neural networks – has become the standard approach for learning to recognize patterns in these complex, noisy feature spaces.

However, working with underwater signals comes with its own unique set of challenges. For one, the background noise in the ocean does not behave the way it does in electronic circuits; it often follows patterns that are much trickier to predict. On top of that, the conditions underwater can change quickly; currents shift, temperatures fluctuate, and the way sound travels can be different from one moment to the next. This means that signal processing techniques have to be flexible and able to adapt on the run. Another challenge is the limited space for information; underwater communication does not support high-speed data transmission that we see with radio waves in the air. Despite these obstacles, researchers are making exciting progress. New methods like compressed sensing are helping to pick out useful information from sparse data, while cognitive SONAR systems are learning to adjust and "think" for themselves, opening new possibilities for underwater listening and communication.

16.1.5 Applications

The science of how sound travels underwater is not just limited to research labs; it plays a significant part in both military activities and everyday civilian life. On the military side, navies around the world depend on SONAR technology for a range of critical tasks, from trying to spot stealthy submarines hiding below the surface to detecting underwater mines and keeping an eye on wide stretches of ocean. The demands of modern operations mean these systems need to perform even in noisy, unpredictable conditions, often while avoiding being detected themselves. It is becoming more common to send robotic submarines, known as autonomous underwater vehicles (AUVs), on missions of their own, using their advanced SONAR systems to scout out the seabed or gather information in places that are hard for humans to reach. The application of underwater acoustics extends far beyond military uses. Today, scientists are using acoustic tools to study the mysteries of the deep ocean, track the movements of marine animals, investigate the impacts of climate change, or learn about the geology hidden below the waves. In commercial fishing, SONAR helps boats pinpoint schools of fish, making operations more efficient and sustainable. The energy sector, too, benefits from these technologies, with offshore oil rigs and wind farms using underwater sound monitoring to keep their infrastructure safe. And when accidents happen, search and rescue teams often depend on acoustic equipment to find objects or wrecks below the water's surface. With rapid advances in technology, AI is starting to work hand in hand with underwater acoustics. This partnership is leading to impressive new possibilities, such as real-time environmental monitoring and smart systems that can communicate and adapt beneath the waves, helping us understand and protect our oceans like never before [3].

16.1.6 Outcome

The world beneath the waves is a place full of both difficulties and possibilities. Because sound travels in such unusual ways underwater, making sense of it requires some pretty advanced processing methods. At the same time, the ocean remains one of Earth's least explored frontiers, offering a wealth of opportunities for new discoveries and cutting-edge technology. Sonar, whether it is sending out signals or simply listening, has become essential for finding our way underwater, helping out everywhere from naval defense to industrial operations and scientific research. As our computational tools keep improving, especially with the rise of deep learning and AI, we are gaining an even deeper ability to study, understand, and make use of the mysteries that the underwater world still holds.

16.2 Deep learning for underwater acoustic analysis

In recent times, the oceans of the world have gained the attention of the scientists, industries and military organizations, to explore underwater archeology [19], perform maritime explorations [20], understand transportation logistics [21], take renewable energy initiatives [22] and for military applications [23]. These applications have a common challenge, that is to conduct safe surveys, data collection and exploring the unpredictable environment of underwater. One of the most viable solution of data collection is provided by SONAR sensors, that rely on sound waves for underwater sensing and navigation. Combining the SONAR based data collection and deep learning in underwater acoustic analysis has truly changed the way the underwater sounds are listened to, interpreted, and used. This shift in technology is helping to tackle some of the trickiest challenges in understanding the underwater world, thanks to neural networks specially tuned for these soundscapes.

New systems powered by deep learning using tools like neural networks excel at picking out everything from the haunting songs of whales to the tell-tale pulses of ship engines, even when there's a lot of background noise or overlapping sounds. These smart models are always learning what "normal" ocean noise sounds like so they can quickly spot anything out of the ordinary, whether that is a rare weather event or signs of human activity.

When it comes to classifying what is making those sounds, deep learning goes far beyond what traditional methods could do. Modern networks, can now tell apart different species of marine mammals by recognizing subtle differences in their calls that would be almost impossible for a human ear or even older algorithms to distinguish. For sounds created by people, like ship traffic [24], these neural networks sort through complex harmonics to identify not just the presence of a ship, but even to guess its type or engine specs with remarkable accuracy. And when mapping the seafloor, deep learning helps us pull detailed information from acoustic data, giving us clearer pictures of underwater landscapes than before [25].

Locating where a sound is coming from underwater has also become much more precise with these techniques. By learning how local conditions affect how sound travels, new deep learning models do a better job than traditional approaches, especially in places where echoes and reflections would normally cause confusion.

Some networks even build a kind of "acoustic map" of the ocean, helping pinpoint sounds in three dimensions and under changing conditions.

Deep learning is not just about hearing things better it is about understanding the whole scene. It can combine audio with extra information like temperature or salinity readings to create rich maps that reveal how ocean environments change over time. Today's smart networks can analyze the echoes from the seabed to figure out what is down there, or even detect important features like thermoclines simply from the way they change the sound.

There are different types of SONARs including forward-looking SONAR (FLS), side-scan SONAR (SSS), and multibeam echo sounders (MBES), where each SONAR is designed for specific tasks [26]. The SONAR signals can be used for automatic target recognition (ATR) tasks including feature extraction, classification, detection, and segmentation. However, SONAR signals do not have plenty of information on detailed content of the target [27]. Deep learning is an effective tool that can be used to analyze and process these SONAR signals for ATR tasks. The next section will explain some commonly used deep learning models.

What sets these AI-driven systems apart is their ability to keep learning and adapting on their own, handle complicated types of noise, and pick up on connections buried deep in the data. Because they are built with a foundation in the physics of sound, but enhanced by the flexibility of modern machine learning, these methods are setting new standards for research, environmental monitoring, and navigation beneath the waves. Blending acoustic know-how with advanced AI means scientists and engineers can now solve problems that once seemed out of reach and explore the underwater world with fresh eyes and ears.

16.2.1 Deep learning algorithms

In recent years, the landscape of underwater acoustic signal analysis has been profoundly reshaped by the integration of deep learning, with several foundational neural architectures now underpinning state-of-the-art systems. Among these, CNNs [28] have become a mainstay in many underwater acoustic research applications due to their effectiveness in handling grid-structured data such as spectrograms and time–frequency representations. Their utility extends from detecting and classifying marine targets to identifying fine-grained patterns in SONAR imagery.

A typical CNN architecture is composed of an input layer responsible for ingesting either raw acoustic inputs or preprocessed spectrograms followed by successive convolutional layers that apply a set of learned filters to highlight salient features. Pooling layers help to distill the information by reducing spatial dimensions, while retaining the key characteristics necessary for meaningful interpretation. Finally, one or more fully connected layers perform classification or regression tasks to yield the network's predictions. A defining quality of CNNs is their use of weight sharing in convolutional layers, conferring shift-invariance, which facilitates robust pattern detection even in the presence of time–frequency shifts, a common occurrence in underwater channels. Beyond underwater acoustics, CNNs are recognized for their versatility in diverse fields including image classification, video analysis, and language processing.

The primary strengths of CNNs lie in their ability to autonomously learn hierarchical features, largely eliminating the need for handcrafted feature engineering. Their inherent translation invariance ensures resilience against Doppler effects and temporal shifts. However, these networks are data-intensive, often necessitating large datasets to achieve generalizable performance. Fixed input dimension requirements can also restrict their applicability to situations with variable-length signals.

For the analysis of inherently sequential data, RNNs [29] and in particular, the long short-term memory (LSTM) [30] variant has demonstrated substantial value. The basic LSTM cell incorporates input, forget, and output gates, as well as a memory cell, enabling the network to capture dependencies over extended temporal spans. This makes LSTMs particularly effective for tracking temporal structure within acoustic sequences. They excel at adapting to variable-length inputs and maintain context over long episodes, while their architecture helps to mitigate the vanishing gradient problem historically associated with deep sequential models. Nevertheless, LSTM networks process data sequentially, limiting opportunities for parallelization and resulting in increased training times compared to their convolutional counterparts. Memory constraints may also become significant when handling long-duration signals, presenting challenges for deployment in real-time systems with limited resources.

The advent of the transformer architecture [31] has introduced a paradigm shift in sequential data modeling. By leveraging mechanisms such as multi-head self-attention, positional encoding, feed-forward sub-layers, and robust normalization, transformers facilitate the efficient processing of long sequences. This architecture allows for parallelized operations and excels at modeling long-range dependencies capabilities especially beneficial for interpreting complex and extended underwater acoustic events. Yet, transformers can be computationally demanding, with memory and processing requirements that scale quadratically with input length, and they remain bounded by fixed positional encoding constraints, which may pose limitations for the analysis of exceedingly long acoustic records.

Autoencoders [32] represent another valuable class of models within the underwater acoustics domain, particularly in scenarios requiring noise reduction and compact feature learning. Comprised of an encoder that compresses input data, a bottleneck layer for latent representation, and a decoder for reconstructing the original signal, autoencoders leverage unsupervised learning techniques to distill essential features and mitigate noise. Their utility in dimensionality reduction not only enhances computational efficiency but also aids in anomaly detection and data compression. As with any compression technique, however, autoencoders may introduce reconstruction errors and require a careful balance between representing the essential structure of the input and preserving fine detail. Robust training of autoencoders is often sensitive to model architecture and demands careful hyperparameter optimization.

Siamese networks [33] have gained traction for their role in comparative analysis and tracking within underwater acoustic contexts with small sample sizes. Characterized by parallel network branches with shared parameters, a similarity metric, and a contrastive loss, these models excel at identifying relationships between pairs of input signals. Their ability for one-shot or few-shot learning supports robust matching and improved generalization, even in scenarios with limited labeled data. However, training these networks can be computationally intensive, as they necessitate curated positive and negative input pairs. Their effectiveness is also closely linked to the appropriateness of the chosen distance metric, which may present difficulties when subtle or complex similarities are involved.

In summary, these deep learning architectures CNNs, LSTMs, transformers, autoencoders, and Siamese networks collectively enable more nuanced, adaptive, and robust analysis of underwater acoustic data. Nevertheless, each comes with inherent trade-offs in terms of data requirements, computational cost, and model design. Ongoing advances in network architecture, optimization, and hybridization of these models continue to expand the frontier of what is possible in underwater acoustic signal processing.

16.2.2 Acoustic signal analysis based on deep learning

Acoustic signal analysis involves a range of tasks that aim at improving the quality and interpretability of SONAR data, as well as extracting the useful information. The underwater acoustics can be categorized as remote sensing. When an object, condition, or any phenomenon of interest cannot be directly observed and information is acquired indirectly using the data, it comes under remote sensing. Sonar systems are also a form of remote sensing. As described in Section 16.1.3, the SONAR systems act as both active and passive SONARs. In acoustic signals processing, SONARs are used as remote sensors for the tasks including but not limited to detection/identification of underwater sound sources and targets and their localization [23,34–36].

Underwater acoustic environments present significant opportunities and challenges for deep learning methodologies. The critical need for robust underwater surveillance, communication, and environmental monitoring systems has propelled research into advanced machine-learning techniques for acoustic signal processing. This literature review critically examines recent developments in applying deep learning techniques, specifically focusing on target detection, tracking, recognition, and classification in underwater acoustics. It aims to analyze existing methodologies, their theoretical underpinnings, and operational implications, while identifying significant research gaps to guide future investigations. Deep learning, a subset of machine learning, employs neural networks structured in multiple layers to model complex patterns in data. Deep learning architectures like CNNs, RNNs, and transformers are particularly adept at feature extraction, classification, and temporal pattern recognition. In underwater acoustics, traditional signal processing methods such as matched-field processing (MFP), FT, and wavelet analysis have historically dominated. However, their performance deteriorates under environmental mismatches and complex acoustic conditions, thus opening avenues for deep learning techniques.

Recent studies have highlighted significant methodological innovations in deep learning applications to underwater acoustics. [37] reviewed a range of techniques, including CNNs, multi-task learning (MTL), and gated feedback recurrent units (GFGRU), emphasizing their effectiveness in source localization and recognition. They demonstrated that deep learning methods significantly outperform traditional

MFP under real-world conditions, especially regarding robustness against environmental variability and noise interference.

[5] categorized deep learning methodologies for SONAR data classification into five primary strategies: CNN-based direct approaches, biologically inspired preprocessing (e.g., Gabor filters), time–frequency domain analysis (spectrograms, cepstrograms), ML-based feature extraction, and transfer learning. They highlighted high detection accuracies in controlled shoreline surveillance contexts, using active and passive SONAR data.

[38] proposed integrating advanced terrestrial network technologies, such as software-defined networks (SDN), network function virtualization (NFV), fog/cloud computing, and the Internet of Underwater Things (IoUT), into underwater acoustic networks. This architectural shift aims to enhance flexibility, scalability, and robustness of underwater communication and target tracking infrastructures, emphasizing the necessity of accurate sound propagation modeling.

[39] noted severe limitations posed by the scarcity of annotated acoustic datasets, proposing extensive use of data augmentation techniques and transfer learning to mitigate these challenges. Their findings highlight that deep learning models, while showing superior performance under controlled conditions, significantly degrade in more complex or low SNR scenarios, especially in shallow, cluttered waters.

Deep learning methods have undeniably advanced underwater acoustic signal processing, demonstrating superior performance over traditional techniques. The structured taxonomy by [5] and [39] clarifies methodological approaches and reveals significant performance improvements achievable through biologically inspired preprocessing and transfer learning. These methodologies, particularly CNN-based models, exhibit robust generalization capabilities when trained appropriately. However, critical analysis reveals consistent challenges regarding real-world applicability. Most studies demonstrate high performance in controlled environments or simulations but face significant accuracy drops in realistic, variable oceanic conditions. For instance, [37] demonstrated enhanced robustness using numerical simulations but acknowledged challenges in operational scenarios lacking extensive labeled data. Similarly, [39] confirmed that advanced deep learning methods' real-world applicability remains limited without significant environmental-specific fine-tuning. Integrating emerging terrestrial technologies, as discussed by [38], offers theoretical advantages but lacks empirical validation. The absence of standardized evaluation frameworks further complicates meaningful comparisons across studies, weakening the evidence base for practical deployments.

16.2.2.1 Source localization

The sources of sound are located in different locations in underwater environments as shown in Figure 16.1. The received sound varies as a result and this can be used for passive sound source localization. The process of localization involves identifying a mapping relationship between the received sound and the location of the source. The source localization can be divided into active location and passive location. The passive positioning involves sensor arrays that extract helpful information from the received signals and determine the target location by different methods. The passive positioning is more commonly used due to its benefits including low cost, better camouflage, high security and long range operation [40].

The traditional method used for passive source localization since 1990s is known as MFP. The MFP algorithm was based on underwater physical parameters of sound and environment. It is basically the extension of beamforming algorithm in the marine environment. For a long time, scientists have worked on the development of MFP technology by making it robust, adaptive and high resolution [41–43]. However, there are still many issues in actual implementation of MFP algorithms such as extensive computation requirements, complex real-time processing and environmental mismatch problems [37].

ML methods can solve the optimization problems and they are in use for sound source localization since 1991 [44]. This was a pioneer study to employ neural networks to localize an acoustic point source within a homogenous medium. Shortly after that in 1991, [45] applied associative feedforward neural networks with no hidden layers to estimate the range and depth using a vertical array of sensors. These ML algorithms failed at that time due to limited hardware resources and algorithms. Later in 2017, the rapid advancement of computer hardware and ML algorithms paved the way for ML related methods in underwater acoustics. The study by [46] demonstrated the localization performance of a nonlinear regression algorithm in fluctuating environment of ocean by using the data from water tank experiments. The results displayed the advantages and the potential of ML algorithms for using in underwater acoustic source localization. In 2017, scientists [47,48] developed a practical class of ML-based underwater acoustic source localization methods. They analyzed Noisy09 experiment dataset using three ML methods including feedforward neural network (FNN), support vector machines (SVM), and random forest (RF). This work was known as a significant milestone in employing ML for underwater source ranging using sea trial data. They also succeeded in alleviating the environmental mismatch problem.

Deep learning, with its powerful self-learning and adaptive capabilities, has become a cornerstone for addressing underwater acoustic ranging challenges provided one has suitable training data. A research in 2018 [49] showed that a CNN could estimate the range of broadband noise sources in shallow seas more reliably than traditional approaches that require precise environmental models. Building on this, Huang combined CNN and FNN to create a hybrid CNN-FNN architecture that markedly improved performance under low signal-to-noise conditions [49]. In 2019, Niu in [50] demonstrated the advantages of depth by employing a 50-layer residual neural network to jointly estimate both range and depth of an acoustic source, outperforming MFP. Around the same time, Liu [51] leveraged gradient boosting decision trees (GBDT) with South China sea data to achieve accurate deep-sea source ranging, and Wang in [52] introduced transfer learning to adapt models trained in one environment for use in another, yielding superior deep-sea performance.

By 2020, both Liu and Wang had applied various CNN designs to extend ranging capabilities to long-distance, deep-sea scenarios [53,54]. Huang also explored multi-source environments by integrating seed-array beamforming with a two-stage deep neural network (DNN): The first DNN determined source count and arrival directions, and a subsequent RNN refined the actual ranging [55].

More recently, Jin *et al.* [56] combined self-attention and semi-supervised learning within a convolutional autoencoder and multi-layer perceptron to further boost ranging accuracy. A latest work in 2024, combines the local feature acquisition capability of CNN and the global feature capture capability of the transformer to improve the ranging accuracy. The results of sea trial experiment demonstrated that the method based on CNN and transformer model has the best performance, compared with the FNN and CNN neural networks [40].

16.2.2.2 Target detection/identification

Underwater acoustic target recognition (UATR) methods are commonly used in the fields of marine biodiversity detection, marine search and rescue, and seabed mapping, thus enabling the human marine economic and military activities [39]. The primary objective of target recognition is to identify the underwater targets by analyzing their emitting sounds [57]. UATR mostly focuses on the radiated sounds from passive SONARs. UATR is a challenging task that comes along with many practical hurdles [58–60]. These hurdles are due to various factors that include ever changing underwater environments, unpredictable transmission channels and the volatile motion states of vessels. People can spend countless hours manually listening to and identifying underwater sounds and targets, which is not practical for real-world applications [59]. On top of that, the ocean's acoustic data often hides subtle patterns that our ears and even our brains miss entirely [61]. That is why a clear shift has been noted toward using machines do the job of recognizing underwater targets. A modern automatic recognition system for underwater acoustics works a lot like other pattern-recognition tools, following three main steps:

1. Pre-processing strategies are applied on the data collected using different kind of sensors in order to amplify the target signals and mitigate the irrelevant interference.
2. Feature extraction methods transform these processed signals into informative and low-dimensional acoustic features.
3. Recognition models use the statistical methods, linear or non-linear classifiers, or neural networks to extract knowledge from these input features and predict the potential underwater source/target.

A typical pipeline of acoustic source identification process is shown in Figure 16.6.

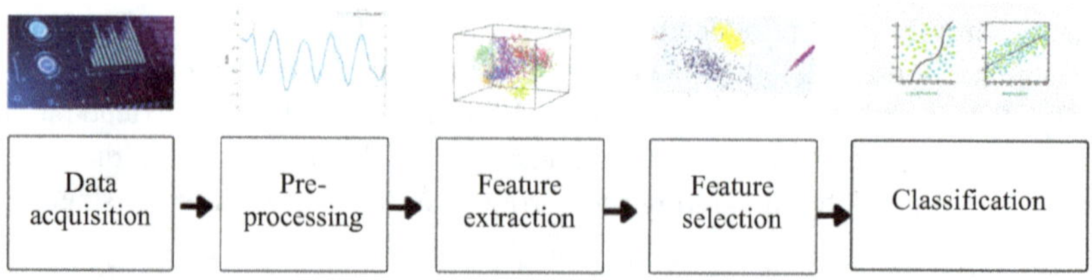

Figure 16.6 Acoustic source identification process

Deep learning has been a game-changer here. It has helped to automate each of these stages, build smarter features, and dramatically improve how accurately and robustly the systems recognize sounds beneath the waves. In recent times, the development of new and advanced algorithms and ever-growing libraries of underwater data, has paved the way to make these systems more reliable. The research in this field can be categorized into multiple directions. Some of the studies have focused to optimize pre-processing algorithms to tackle background noise, signal interferences, low SNR, and even small datasets [58,62]. Others are inventing feature-extraction methods that are specifically designed according to the unique characteristics of underwater acoustics [63,64]. A third kind of studies work on building adaptive and accurate recognition models that are capable of discerning the underwater signals effectively [60]. And lately, a few studies have even started looking past the source of the sounds and instead focusing on the characteristics of acoustic fields [65,66].

16.2.3 Public databases and benchmarks

The underwater acoustic signal acquisition has many challenges and high costs [32,67] and in addition to that the limited data availability and restrictions due to security concerns of military organizations, have all contributed to the scarcity of real-world underwater acoustic signal data. Many works have reported relied on simulated data with predetermined characteristics such as speed, direction and distance. However they failed to simulate the complex interference factors in the underwater environments and even doing so is not practical. This disparity between simulated data and real-world scenarios often leads to reduction in generalization of recognition systems. In recent times, the acquisition technology has improved in response to the increased demands by the research community and thus two real-world underwater acoustic data-bases have been released publicly. These databases have become the fundamental resources for recent work in this domain. These databases are comprised of comprehensive, authentic and diverse datasets that mirror actual underwater scenarios. The details of these two databases are provided in Table 16.1. Many recent research works are based on these publicly available databases and have yielded promising results [59,67–71].

Table 16.1 Benchmark databases information [37]

Dataset	**Location**	**Record**	**Categories**	**Duration (h)**	**Sample Rate (Hz)**	**Extra Info**
ShipsEar	Georgia Delta node near Canada	90	12	2.94	52,734	Yes
DeepShip coast in Northwest Spain	Spanish Atlantic	265	5	47.07	32,000	No

- ShipsEar [67] is comprised of 90 records of sounds of boats and ships from 12 different types (dredgers, fishing boats, mussel boats, trawlers, motorboats, tugboats, sailboats, pilot boats, passenger ferries, Ro-Ro vessels, ocean liners, and background noise recordings). This makes total 2.94 h of recordings. In addition to sound records, ShipsEar also has some supplementary information like the target images, localization data, acquisition time, channel depth, distance, wind conditions, atmospheric and oceanographic data and notes. This additional information is helpful for making a more comprehensive and detailed acoustic data analysis.
- DeepShip [32], comprises of 47.07 h of real-world acoustic data. The data has recordings of 256 ships that are divided in four categories (cargo, passenger boats, tug, and tanker). The DeepShip is an extensive data that enables it to be used in ML and deep learning algorithms. These two databases have appeared to be a valuable benchmark for underwater acoustic analysis. These two databases can be used for training, validation and testing for recognition tasks.

16.2.4 Strategies for efficient deep learning implementation

The effective deployment of deep learning architectures for underwater acoustic analysis increasingly calls for integration strategies that maximize the unique benefits of different models. A popular and promising approach is to employ hybrid methodologies, where the complementary strengths of multiple architectures are harnessed within a unified system. For example, by combining CNNs, well-suited for extracting spatial features from spectrograms with RNNs, which excel at modeling sequential dependencies, researchers can construct frameworks capable of capturing both the spatial and temporal complexities present in underwater acoustic signals [72]. These hybrid architectures are often further refined with task-specific modifications, allowing adaptation of certain components for improved target identification, signal enhancement, or noise suppression, depending on the requirements of a given application [73]. Furthermore, adapting processing pipelines in real time, based on the unique features of incoming data, ensures that such systems remain robust and responsive to diverse and fluctuating underwater conditions.

Another encouraging trend is the refinement of attention mechanisms, which now enable neural networks to selectively pinpoint the most informative segments within complex acoustic inputs. This targeted focus not only supports clearer model interpretation but also often streamlines computational demands, contributing to more efficient yet still accurate performance [74].

At the same time, the field is witnessing a growing emphasis on lightweight neural networks. These compact models retain much of the predictive power of larger architectures while demanding far less from hardware, facilitating reliable operations on devices with limited onboard resources, such as ocean-deployed sensors and AUVs [35]. Another innovative direction involves the emergence of physics-informed neural networks, which weave well-established physical laws and domain expertise directly into their learning processes. By rooting their predictions in sound physical principles, these models can achieve higher fidelity and greater alignment with the realities of underwater environments [75].

Alongside these architectural shifts, significant strides are being made in implementation techniques. Optimizing models for edge computing, for example, allows data to be processed directly where it is collected, substantially reducing both latency and reliance on wide-band communication links – a clear advantage for time-sensitive operations below the surface [35]. The use of hardware acceleration, through technologies like GPUs and domain-specific co-processors, is another key development. This enables resource-intensive architectures to be run more efficiently, bridging the gap between experimental models and their practical deployment [76]. Moreover, distributed processing approaches, which harness the computational power of multiple interconnected devices, offer impressive scalability-making it possible to analyze acoustic data in real time across vast expanses of the ocean.

Adapting models for differing environmental contexts is equally crucial. Transfer learning offers a practical route for accelerating model deployment, as existing pre-trained networks can serve as a starting point for new tasks with related datasets. By fine-tuning these models using data from the target environment, practitioners can significantly reduce training overhead while still leveraging established feature representations relevant to underwater acoustics [76]. Beyond transfer learning, domain adaptation techniques are indispensable for addressing mismatches between source and target distributions, further enhancing the ability of models to generalize across a variety of marine environments. Moreover, the integration of multi-task learning, where a single network is trained to address multiple related objectives at once, has demonstrated notable efficiency advantages. Multi-task settings promote the sharing of useful representations between tasks, leading to improved learning efficiency and performance when data are sparse or when diverse classification and regression goals must be met simultaneously.

Collectively, these trends are paving the way for a new generation of underwater acoustic systems that are not only more accurate, agile, scalable, and accessible but also practical for the vastly variable and complex conditions characteristic of underwater acoustics. As these innovations continue to gain traction, they are well-positioned to transform how the scientific community and industry alike address the multifaceted challenges inherent to the underwater acoustic domain. The careful combination of hybrid modeling, environmental adaptation, and task-sharing reflects a broader trend within the field recognizing the importance of flexibility and contextual awareness in meeting the evolving challenges of real-world acoustic data analysis.

16.2.5 Challenges and future prospects in underwater acoustic analysis

A consistent and pressing gap across reviewed literature is the absence of large-scale, standardized, and publicly available acoustic datasets. Collecting and labeling such datasets is time-consuming and expensive. As a result, much work relies on small custom datasets or attempts to adapt data from other domains. The scarcity severely constrains deep learning model training and robust performance evaluations under varied environmental conditions. Strategies like transfer learning, data augmentation, and integration of physics-based simulations have been used to mitigate this problem [77]. But developing benchmark datasets for underwater sensing remains a key priority to enable more rigorous evaluation and comparison of models.

Many studies implicitly overlook the comprehensive evaluation of deep learning models across diverse oceanographic scenarios (e.g., seasonal variations, geographic differences). Robustness under varying SNR conditions and environmental mismatches remains inadequately addressed. While deep learning models achieve high performance, the ability to interpret the decisions remains limited, posing significant barriers for critical and defense-related applications, as highlighted explicitly by [39].

Despite recognizing the importance of accurate acoustic propagation emulation [38], integration of detailed sound propagation physics into deep learning training and evaluation frameworks is insufficiently explored, presenting an implicit yet significant gap.

Other key priorities for future work include improving the computational efficiency of models for real-time embedded deployment, handling occlusion and camouflage, and leveraging multi-modal sensor fusion [78]. Sonar can be combined with inertial, magnetic, and chemical sensors to provide a richer understanding of the environment. Adapting models to use bio-inspired attention mechanisms and cross-modal representations is a promising direction.

In summary, deep learning is driving rapid progress in underwater acoustic perception across a range of applications and modalities. While significant challenges remain, particularly in terms of data availability and model robustness, there is a clear trajectory toward more intelligent, adaptive, and autonomous underwater systems. Continued research integrating innovations in deep learning with domain knowledge from marine science, ocean engineering, and underwater robotics will be essential to realizing this vision.

A thorough grasp of these foundational deep learning architectures is essential, as they serve as the groundwork upon which more sophisticated approaches to underwater acoustic analysis are constructed. Familiarity with their core operating principles, as well as their respective strengths and constraints, is indispensable when tailoring solutions to address distinct challenges within underwater acoustic environments. Selecting the most appropriate architecture is ultimately a context-driven decision, one that must account for the unique goals of the application, the computational resources at hand, and the particular characteristics of the acoustic data involved. Thoughtful alignment of these elements enables researchers and practitioners to maximize both the efficiency and accuracy of their analysis in the demanding setting of underwater signal processing.

References

[1] Hodges RP. *Underwater acoustics: Analysis, design and performance of SONAR*. John Wiley & Sons; 2011.

[2] Kuperman WA and Roux P. *Underwater acoustics. Springer Handbook of Acoustics*. 2014; pp. 157–212.

[3] Lurton X. *An introduction to underwater acoustics: Principles and applications*. Springer Science & Business Media; 2002.

[4] Nielsen RO. *Sonar signal processing*. Artech House, Inc.; 1991.

[5] Domingos LC, Santos PE, Skelton PS, *et al.* A survey of underwater acoustic data classification methods using deep learning for shoreline surveillance. *Sensors*, 2022;22(6):2181.

[6] Boashash B. *Time–frequency signal analysis and processing: A comprehensive reference.* Academic Press; 2015.

[7] Tsai WH and Lin HP. Background music removal based on cepstrum transformation for popular singer identification. *IEEE Transactions on Audio, Speech, and Language Processing*, 2010;19(5):1196–1205.

[8] Lee CH, Shih JL, Yu KM, *et al.* Automatic music genre classification based on modulation spectral analysis of spectral and cepstral features. *IEEE Transactions on Multimedia*, 2009;11(4):670–682.

[9] Oppenheim AV and Schafer RW. From frequency to quefrency: A history of the cepstrum. *IEEE Signal Processing Magazine*, 2004;21(5):95–106.

[10] Bogert BP. The quefrency alanysis of time series for echoes: Cepstrum, pseudo-autocovariance, cross-cepstrum and saphe cracking. In *Proceedings of the Symposium Time Series Analysis*, 1963; pp. 209–243.

[11] Oxenham AJ. How we hear: The perception and neural coding of sound. *Annual Review of Psychology*, 2018;69(1):27–50.

[12] Stevens SS, Volkmann J, and Newman EB. A scale for the measurement of the psychological magnitude pitch. *The Journal of the Acoustical Society of America*, 1937;8(3):185–190.

[13] Logan B. Mel frequency cepstral coefficients for music modeling. In *International Society for Music Information Retrieval Conference*. vol. 270. Plymouth, MA; 2000. p. 11. https://api.semanticscholar.org/CorpusID:17454278

[14] Donald DA, Everingham Y, McKinna L, *et al. Feature selection in the wavelet domain: Adaptive wavelets*. Elsevier; 2009.

[15] Tzanetakis G, Essl G, and Cook P. Audio analysis using the discrete wavelet transform. In *Proceedings of the Conference in Acoustics and Music Theory Applications*, vol. 66. Citeseer; 2001.

[16] Chuan CH, Vasana S, and Asaithambi A. Using wavelets and gaussian mixture models for audio classification. In *2012 IEEE International Symposium on Multimedia*. IEEE; 2012. pp. 421–426.

[17] Jones JP and Palmer LA. An evaluation of the two-dimensional Gabor filter model of simple receptive fields in cat striate cortex. *Journal of Neurophysiology*, 1987;58(6):1233–1258.

[18] Costa Y, Oliveira L, Koerich A, *et al.* Music genre recognition using Gabor filters and LPQ texture descriptors. In *Progress in Pattern Recognition, Image Analysis, Computer Vision, and Applications: 18th Iberoamerican Congress, CIARP 2013, Havana, Cuba, November 20–23, 2013, Proceedings, Part II 18*. Springer; 2013, pp. 67–74.

[19] Wu J, Bingham RC, Ting S, *et al.* Multi-AUV motion planning for archeological site mapping and photogrammetric reconstruction. *Journal of Field Robotics*, 2019;36(7):1250–1269.

[20] Nakatani T, Ura T, Ito Y, *et al.* AUV "TUNA-SAND" and its Exploration of hydrothermal vents at Kagoshima Bay. In *OCEANS 2008-MTS/IEEE Kobe Techno-Ocean*. IEEE; 2008. pp. 1–5.

[21] Wright M, Gorma W, Luo Y, *et al.* Multi-actuated AUV body for wind-farm inspection: Lessons from the bio-inspired robofish field trials. In *2020 IEEE/OES Autonomous Underwater Vehicles Symposium (AUV)*. IEEE; 2020. pp. 1–6.

[22] Hagen PE, Storkersen N, Vestgard K, *et al.* The HUGIN 1000 autonomous underwater vehicle for military applications. In *Oceans 2003. Celebrating the Past... Teaming Toward the Future (IEEE Cat. No. 03CH37492)*. vol. 2. IEEE; 2003. pp. 1141–1145.

[23] Aubard M, Madureira A, Teixeira L, *et al.* Sonar-based deep learning in underwater robotics: Overview, robustness, and challenges. *IEEE Journal of Oceanic Engineering*, 2025;50(3):1866–1884.

[24] Character L, Ortiz Jr A, Beach T, *et al.* Archaeologic machine learning for shipwreck detection using LiDAR and SONAR. *Remote Sensing*, 2021;13(9):1759.

[25] Krogstad TR, Wiig M, Cleophas P, *et al.* Autonomous object identification in mine countermeasure missions. In *OCEANS'11 MTS/IEEE KONA*. IEEE; 2011. pp. 1–9.

[26] Wang X, Fan X, Shi P, *et al.* An overview of key SLAM technologies for underwater scenes. *Remote Sensing*, 2023;15(10):2496.

[27] Christensen JH, Mogensen LV, and Ravn O. Deep learning based segmentation of fish in noisy forward looking MBES images. *IFAC-PapersOnLine*, 2020;53(2):14546–14551.

[28] LeCun Y, Bengio Y, *et al.* Convolutional networks for images, speech, and time series. *The Handbook of Brain Theory and Neural Networks*, 1995;3361(10):1995.

[29] Lipton ZC, Berkowitz J, and Elkan C. A critical review of recurrent neural networks for sequence learning. *arXiv preprint arXiv:150600019*, 2015.

[30] Kamal S, Chandran CS, and Supriya M. Passive SONAR automated target classifier for shallow waters using end-to-end learnable deep convolutional LSTMs. *Engineering Science and Technology, An International Journal*, 2021;24(4):860–871.

[31] Feng S and Zhu X. A transformer-based deep learning network for underwater acoustic target recognition. *IEEE Geoscience and Remote Sensing Letters*, 2022;19:1–5.

[32] Irfan M, Jiangbin Z, Ali S, *et al.* DeepShip: An underwater acoustic benchmark dataset and a separable convolution based autoencoder for classification. *Expert Systems with Applications*, 2021;183:115270.

[33] Liu D, Shen W, Cao W, *et al.* Design of Siamese network for underwater target recognition with small sample size. *Applied Sciences*, 2022;12(20):10659.

[34] Zaheer R, Ahmad I, Habibi D, *et al.* A survey on artificial intelligence-based acoustic source identification. *IEEE Access*, 2023;11:60078–60108.

[35] Yang H, Lee K, Choo Y, *et al.* Underwater acoustic research trends with machine learning: Passive SONAR applications. *Journal of Ocean Engineering and Technology*, 2020;34(3):227–236.

[36] Yang H, Byun SH, Lee K, *et al.* Underwater acoustic research trends with machine learning: Active SONAR applications. *Journal of Ocean Engineering and Technology*, 2020;34(4):277–284.

[37] Niu H, Li X, Zhang Y, *et al.* Advances and applications of machine learning in underwater acoustics. *Intelligent Marine Technology and Systems*, 2023;1(1):8.

[38] Ghafoor H and Noh Y. An overview of next-generation underwater target detection and tracking: An integrated underwater architecture. *IEEE Access*, 2019;7:98841–98853.

[39] Luo X, Chen L, Zhou H, *et al.* A survey of underwater acoustic target recognition methods based on machine learning. *Journal of Marine Science and Engineering*, 2023;11(2):384.

[40] Qu Y, Huang Y, Ren X, *et al.* Underwater sound source range estimation based on deep learning. In *OCEANS 2024 – Singapore*. IEEE, 2024. pp. 1–8.

[41] Czenszak SP and Krolik JL. Robust wideband matched-field processing with a short vertical array. *The Journal of the Acoustical Society of America*, 1997;101(2):749–759.

[42] Orris GJ, Nicholas M, and Perkins JS. The matched-phase coherent multi-frequency matched-field processor. *The Journal of the Acoustical Society of America*, 2000;107(5):2563–2575.

[43] Debever C and Kuperman W. Robust matched-field processing using a coherent broadband white noise constraint processor. *The Journal of the Acoustical Society of America*, 2007;122(4):1979–1986.

[44] Steinberg BZ, Beran MJ, Chin SH, *et al.* A neural network approach to source localization. *The Journal of the Acoustical Society of America*, 1991;90(4):2081–2090.

[45] Ozard JM, Zakarauskas P, and Ko P. An artificial neural network for range and depth discrimination in matched field processing. *The Journal of the Acoustical Society of America*, 1991;90(5):2658–2663.

[46] Lefort R, Real G, and Drémeau A. Direct regressions for underwater acoustic source localization in fluctuating oceans. *Applied Acoustics*, 2017;116: 303–310.

[47] Niu H, Ozanich E, and Gerstoft P. Ship localization in Santa Barbara Channel using machine learning classifiers. *The Journal of the Acoustical Society of America*, 2017;142(5):EL455–EL460.

[48] Niu H, Reeves E, and Gerstoft P. Source localization in an ocean waveguide using supervised machine learning. *The Journal of the Acoustical Society of America*, 2017;142(3):1176–1188.

[49] Ferguson EL, Williams SB, and Jin CT. Sound source localization in a multipath environment using convolutional neural networks. In *2018 IEEE International Conference on Acoustics, Speech and Signal Processing (ICASSP)*. IEEE; 2018. pp. 2386–2390.

[50] Niu H, Gong Z, Ozanich E, *et al.* Deep-learning source localization using multi-frequency magnitude-only data. *The Journal of the Acoustical Society of America*, 2019;146(1):211–222.

[51] Liu Y, Niu H, Li Z, *et al.* Source localization using gradient boosting decision tree with a single hydrophone in deep ocean. In *Global Oceans 2020: Singapore–US Gulf Coast*, IEEE; 2020. pp. 1–5.

[52] Wang W, Ni H, Su L, *et al.* Deep transfer learning for source ranging: Deep-sea experiment results. *The Journal of the Acoustical Society of America*, 2019;146(4):EL317–EL322.

[53] Liu Y, Niu H, and Li Z. A multi-task learning convolutional neural network for source localization in deep ocean. *The Journal of the Acoustical Society of America*, 2020;148(2):873–883.

[54] Wang W, Wang Z, Su L, *et al.* Source depth estimation using spectral transformations and convolutional neural network in a deep-sea environment. *The Journal of the Acoustical Society of America*, 2020;148(6):3633–3644.

[55] Huang Z, Xu J, Gong Z, *et al.* Source localization using deep neural networks in a shallow water environment. *The Journal of the Acoustical Society of America*, 2018;143(5):2922–2932.

[56] Jin P, Wang B, Li L, *et al.* Semi-supervised underwater acoustic source localization based on residual convolutional autoencoder. *EURASIP Journal on Advances in Signal Processing*, 2022;2022(1):107.

[57] Yang H, Lee K, Choo Y, *et al.* Underwater acoustic research trends with machine learning: General background. *Journal of Ocean Engineering and Technology*, 2020;34(2):147–154.

[58] Dong Y, Shen X, Jiang Z, *et al.* Recognition of imbalanced underwater acoustic datasets with exponentially weighted cross-entropy loss. *Applied Acoustics*, 2021;174:107740.

[59] Xie Y, Ren J, and Xu J. Underwater-art: Expanding information perspectives with text templates for underwater acoustic target recognition. *The Journal of the Acoustical Society of America*, 2022;152(5):2641–2651.

[60] Zhang W, Yang X, Leng C, *et al.* Modulation recognition of underwater acoustic signals using deep hybrid neural networks. *IEEE Transactions on Wireless Communications*, 2022;21(8):5977–5988.

[61] Bianco MJ, Gerstoft P, Traer J, *et al.* Machine learning in acoustics: Theory and applications. *The Journal of the Acoustical Society of America*, 2019;146(5):3590–3628.

[62] Zhou X and Yang K. A denoising representation framework for underwater acoustic signal recognition. *The Journal of the Acoustical Society of America*, 2020;147(4):EL377–EL383.

[63] Jiang J, Shi T, Huang M, *et al.* Multi-scale spectral feature extraction for underwater acoustic target recognition. *Measurement*, 2020;166:108227.

[64] Jiang J, Wu Z, Lu J, *et al.* Interpretable features for underwater acoustic target recognition. *Measurement*, 2021;173:108586.

[65] Zhang W, Wu Y, Shi J, *et al.* Surface and underwater acoustic source discrimination based on machine learning using a single hydrophone. *Journal of Marine Science and Engineering*, 2022;10(3):321.

[66] Yu Q, Zhu M, Zhang W, *et al.* Surface and underwater acoustic source recognition using multi-channel joint detection method based on machine learning. *Journal of Marine Science and Engineering*, 2023;11(8):1587.

[67] Santos-Domínguez D, Torres-Guijarro S, Cardenal-López A, *et al.* ShipsEar: An underwater vessel noise database. *Applied Acoustics*, 2016;113:64–69.

[68] Ke X, Yuan F, and Cheng E. Integrated optimization of underwater acoustic ship-radiated noise recognition based on two-dimensional feature fusion. *Applied Acoustics*, 2020;159:107057.

[69] Khishe M. DRW-AE: A deep recurrent-wavelet autoencoder for underwater target recognition. *IEEE Journal of Oceanic Engineering*, 2022;47(4): 1083–1098.

[70] Ren J, Xie Y, Zhang X, *et al.* UALF: A learnable front-end for intelligent underwater acoustic classification system. *Ocean Engineering*, 2022;264: 112394.

[71] Xie Y, Ren J, and Xu J. Adaptive ship-radiated noise recognition with learnable fine-grained wavelet transform. *Ocean Engineering*, 2022;265:112626.

[72] Doan VS, Huynh-The T, and Kim DS. Underwater acoustic target classification based on dense convolutional neural network. *IEEE Geoscience and Remote Sensing Letters*, 2020;19:1–5.

[73] Hu G, Wang K, Peng Y, *et al.* Deep learning methods for underwater target feature extraction and recognition. *Computational Intelligence and Neuroscience*, 2018;2018(1):1214301.

[74] Wu H, Song Q, and Jin G. Deep learning based framework for underwater acoustic signal recognition and classification. In *Proceedings of the 2018 2nd International Conference on Computer Science and Artificial Intelligence*; 2018. pp. 385–388.

[75] Khan A, Sohail A, Zahoora U, *et al.* A survey of the recent architectures of deep convolutional neural networks. *Artificial Intelligence Review*, 2020;53:5455–5516.

[76] Zhang Q, Da L, Zhang Y, *et al.* Integrated neural networks based on feature fusion for underwater target recognition. *Applied Acoustics*, 2021;182: 108261.

[77] Liu W, Bai K, He X, *et al.* FishGym: A high-performance physics-based simulation framework for underwater robot learning. In *2022 International Conference on Robotics and Automation (ICRA)*. IEEE; 2022. pp. 6268–6275.

[78] Elmezain M, Saoud LS, Sultan A, *et al. Advancing underwater vision: A survey of deep learning models for underwater object recognition and tracking.* IEEE Access, 2025.

Chapter 17
Deep learning techniques for optimizing underwater acoustic communication

Asiya Ali Khan[1], Minahel Ahsan Khan[1] and Ayesha Kanwal[1]

Underwater acoustic communication (UAC) is one of the most challenging forms of communication. Conventional terrestrial methods fail underwater due to several precarious conditions. Several studies have been conducted to make this task more efficient and accurate. However, deep learning (DL)-driven techniques help sensor nodes address these limitations. It has revolutionized UAC by enabling advancements in channel modeling and prediction, signal detection and classification, adaptive channel equalization, efficient modeling and coding schemes, and intelligent acoustic learning due to their ability to learn intricate patterns and adapt to different environments. It promises high-performance solutions for wireless networks that contribute to the applications of UACs from environmental monitoring to defense and underwater robotics. This chapter aims to highlight the transformative impact of DL on UAC, showcasing the contributions of various model implementations. This development of different models actively bridges various existing gaps, fostering the creation of powerful and resilient communication networks.

17.1 Introduction

UAC is a form of communication in the ocean based on acoustic waves, since radio waves are not preferred because they do not perform well within short distances and are absorbed in water. It is basically how underwater systems, such as submarines or sensor nodes deployed to collect data for various underwater applications, transmit data. UAC has a great bunch of applications around, such as environmental monitoring, offshore industry, scientific research, defense and security, disaster prevention, etc. [1]. Despite the fact that the speed of sound is faster in water than in air, there are loads of challenges that UACs experience. The speed of sound is proportional to the water temperature, salinity, and depth of water; therefore, it is affected by all kinds of

[1]School of Electrical Engineering and Computer Science (SEECS), National University of Sciences and Technology (NUST), Pakistan

distortion. Moreover, UACs face challenges like Doppler effects, multipath propagation, high latency, and noise. The bandwidth for acoustic communication is also much narrower than that for radio frequency communication. There are many schemes discussed in the literature [1] that cater to these challenges, such as modulation schemes, frequency shift keying (FSK), phase shift keying (PSK), orthogonal frequency division multiplexing (OFDM), spread spectrum (SS), etc., that help encode data in a way that it is easier to transmit over channel conditions.

Techniques such as equalization, Doppler compensation, and error correction coding compensate for distortions such as multipath propagation, frequency shifts, and correction of errors during transmission [2]. Likewise, to improve UAC reliability, adaptive modulation, machine learning (ML), DL, and delay-tolerant networking are used. Mutual communication between sensor nodes strengthens information transmission over longer distances. Currently, researchers are focusing on combining acoustic, optical, and radio approaches to take advantage of each communication technology and improve the overall network performance of underwater networks. However, due to the use of different communication technologies, challenges such as temperature changes, salinity, and ocean currents can still affect communication between underwater nodes. Eventually, UAC is developing rapidly and is key to future underwater observation. Huang *et al.* [3] put an emphasis on the importance of underwater resources, but due to adverse and changing conditions in the ocean, the nodes are unable to communicate smoothly. The solution offered in Huang *et al.* [3] was to apply ML methods to UAC to develop an intelligent technology that adapts to its environment and provides the desired results with fewer losses by improving adaptive modulation and coding (AMC), which helped in real-time data processing by dynamically adjusting the modulation scheme and error correction coding rate in the changing underwater environment. It helped in efficiently exchange data by balancing data rate, robustness, and energy.

Even simpler techniques, such as K-nearest neighbors (KNN), delivered impressive results for AMC, offering preliminary guidelines for integrating ML into UAC systems and highlighting the potential of ML across various layers of underwater networks. But the solution they provided also came with huge challenges, such as the high quality of underwater data for training, the huge computational cost for underwater sensor nodes, and a diverse and complex study of the ocean environment. Furthermore, integrating ML across network layers and security aspects was also a huge problem. Building on the potential of ML to improve UAC, many important obstacles needed to be addressed. These obstacles arose from the distinctive and severe conditions of underwater networks. Among the leading issues was the data unavailability for training ML models, as high-quality labeled data in the underwater domain was expensive and complicated to understand. In addition, a high computational cost was required for underwater nodes due to their limited processing capabilities and the lack of energy sources. Another layer of complexity was added by the volatile behavior of the ocean environment, characterized by erratic water currents, noise sources, and node mobility. Despite difficulties, the integration of underwater networks with DL models, a special subset of ML, offers great opportunities to optimize communication by responding to real-time environments. It can improve routing stability, energy management, and resource use. The next section will discuss the challenges and possible solutions to improve UAC performance with DL.

17.2 How deep learning can improve UAC?

DL methods utilize a purified signal for a layered network to look for significant features, where the representation of the data is determined. Such methods that use DL techniques are considered far more effective than manually crafting filters, as algorithm-based processing is faster and smarter, capable of learning and predicting results simultaneously (see Figure 17.1). Among many proposed solutions for UAC, the authors in Lee-Leon *et al.* [2] provided quite a qualitative and quantitative study using both simulations and real-time sea trials and proposed solutions using the DL technique, the deep belief network (DBN), for de-noising and classifying the received signals. The utilization of a DBN-based denoising technique has been proposed to mitigate distortions such as Doppler effects and multi-path propagation. This effectiveness can be seen through results under various channel conditions, including additive white gaussian noise (AWGN), multi-path propagation, and Doppler effect, showing great improvement as compared to other methods like conventional maximum likelihood estimation (MLE) and denoising autoencoders, especially in high noise and multi-path environments. One of the prime objectives of UAC is mapping and detecting underwater objects using sonar and sound wave reflections. This process comes with loads of complex integrations. [4] discussed the difficulties faced in underwater object detection and how visual data could be intercepted.

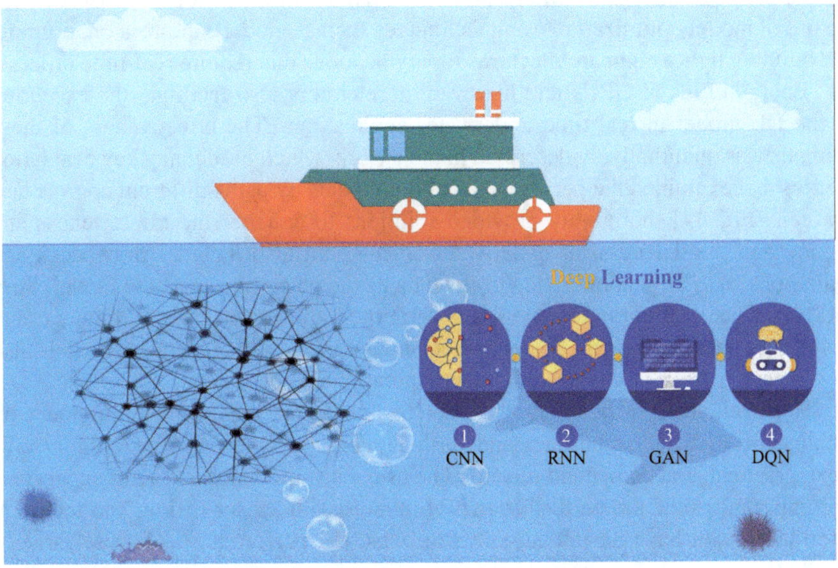

Figure 17.1 Deep learning-based underwater acoustic communications

Image deterioration can be caused by various factors such as light absorption, scattering, color distortion, turbidity, and low visibility, which result in corruption of the image quality and make the traditional image processing method ineffective. Accurate and effective object detection and image processing are quite pivotal for different applications such as deep-sea research, environmental monitoring, autonomous underwater vehicles (AUV), underwater archeology, inspections, and analysis. The objective of their work was to introduce cutting-edge technologies and the role of DL models that can easily overcome these hindrances. Convolutional neural networks (CNN) were introduced as one of the most powerful tools in [4] that had the ability to learn hierarchical features from images of patterns and details. Methods used previously in the literature, such as optical flow, edge detection, and histogram equalization, are believed to be weak in real-world detection due to noise and distortion. The models explained in their work were used for object identification in different domains, such as faster region-based (R)-CNN, YOLO, and single-shot multi-box detector (SSD). In hybrid architecture, these models were used with other DL architectures such as attention mechanisms and recurrent neural networks (RNN). Transfer learning is also an important technique that helps with the scarcity of datasets, since it obtains data from large datasets such as ImageNet or COCO. Pre-trained architectures such as residual neural network (ResNet) and VGG were adapted for fine-tuning, which increased training and improved model performance.

Data augmentation and synthetic data generation are also considered viable options for further dealing with this problem. In addition, to improve the quality of the underwater water, they used the preprocessing methods before feeding them as input into the models. This method enhanced images in low-light and high-light turbidity conditions, which further enhanced the performance of the models by using techniques that assisted in color correction, noise reduction, dehazing, and contrast-limited adaptive histogram equalization (CLAHE). Furthermore, they also devised the use of model optimization using techniques such as model compression, pruning, and building lightweight architectures for applications that require real-time processing. Energy-efficient GPUs and hardware accelerators also facilitate the execution of the DL model in real-time underwater applications. The effectiveness of these methods was quantitatively described in their work, which used many key evaluation metrics that examined the performance of the models by using different approaches. Praveen *et al.* [4] thoroughly provided solutions to enhance the effectiveness and accuracy in the detection of underwater objects, but also includes drawbacks and limitations such as generalization of the data as the performance of the models vary and might become unpredictable for uncertain conditions of the environment, and this further changes the visual appearance of the objects. These complex models also have a greater price to pay in terms of computational expense.

However, their work opens the door to further research and development by making use of these techniques and further integrating them to make highly efficient, practical, and competent models. The use of unsupervised and semi-supervised learning, multi-modal data fusion, hybrid models with sensor fusion, and temporal information can have an impact that can boost the performance of these models. In addition, the use of refined simulations that can depict underwater real-time conditions can greatly improve the quality of these models. UAC networks are inherently limited by the unfriendly nature of the aquatic channel, with long delay spreads, time-varying multipath propagation, and extremely severe Doppler effects, as previously stated. The orthogonal time-frequency space (OTFS) modulation has recently drawn significant attention as a promising method to overcome these impairments due to its resilience in doubly selective channels. Nevertheless, traditional OTFS receivers that take advantage of the sparsity of the channel matrix in the delay-Doppler (DD) domain rarely work well under the Doppler squint effect (DSE), which ruins this sparsity and causes channel estimation and signal detection performance to be poor. Facing these difficulties, recent studies suggested a DL-based OTFS receiver framework for UAC networks [5]. They used a pilot-assisted OTFS frame structure in which the pilot was placed at a specific DD grid position and surrounded by guard symbols. Data transmission was performed at the other grid positions. The pilot and data signals were demultiplexed and treated separately at the receiver.

The pilot data was fed through a channel feature extraction subnet made up of convolutional layers, which learned nonlinear representations of the underwater acoustic channel. The learned features were concatenated with the data signals, following zero padding to align, and input into an enhanced ResNet specifically tailored for signal detection. Their suggested ResNet structure avoids using pooling layers in order to avoid information loss and consists of a preprocessing convolution layer, four residual blocks, and a fully connected layer. All residual blocks had two convolution layers, batch normalization, and activation functions, while the number of kernels was growing progressively across the blocks. With their architecture, the ability of the network to detect more complex propagation effects was increased without adding redundant computation overhead. The model was trained on a dataset of 500,000 OTFS frames simulated under the DSE conditions using the mean squared error (MSE) loss function and was optimized with the Adam optimizer along with a learning rate decay strategy to ensure convergence stability. Their results showed that the proposed DL-based OTFS receiver performed better than current solutions in detection accuracy and resilience. Their findings emphasized the insufficiency of conventional sparse assumption-based receivers in real-world underwater networks and the benefit of using learned channel features in the detection process.

Such observations are consistent with the overall trend of incorporating data-driven methods into physical layer design, with deep neural networks becoming more frequently applied to substitute or complement conventional signal processing methods. The flexibility of DL models to learn in a self-adaptive way under complex, nonlinear channel behaviors without making strict prior assumptions is particularly beneficial in underwater networks, in which channel behaviors are hard to predict and extremely dynamic. This development represents a major step toward intelligent communication systems that can adapt autonomously and are more resilient in harsh environments. To assess the performance of the suggested ResNet receiver under actual underwater conditions, extensive simulation experiments with dynamic system parameters were performed. Even with the encouraging performance of their proposed ResNet receiver under simulated DSE conditions, one of the key limitations of the work by Zhang *et al.* [5] was its exclusive use of simulation-based testing.

Although comprehensive, simulations cannot emulate the capricious and often more intricate dynamics of real underwater environments. Consequently, the absence of verification through field experiments or even real-field deployment instills fear regarding the model's generalizability and robustness in real field operating environments. In the following sections, we will examine both CNN and RNN models separately to see their impact on UAV networks.

17.2.1 CNN-based models to improve UAC networks

UAC networks are faced with drastic impairment from highly time- and frequency-varying underwater channels. Doubly selective characteristics render traditional estimation methods like least squares (LS) and minimum MSE unsuitable in low SNR and nonstationary conditions. Consequently, CNN-based models such as DenseNet [6] have been introduced as strong alternatives to channel estimation. DenseNet, through the connectivity of dense layers, supports feature reuse and effective gradient propagation, especially important when characterizing UAC channels' complex and irregular nature. The model was trained on the watermark dataset with three different underwater channel environments, BCH, KAU1, and KAU2, which offered diverse channel impulse responses. The diversity enabled DenseNet to generalize well under different channel conditions. They divided the data into 75% for training and 25% for testing to reduce overfitting. The model was also tested under both binary PSK and quadrature PSK modulations, which added to its flexibility. To our understanding, the choice to use a thoroughly vetted dataset and test on several modulations shows the model's viability in real-world applications. The flexibility of DenseNet under different conditions holds promise for future use in mission-critical underwater environments. This renders it an attractive option for embedded use in AUVs. The combination of strength and flexibility is a significant step towards underwater communication solutions.

DenseNet was comprehensively compared with LS, minimum MSE, and fully connected neural network (FC-NN) models. LS, while computationally cheap, performed poorly in noisy and dynamic environments, while minimum MSE, while more accurate, was marred by cubic computational complexity and required noise statistics expertise. FC-NNs were more expressive but over-fitted, particularly with limited data. DenseNet, however, performed better than these methods at all times in the MSE measures. It demonstrated good generalization with little performance degradation in test environments. Above all, when used in combination with frequency-domain equalization, DenseNet's output led to improved signal reconstruction. Even at 0 dB SNR, the performance of loss of training was below 7% and only increased marginally with increasing SNR. Such robustness is rare in UAC systems and, in our opinion, a reflection of the architecture's ability to function under real-world, uncertain underwater conditions. Although minimum MSE's high complexity or FC-NN's instability are undesirable traits, DenseNet maintains learning capacity and usability in perfect balance. Its consistency across conditions also makes it less necessary to retrain frequently. Overall, DenseNet provides an energy-efficient and stable solution in harsh communication environments. Also, its lightweight inference makes it ideal for embedded deployment in energy-limited applications. Such qualities increase its viability for long-duration underwater operations, where model consistency is paramount to ensuring reliable communication in challenging and unpredictable environments.

The model's performance was also compared with pilot strategies and modulation types. Binary PSK always yielded superior bit error rate (BER) values over quadrature PSK, with BER values of down to 0.0004 at 20 dB SNR, while those of QPSK were 0.0009. This conforms to theory, since binary PSK inherently enjoys greater resistance to noisy channels. DenseNet also supported comb-type and block-type pilot structures, so it could work in any fading environment without unnecessary pilot overhead. By equating pilot spacing to coherence bandwidth in the BCH and KAU channels, the model was successful in maintaining low BER even with minimal pilot density. We support this dual compatibility the most because it is compatible with bandwidth and energy constraints common in UAC systems. The combination of DenseNet with an LS equalizer also demonstrates that hybrid approaches are able to provide high accuracy without entirely discarding traditional approaches. With these advantages in mind, we believe that DenseNet is a good compromise between high-performance ML models and traditional low-complexity approaches, and thus is appropriate for future UAC deployments. Such versatility may make DenseNet a cornerstone of the next-generation underwater communication infrastructure. Its ability to optimize performance without adding to system expense is another major advantage, making it an ideal solution for resource-limited and cost-sensitive underwater communication systems. Likewise, Rathor [7] emphasizes how determining the atmosphere in acoustic communication is revolutionary in terms of applications of different fields such as security and health monitoring, by using DL techniques, such as CNN.

CNN improves the accuracy and reliability of ambiance recognition due to its effectiveness in capturing spatial and temporal features. Such models can improve human–computer interaction by creating a smart environment in which computers can act more naturally. This aims to make technology more sensitive to changing factors in human behavior and act accordingly, which requires more advanced feature extractions and techniques that are able to identify different patterns in the audio to relate to different possibilities. CNN-based models play a pivotal role in achieving this goal. The data is further used as an input to a multi-level CNN architecture that reduces the matrix size and combines different output layers; this enhances the computational efficiency, which is then processed through fully connected layers, producing probabilities that the data belong to. They provided concrete data and experimental backing to illustrate how their model can effectively achieve the goal. Their evaluation was based on training and testing datasets, considering the number of parameters, confusion matrices, and quantitative measures, such as precision, recall, and F-score. Parameters are categorical values that a model learns from the data. This directly relates to showcasing the complexity of the model; the metrics show lower parameter, which makes this model highly accurate, indicating its computationally efficient. Their data also lack detailed error analysis, which limits our understanding of the weakness of the model and where it underperforms. They provided an exciting revolution in the field of acoustic communication analysis.

Their model features commendable and efficient results that outperform all traditional methods. They identified the gap between existing methodologies in the literature and the way in which they overlooked the context of acoustic communication. They effectively suggested future research in this area of communication. Their model architecture provided a clear understanding of the technical approach that this model employs, helping it achieve effectiveness and great efficiency. However, with its various strengths, there is always room for improvement. The overreliance of this model on textual embedding derived from acoustic communication risks discarding crucial preverbal cues such as tone, pitch, rhythm, and non-verbal sound present in the raw audio signal; this limits the model from fully capturing the spectrum and ambient nuances. Implementing the model on more diverse, real-world, and noisy datasheet can configure more accurate results. A comprehensive ablation study for each architecture component can help us identify decision-driven performance gains for future replication, optimization, or simplification of the model. Future work should be directed more toward finding methodologies that process raw audio features to fully instill the important specifications of the data. In the transition of their method from academics to real-world impact, the model should be able to overcome limitations involving data representation, environmental variability, architecture transparency, and comparative rigor through research, which can further pave the way for highly effective, efficient, and robust systems for functioning effectively across a wide range of human-centric applications.

17.2.2 RNN-based models to improve UAC networks

RNNs, especially long short-term memory (LSTM) models, are effective for sequence prediction tasks in nonstationary networks, such as UAC networks. In contrast with static classifiers that use independently and identically distributed feature distributions, LSTMs take advantage of temporal coherence in the trajectory of physical layer metrics such as angle of arrival, time difference of arrival, received signal strength (RSS), and channel impulse response (CIR) [8]. In the context of UAC, transmitter mobility and multipath effects create a non-Gaussian and time-varying structure of CSI. Classic detectors, learned from static features, do not work when the distribution drifts. In contrast, LSTMs learn the time evolution, allowing prediction and anomaly detection even for unseen paths. Receivers use these time-series predictions to detect spoof attacks based on the norm of residuals. Their single-node decisions failed because of poor separability, which led to the generalized likelihood ratio test (GLRT) to centralized aggregation over receivers. This combines distributed temporal forecasts, elevating detection performance. Additionally, temporal coherence in LSTM outputs mirrors the physical realism of feature sequences, which attackers have difficulty duplicating exactly. Our perspective is that this change from spatial snapshots to temporal structure through RNNs constitutes a paradigm shift for physical layer authentication in UAC systems, where environmental instability otherwise constrains traditional DL performance. Their model capitalized not on the values of individual features, but on their changes, that is, the derivative of channel behavior. This validates that LSTMs acquire feature dynamics, not trajectories. The channel's higher randomness at high velocities adds noise in prediction as well as spoofing. Another strength of LSTM-based systems in UAC networks is their ability to generalize under limited training data. Collecting real underwater channel data is expensive and logistically complex.

However, models trained offline on synthetic or approximate trajectories generalize well in practice. Bellhop simulation enables the generation of trajectory-conditioned CIRs, which serve as training input for neural networks. These networks can be pre-trained and later fine-tuned during short online calibration phases post-deployment. This is key in resource-constrained deployments, where retraining on site is impractical. Moreover, centralized GLRT avoids sensor-level computation overload by leveraging the statistical structure of prediction errors between nodes. From our perspective, the low resource demands, generalization to new paths, and resilience to high mobility variations make RNN-based architectures indispensable for practical and scalable physical authentication in future UAC network deployments, ultimately enhancing the security and reliability of underwater communication systems. Likewise, Wei *et al.* [9] identifies two challenging problems: long delay in acoustic communication due to low speed of sound in water and high data loss rate due to different underwater factors that greatly affect the monitoring of the underwater environment using wireless sensors. The slow propagation results in end-to-end transmission delays that are long, particularly in multi-hop networks where packets travel through a number of relay nodes. As mentioned above, underwater communication is quite challenging and complex due to high ambient noise, packet collisions, and multipath propagation, all of which cause a high rate of data loss. Conventional MAC protocols that use automatic repeat request (ARQ) schemes to retransmit lost packets increase both the delay and waste energy on battery-restricted nodes. To address this, they present two solutions. Firstly, they propose a delay-guaranteed transmission approach through protocol modifications such as slotted floor acquisition multiple access (FAMA) with retransmission removal to stop latency and power consumption.

In doing so, packet loss grows, but they counter this using their second innovation: an RNN-based framework for data imputation known as LSTM-decay. This goes beyond vanilla LSTM by including time decay functions based on how long a data point has been unavailable. The decay weights are learned in conjunction with the conventional model parameters, so that both the temporal patterns and the missing data modes are learned by the network. Positional correlations and multivariate time series can be handled with robust predictions through LSTM-decay even with strong missing-data scenarios. Through remote execution via the cloud, the model can be run computationally efficiently without loading sensor nodes. Their model was also tested on actual datasets from the ocean networks of Canada and showed better delay reduction, energy efficiency, and imputation accuracy compared to the traditional methods of KNN and support vector regression (SVR). The solutions they provided come with a load of benefits and a few limitations of their own that deserve consideration. The delay-guaranteed approach to transmission by eliminating the auto-retransmit feature in customary MAC protocols, such as slotted FAMA, hugely lowers end-to-end latency and saves power, which is important where underwater sensor nodes are powered through batteries. However, their computational intensity was much higher than that of other conventional approaches, and hence, this was more fitting for implementation over cloud-based transmission. In summary, for the transmission strategy, an adaptive algorithm that selectively initiates retransmissions depending on actual network conditions, such as node congestion or link quality, might provide an improved trade-off between delay and data quality. Other hybrid approaches may also be established to enable retransmission only for critical data packets. Adding attention mechanisms or transformer-based models might further enhance performance by more accurately modeling long-range dependencies. Real-time environmental feedback might be employed to dynamically fine-tune predictions, making the system more responsive to change. Overall, with these improvements, the framework might be more scalable and usable for a larger number of underwater monitoring applications.

17.3 Conclusion

This chapter illustrates how acoustic systems can become more intuitive and effective through the inclusion of temporal information, behavioral patterns, and multi-modal information, leading to improved understanding of intricate auditory environments. These developments make systems stronger in noisy or unstable settings, allowing for more accurate responses to varying human and environmental inputs. Sequence-aware learning enables systems to track changing patterns, critical in underwater communication or affective recognition. The use of dynamic and temporal contextual cues increases system adaptability, decreases false positives, and enhances contextual knowledge. One of the central themes throughout the studies discussed in this chapter is optimizing data processing efficiency while preserving depth and accuracy. These studies indicate that simple classification is no longer adequate; audio analysis needs to decipher intricate, layered signals. This makes the system more flexible in applications ranging from surveillance to healthcare and immersive technologies. Interestingly, most models are effective with small training sets, indicating robustness and generalizability. Although promising performance is shown in the models discussed in this chapter, there are still challenges, such as domain shift, class overlap, and insufficiently diverse datasets, that can be addressed as future work. More real-world variability and preverbal cues must be integrated to increase accuracy. These results support more comprehensive, human-centric acoustic systems that can contextualize, predict, and adapt. This work sets the stage for more adaptive systems that develop over human interaction, defining the future of intelligent acoustic communications.

References

[1] Zia, M.Y.I., Poncela, J., and Otero, P. "State-of-the-art underwater acoustic communication modems: Classifications, analyses and design challenges," *Wireless Personal Communication*, vol. 116, pp. 1325–1360, 2021.

[2] A. Lee-Leon, C. Yuen, and D. Herremans, "Underwater acoustic communication receiver using deep belief network," *IEEE Transactions on Communications*, vol. 69, no. 6, pp. 3698–3708, 2021.

[3] L. Huang, Y. Wang, Q. Zhang, J. Han, W. Tan, and Z. Tian, "Machine learning for underwater acoustic communications," *IEEE Wireless Communications*, vol. 29, no. 3, pp. 102–108, 2022.

[4] R. Praveen, M. Maindola, A. R J, N. Thandra, S. Bansal, and V. Sharma, "Deep learning techniques for object detection in underwater environments," in *2025 International Conference on Computational, Communication and Information Technology (ICCCIT)*, 2025, pp. 595–600.

[5] Y. Zhang, Y. Wang, Y. Liu, L. Shi, and Y. Zang, "A deep learning receiver for underwater acoustic OTFS communications with doppler squint effect," *IEEE Wireless Communications Letters*, vol. 14, no. 4, pp. 1179–1183, 2025.

[6] S. Liu, M. Adil, L. Ma, S. Mazhar, and G. Qiao, "DenseNet-based robust channel estimation in OFDM for improving underwater acoustic communication," *IEEE Journal of Oceanic Engineering*, vol. 50, no. 2, pp. 1518–1537, 2025.

[7] S. Rathor, "A sturdy CNN-based model for ambience recognition of acoustic communication," *Neural Computing & Applications*, vol. 37, pp. 8261–8271, 2025.

[8] F. Ardizzon, P. Casari, and S. Tomasin, "A RNN-based approach to physical layer authentication in underwater acoustic networks with mobile devices," *Computer Networks*, vol. 243, p. 110311, 2024. https://www.sciencedirect.com/science/article/pii/S1389128624001439

[9] X. Wei, Y. Liu, S. Gao, X. Wang, and H. Yue, "An RNN-based delay-guaranteed monitoring framework in underwater wireless sensor networks," *IEEE Access*, vol. 7, pp. 25959–25971, 2019.

Chapter 18

Underwater acoustic communications powered by hybrid intelligent models

Muneeb ur Rehman[1], Ahmed Jamil[1] and Pham Duy Thanh[2]

Underwater acoustic communication faces peculiar challenges, including dynamic multipath transmission, bandwidth limitations, and rapidly varying data that make traditional physics-based or purely data-driven approaches obsolete on their own. By fusing first-principle acoustic modeling with modern machine learning (ML) techniques, this chapter presents a unified hybrid framework to deliver strong, adaptive, and easy-to-understand communication solutions for underwater environments. We start with looking back to already established methods, including matched field processing, channel equalization, and routing protocols, characterizing key channel impairments such as attenuation, Doppler spread, and multipath scattering, and pointing out their sensitivity to rapidly changing environments and inadequate data. Then, we explore hybrid physics-guided architectures, including physics-based preprocessing of time delay and Doppler features, as well as integrating mechanistic channel models directly within neural network layers and ensuring acoustic propagation invariants through physics-derived loss functions. We then proceed to the real-world testing of models in the literature, where we show that the hybrid models proposed in the literature have an edge compared with standalone baselines in terms of bit error rate performance, faster convergence with limited labeled data, and improved generalization across rapidly changing underwater conditions. In addition, these models have uncertainty estimates via Bayesian dropout, and easy-to-understand parameters show model confidence and give the researchers compatible protocol decisions in real time. Instead of going the traditional route and retrofitting ML onto existing stacks, we have discussed co-designed communication protocols with hybrid estimators through which we can unlock energy-efficient, scalable, and durable underwater acoustic networks. This work sets out a clear research road map including multi-fidelity modeling, graph neural network topologies for node cooperation, and physics-aware active learning, a step toward the next generation of intelligent, ocean-scale physical systems.

[1]School of Electrical Engineering and Computer Science (SEECS), National University of Sciences and Technology (NUST), Pakistan
[2]Big Data Integration Research Center, National Institute of Information and Communications Technology (NICT), Japan

18.1 Introduction

Communicating underwater is not as straightforward as communicating in an air medium; radio waves quickly attenuate, and light barely penetrates the dark layers. However, sound waves travel farther, making acoustics the backbone of underwater communication. In addition, even sound propagation has challenges; the environment is dynamic, complex, noisy, and unpredictable. Thus, underwater communication systems need robust adaptive technologies. In underwater acoustics networks, communication relies primarily on the low-frequency acoustic band, since at these frequencies the sound attenuates to a minimum. This limited available bandwidth limits data rates, making it difficult to support high-speed communication. In addition, multipath effects, noise, and dynamic channel variations also make reliable transmission more complex within this narrow band. Machine learning (ML) does not expand the acoustic band; instead, it makes communication and signal detection within that limited band more reliable and adaptable. ML helps systems adapt to complex and dynamic underwater conditions using very limited information by recognizing patterns such as sound frequency, timing amplitude, or signal distortion. These patterns may be needed for ML models for specific tasks such as source localization, target recognition, signal equalization, and geoacoustic inversion.

Although traditional models, such as matched field processing (MFP), are usually used for source localization, they might suffer if there is a surprise scenario underwater, such as a sudden change in ocean temperature or an unexpected vessel entering the area. ML provides flexibility, robustness, and real-time adaptability, which are essential for reliable underwater communication. ML has already come a long way in underwater acoustic communication, from early models struggling with tracking underwater sounds to now advanced ML models handling real-time underwater tracking. ML has also improved underwater object recognition, communication, and mapping the ocean floor much faster and better than before. However, there are some challenges. Labeled underwater data are hard to obtain, and models that work in one condition might fail somewhere else. The ocean is super noisy and unpredictable, and on top of that, ML models can be total black boxes, not ideal when the stakes are high. Possible solutions could be the generation of synthetic datasets with generative adversarial networks (GAN), the application of physics-informed neural networks (PINN) to embed real-world limitations, and also the establishment of global standards for the creation and sharing of datasets [1]. Hybrid intelligent models (physics-based models married to ML) could give us the best of the two worlds. In this chapter, our focus is on hybrid models that improve underwater performance by addressing the challenges discussed above. The rest of the chapter is arranged in the following manner: Section 18.2 discusses related work, Section 18.3 presents solutions to challenges with the aid of hybrid models, and Section 18.4 concludes the chapter.

18.2 Related work

When it comes to underwater systems, purely physics-based or purely data-driven models become rather insufficient due to the exhibition of extreme non-linearity and uncertainty. PINN is introduced by Amer *et al.* [2] to tackle this dilemma, in which we get the best of both worlds, that is, a hybrid approach in which we can integrate known physical laws into ML models. As PINN uses physics as a regularizer, it helps to improve generalization and sample efficiency, ultimately generating outputs that are within the realms of possibility. The key technology in making all of this possible is a specialized physics informed neural network (PINN) with control, which through data loss, and physics-based loss understood the dynamics while taking in the vehicle's state, control input, and time input. Similarly, future states can be predicted by this model, which is not only consistent with the underlying motion equations, but a PINC can make more accurate long-horizon predictions than a conventional neural network during simulation stages of an underwater vehicle. Pure ML models' outputs usually have a high proportion of instability and drift, but by constraining the learning with physics, those effects are dwindled.

Their work reiterates the limitations mentioned earlier, the unpredictable underwater environment and data scarcity are handled by the model's physics priors, which reduces the set of data needed for training while ensuring the results beyond the seen data. Work has been done that includes, but is not limited to, the PINC framework, where PINC models have been designed for simplified remotely operated vehicle (ROV) dynamics [2]. Physics-augmented training, by combining a physics-based loss with network enhancements helps improve model fidelity. And on top of it, they provide an open source implementation and simulate the ROV model so that other researchers can continue and build upon the results of the previous ones. The results are a testimonial of how PINC can be used to model complex underwater dynamics, a concrete example of the hybrid modeling strategy mentioned in our introduction section of this chapter. On the other side of the coin, despite these successes, they still have some unresolved issues. On top of which is all the work they have conducted is purely in simulation and there is a need to apply PINC to real world ROV data to judge its reliability.

Moreover, the current version handles only simplified planar motions, and extending to full six-degrees-of-freedom (6-DOF) dynamics is needed to have results in alignment with the real world. In short, PINC is invaluable when it comes to underwater communication and control environment, but it is just a start for now, and further work is required to test and extend it in a fully operational form. Although traditional physics-based and data-driven models individually fall short in underwater domains, hybrid intelligent models, however, particularly those that combine physics and ML, have done a great job of closing that gap. An impressive and strong demonstration of such a hybrid model is the use of PINNs to enable safe navigation of autonomous underwater vehicles (AUVs) in messy, ever-changing ocean settings. One such innovative framework is the underwater sense and avoidance PINN (USAAPINN), designed specifically to address the challenges of obstacle avoidance in messy and unpredictable underwater settings [3]. Traditional methods, such as fuzzy logic, artificial potential fields (APFs), and model predictive control (MPC), have attempted to offer solutions, but usually do not offer solid safety assurances or run efficiently enough for real-time use.

USAAPINN fills this gap by embedding control barrier functions (CBFs) into the neural network architecture, which mathematically define safety boundaries. Their resulting model does not simply react to detected obstacles; it proactively plans control inputs that maintain safety even in the face of uncertainty. The USAAPINN approach deals with a quadratic programming (QP) problem based on safety constraints, which is then transformed into a Hamilton–Jacobi–Bellman (HJB) partial differential equation. The equation is then solved using a PINN, which uses the physical laws of AUV motion as a regularizer. This hybrid approach reduces the need for many labeled data, and unlike black-box models, it makes sure that the outputs stay physically realistic. Their simulations show that USAAPINN navigated both convex and non-convex obstacles very efficiently by smoothly switching between a standard PID controller for trajectory tracking and the USAAPINN framework for avoidance maneuvers. These results demonstrate the feasibility of using PINNs not only for prediction but also for real-time control in constrained environments. However, challenges remain.

Most of the experiments are conducted in simulation environments, and the transition to real-world underwater data is essential to validate reliability. Additionally, current models primarily focus on simplified planar motion; extending them to full 6-DOF and bringing them closer to operational reality. To further improve such systems, multiple improvements can be made; real-time integration of sonar feedback, training with synthetic datasets generated through GANs, and expansion into multi-agent AUV scenarios. It is also important to reduce the computational load so that these models can run on embedded systems inside AUVs. In our opinion, hybrid models such as USAAPINN have great promise to handle complex underwater communication and navigation. These hybrid models combine the adaptability of ML with the realness of physics-based modeling, providing reliability even when the data are limited in unpredictable messy underwater environments. As we keep closing the gap between simulation and the real physical world, these models could prove to become the backbone of smart, self-directed underwater missions very soon. The challenges we face are not only limited to the bottom of the water surface but also continue to haunt us above the surface.

18.3 Hybrid intelligent models, their challenges, and possible solutions

Unmanned surface vehicles (USVs) have their own set of challenges above water, including environmental disturbances or traditional control models that make trajectory tracking a personal favorite form of torture. MPC is yet another way to regulate the path of such vehicles, but it is not very reliable due to a very high dependency on high-fidelity mathematical models, which decreases their practicality. If we have a discrepancy between the predicted model and real-world behavior, it can lead to poor performance or system instability. Zheng *et al.* [4] set out to find a solution to this problem and introduces a hybrid physics learning (PL) MPC system, aptly named neural-sailing, where they get the best of both worlds: physics-based modeling is combined with the adaptive flexibility of deep neural networks (DNNs). This hybrid approach uses a physical model that can be acquired through standard parameter identification techniques and a DNN trained on real motion data to predict modeling errors and disturbances, ensuring real-time control in dynamic environments without sacrificing accuracy. To make sure the system is stable, they adopt a quasi-infinite nonlinear MPC framework, and to make sure computational cost does not exceed an acceptable level, they present a novel successive linearization technique.

Overall, this all leads to a significant jump from previous modeling or rigid ML systems with limited reliability. What makes this contribution stand out more is its unique way of dealing with the problem at hand. However, previously, efforts were put mainly in the realm of either improving MPC formulation through disturbance observers or leveraging ML independently of model uncertainty. One of the major advantages of using DNN is that it learns from previous trajectory data to model disturbances directly, no assumptions, just following patterns, while previously letting us take a Gaussian process regressor as an example; they can predict system errors, but they either have a really high computational cost or unrealistic assumptions. By combining these learned dynamics with an interpretable physics-based model, their system can achieve what neither could by being a standalone, real-time, robust control under uncertain conditions. Compared to other approaches such as Koopman-MPC, the hybrid PL-MPC gave higher accuracy and better control smoothness during simulations. However, the challenge of computational cost lies there to ruin the day. The successful linearization strategy saves time by significantly reducing real-time processing requirements while maintaining comparable control performance. It does so because each control sequence from the last iteration is used to linearize the model in the next.

Thus, the balance between adaptability and feasibility is achieved, and hence a significant advancement in the real-world USV trajectory control. As with every research, it has its own limitations. The entire framework, including DNN training, is simulated on a USV platform called CyberShip II. So, they do not have direct real-time confirmation using real-world maritime data. Moreover, modeling is restricted to three DOF in a planar domain, just as researchers previously observed in underwater studies such as PINC and USAAPINN. They also left us guessing about how we can further scale to more complex dynamics or when, in case our computational resources are highly constrained, how it can affect its performance. However, their successive linearization technique shows promise in addressing computational challenges, and further improvements might include the integration of PINNS that are already being explored for underwater communications. Although the need for experimental validation remains a matter of concern, we have a good base that is innovative and apparently feasible. Their research acts as a bridge between traditional physics-based modeling and data-driven control systems. It takes into account the real world problem and the solution it provides is both robust and computationally efficient, which is a rarity in academic proposals.

Still, we can make it better. Starting with going with 6-DOF dynamics instead of planar will make it real-world friendly exponentially. Second, techniques such as GANs could account for data scarcity without risking overfitting. Third, in the future, we can integrate real-time sonar or light detection and range (LIDAR) feedback, further improving the trajectory. If we can adapt the PL-MPC framework into a lightweight version that can be integrated into embedded systems, it could be the core component of future smart marine fleets. Much like we observed in PINC and

USAAPINN and in this hybrid control strategy above water, we come to conclude that integration of physics and learning in modern autonomous systems is of paramount importance. Such research creates a new realm of possibilities where adaptability and safety are not mutually exclusive but a readily available luxury. We have already discussed examples of hybrid models, ML combined with physics, transforming underwater communications, and track-and-trace for surface vessels. However, in dynamic marine systems, there are phenomena that cannot be measured directly at all. These unmeasurable components, for example, actual down-hole weight on bit (DWOB) in offshore drilling, cannot be measured because technology does not exist or is too slow or too costly.

So now we have a new problem: how can we believe in an estimation of something that we cannot measure, in a system governed by flawed laws of physics? The possible solution is proposed by Kaneko *et al.* [5] with an upgraded hybrid framework based not only on data and physics, but also on uncertainty quantification and explainable artificial intelligence (XAI). Unlike conventional physics-based models that assume unknowns to be within tunable parameters, their research approaches structural uncertainty, where even the basic formulas do not fully match how things actually work in real life. Their approach converts quantitative measurements (e.g., weight on bit at surface and heave at crown block) into unobservable objectives (for example, DWOB) and provides redundancy by returning two critical elements, readable parameters, and uncertainty metrics. Their output creates a window on the model's thinking and its confidence, which can be trusted, or questioned, by operators in risky maritime environments. The originality of this study comes from the way that two types of uncertainty are addressed: parameter uncertainty, discrepancy caused by incorrect or unclear model parameters, and structural uncertainty, mismatch caused by oversimplification or incompleteness in models.

Most of the earlier hybrid techniques, including those we have mentioned above for underwater navigation, concentrate primarily on the former. This has pushed the discipline forward by training a hybrid system that translates measurable to unmeasurable variables through a physics-based simulation framework and later tunes it with Monte Carlo dropout, which estimates uncertainty in neural network responses. By comparing the hybrid model with traditional physics-based and gray-box models under both simplified and nonlinear dynamics, their study demonstrates that their model not only deals with noisy settings but also with unforeseen phenomena such as drill bit lifting, when the classical linear spring approximation failed. On top of that, the model not only gives results, but also explains those results by estimating key values. So, for example, it can tell you if something is off because the drill pipe turns out to be longer than expected. This setup, with both an explanation and a result, helps close the gap between real life and the model and makes it less of a mystery box like most AI systems, especially important when the stakes are high. However, as with any progress, there are limitations to this model. Although solid in simulation, its performance against real-world uncertainty must still be proven. Although validated on benchmark data from deep-water drilling systems, the ground truth for DWOB is still beyond reach, so it cannot be directly checked how accurate it is.

Further, although the model does include interpretability, it nonetheless relies on an already trained neural network whose action under unusual edge cases is still not well understood. Their study suggests a few tricks to get around the problem, such as adding future time-step data to get better short-term predictions or using domain randomization to help the model handle different situations more easily. They also illustrate how high-uncertainty outputs function as early warning systems, signals that real-world events are occurring too far beyond what the model has experienced. Their proposed model is particularly valuable for anomaly detection, when abrupt changes in the estimated length of the drill pipe can indicate faults or hazardous operating conditions. In a way, this study remakes hybrid estimation not only as a prediction but as a conversation between data, models, and uncertainty. And as hybrid models continue to evolve, with integrated explainability and safety metrics, they become less tools of estimation and more partners in real-time decision making. In our opinion, this model marks a pivotal transition in hybrid marine modeling. We have moved from what the system predicts to how and why it does so, an evolution that is essential for responsible automation in ocean engineering. This model framework addresses a fundamental tension in high-risk systems, trust.

When predictions are not only accurate but also understandable and aware of uncertainties, operators can make informed decisions with confidence even when the stakes are high. For future systems, integrating this method with other methods, such as GANs to synthesize rare conditions or PINNs to incorporate real-world constraints, may further improve resilience. In addition, extending these models to multiagent systems or integrating them with real-time sonar or visual feedback could make them more autonomous. Like PINC and USAAPINN signaled the advent of underwater vehicle control, this model charters new territory in estimating the invisible. In conclusion, this blended method is not merely an estimator; it is a compass that leads us through the haze of uncertainty that characterizes so much of the oceanic world. Be it underwater sensor networks or ocean surface fleets, one thing common in all of the unmanned maritime systems is that they have complex cyber-physical systems (CPS) that require both correct physics and flexible strategies for extraction of data. Physics-based approaches (model-based, (MB)) and pure ML are two traditional modeling techniques where MB relies on first-principle dynamics to capture both continuous and discrete dynamics, while ML at the risk of violating real-life compatibility, fits high-capacity learners to data [6].

Their survey clarifies one thing as clear as the dawn that neither comes as a standalone approach. On the one hand, MB struggles when faced with non-linearities or a lack of data, while on the other hand, the ML model usually lacks interpretation or might give incorrect results upon extrapolation. This gap is bridged by physics-guided ML, which combines both MB and ML to the ripe benefits of physical consistency while being compatible with real-world data fluctuations. Now, the only question remains: how can we systematically integrate scientific knowledge with data-driven learning to build more robust CPS models? Over a period of trial and error combined with the grit of researchers, hybrid CPS have come a long way, and it is centered around three cornerstones: physics-guided preprocessing, physics-guided network architectures, and physics-informed loss functions. In preprocessing, raw sensor streams, such as sonar returns, are transformed by MB tools into features such as normalized modal energies that make it easy for ML to train. Next, innovative network designs link the MB output directly with neural layers, acting as encoder and decoder modules, which have a physics-informed dynamics block sandwiched between them. Finally, we come to regularization losses, which ensure that conservation laws, invariances or stability constraints are dully followed during ML training, so that what we have learned so far aligns with energy balances.

Collectively, the mechanistic backbone of MB, along with the approximation of ML's function, not only closes the model and data gap but also builds a resilient learning framework that adapts to real-world uncertainties, maintains physical plausibility, and opens the path to generalizing these hybrid approaches to complex dynamic environments such as underwater communication networks, where such synergies are critically needed. Previously, we have seen in underwater acoustic networks that the working of the systems is monitored by phenomena including but not limited to pressure wave propagation and multipath scattering, which are described completely by physics, yet difficult to implement due to the rapidly changing underwater environment. Work on PINNs for underwater trajectory prediction is also encouraged to combine motion equations into learning frameworks. However, this approach is limited to vehicle control and addresses communication channels to a much lesser extent. With hybrid CPS review, the same three cornerstones, that is, preprocessing, such as MB-derived time-delay spreads, network fusion, where for feature extraction, MB channel impulse can be injected into convolutional neural networks, and physics informed losses, for instance, taking violations against known acoustic propagation laws, could exponentially improve ML-based channel estimators and routing algorithms.

Thus, by building up on it, we can reduce the fundamental problem of data variability and insufficient data common to underwater networks, while also enabling more adaptive and interpretable systems that can sustain performance in volatile marine conditions where existing standalone models often fail to generalize or react in real-time. Even as hybrid CPS models have gained ground, there are still issues that remain. First, there are no standardized evaluation metrics: although their research has presented measures for accuracy, generalization, explainability, and computational expense, few researchers implement them all at once. Secondly, how reliable it is in many cases is limited to approximations using Bayesian or dropout techniques, instead of the whole deep dive into random versus what-we-do not-know uncertainties, which is usually needed for high-risk stuff at sea. Third, to handle complex, large-scale systems such as underwater robot swarms or connected surface and underwater networks, we need distributed setups that mix digital and physical control while also keeping everything working together in a physically accurate way across all the parts. Their study points out some promising ideas such as making benchmark tests using real ocean data, using probabilistic models to clearly track uncertainty, and designing hybrid systems that can retrain parts locally without needing to redo the whole setup.

These steps aim to close the gap between what works in the laboratory and what actually works in the ocean. Their study highlights the importance of tools and frameworks to work together. Physics-based codes, such as finite element solvers for ocean modeling, and ML libraries like TensorFlow or PyTorch, are great on their own. But when we combine them for hybrid applications, it might involve a lot of complex "glue code" to make things fit and workable, which might make it harder to reproduce the results. Standardizing data formats for CPS signals, along with APIs that let physics-based solvers easily connect to ML pipelines, and establishing shared community benchmarks, can really accelerate progress in underwater systems. We also need open platforms, like those offered by PINC and USAAPINN, for communication-focused hybrid models. These would make it easier for researchers to share tools like channel simulators, labeled acoustic datasets, and ready-to-use model templates. Ultimately, physics-informed ML can only enable reliable high-speed underwater networks if the entire ecosystem grows and matures together, without friction.

In our opinion, the hybrid model charted out by blending MB's mechanistic certainty with ML's adaptive versatility is needed to address the twin threats of environmental randomness and data paucity. To further enhance underwater network performance, we suggest: (1) multi-fidelity modeling where light MB solvers provide coarse predictions to be refined by ML at runtime; (2) graph neural network models that respect the distributed node topology and consider physics-informed adjacency weights learned from acoustic connectivity; and (3) physics-informed active learning where MB error indicators guide targeted probing of channels to best exploit information gain within energy constraints. By co-designing communication protocols with hybrid models, instead of overlaying ML on top of existing stacks, we can unlock strong, energy-efficient, and scalable underwater communications to support the ocean-scale CPS of the future, laying the groundwork for intelligent, collaborative, and robust systems capable of adapting to the deep sea's ever-changing dynamics while upholding the fundamental physical principles that govern its behavior.

18.4 Conclusion

Underwater communication is difficult because radio waves and light waves do not get transmitted well underwater. Although sound waves work better, they are still subject to noise, multipath interference, and low bandwidth. Low-frequency acoustics can assist, but distortion in the signal is still an issue. ML does not expand bandwidth, but enhances detection through pattern recognition in amplitude, time, or frequency. Classic techniques such as MFP fail in dynamic scenarios, whereas ML is capable of learning online. But ML requires significant labeled datasets and is not interpretable. To solve this, hybrid approaches, such as PINNs, incorporate physical principles in learning, minimizing the need for data, and enhancing generalizability. Derivatives such as PINC have been applied in simulated ROV maneuvers, providing more stable long-term predictions, albeit without real-world validation. USAAPINN, a variation, manages obstacle avoidance through the integration of safety constraints into PINNs. While promising, it is restricted to simulations and not implemented with full 6-DOF. Submerged models such as PL-MPC assist USVs in following trajectories through the use of deep learning and physical modeling and minimizing computational burdens. But they are also not extensively tested in real-world situations. Where unmeasurable critical parameters such as DWOB exist, which is a hybrid model that incorporates explainable AI and uncertainty quantification. It approximates latent variables while exposing model confidence, supporting high-risk decisions. Despite being simulation-based, this contribution highlights the importance of transparency and safety in automation.

References

[1] H. Niu, X. Li, Y. Zhang, and J. Xu, "Advances and applications of machine learning in underwater acoustics," *Intelligent Maritime Technology Systems*, vol. 1, no. 8, pp. 1–23, 2023.

[2] A. Amer, D. Felsager, Y. Brodskiy, and A. Sarabakha, "Modelling of underwater vehicles using physics-informed neural networks with control," 2025. https://arxiv.org/abs/2504.20019.

[3] R. Majumder, R. Makam, P. Mane, Bharathwaj, K. S., and S. Sundaram, "Safe navigation of autonomous underwater vehicles using physics-informed neural networks," in *OCEANS 2024 – Singapore*, pp. 1–6, 2024.

[4] H. Zheng, J. Li, Z. Tian, C. Liu, and W. Wu, "Hybrid physics-learning model based predictive control for trajectory tracking of unmanned surface vehicles," *IEEE Transactions on Intelligent Transportation Systems*, vol. 25, no. 9, pp. 11522–11533, 2024.

[5] T. Kaneko, R. Wada, M. Ozaki, and T. Inoue, "Hybrid physics-based and machine learning model with interpretability and uncertainty for real-time estimation of unmeasurable parts," *Ocean Engineering*, vol. 284, p. 115267, 2023. https://www.sciencedirect.com/science/article/pii/S0029801823016517.

[6] R. Rai and C. K. Sahu, "Driven by data or derived through physics? a review of hybrid physics guided machine learning techniques with cyber-physical system (cps) focus," *IEEE Access*, vol. 8, pp. 71050–71073, 2020.

Chapter 19

AI-based seismic monitoring using underwater acoustic sensing

Khansa Mishqat[1], Faareha Sajjad[1] and Chih-Hsien Hsia[2]

This chapter explores how technologies like distributed acoustic sensing (DAS) study underwater vibrations to detect earthquakes in regions where traditional methods are difficult to operate. These vibrations are converted into measurable signals, and crucial data is extracted from these signals and further analyzed for efficient detection. The chapter further discusses how DAS integrates with artificial intelligence (AI) algorithms to form more precise predictions. The use of AI techniques such as machine learning not only improves the sensitivity and accuracy of the systems but also reduces false alarms. Several works proposed in the literature are discussed that support these notions. Despite some challenges and shortcomings, this approach is fairly reliable for monitoring underwater seismic activity to spot early warnings.

19.1 Introduction

Earthquakes in any region are dangerous, as they can cause significant damage to infrastructure and loss of life. Underwater earthquakes, volcanic eruptions, or land slides can cause high ocean waves, called tsunamis. Early detection of tsunamis and earthquakes can save lives and prevent property loss. The smallest changes in seismic activity underwater are easy to detect and can be detected by underwater sensor nodes using acoustic and optical communication technologies such as the deep-ocean assessment and reporting of tsunamis (DART) system and DAS [1]. Although DART is efficient, it cannot detect high failure rates or the possibility of false alarms. DAS uses scattered light in optical fibers to detect changes along the fibers. Forces such as vibrations and temperature changes stretch or compress the fiber, changing the spacing between scattering centers, which changes the pattern of reflected light. By analyzing these changes, DAS detects disturbances along the fiber. The main issue with DAS is optical fading, which results in noisy data and limited measurement accuracy. Several factors, such as noise interference and lower sensitivity at high

[1]School of Electrical Engineering and Computer Science (SEECS), National University of Sciences and Technology (NUST), Pakistan
[2]Department of Computer Science and Information Engineering, National Ilan University, Taiwan

frequencies, serve as challenges in DAS. Noise mixes with the seismic signal and reduces its clarity and accuracy.

Noise interference can be reduced by using dark fibers in underwater communication [2]. Hollow-core optical fibers, otherwise known as dark fibers, employ a unique design to direct light signals and transmit light through the core of the fiber with minimal loss or interference. In addition, if the cable cannot catch small details carried by high-frequency signals, loss of important information is at risk, and critical events can go unnoticed. Improvements in cable design are required to increase sensitivity. DAS can be further improved by using fiber Bragg gratings (FBG) [3], enhancing signal strength and reliability compared to random scattering centers. Xiao *et al.* [1] test and prove the reliability of DAS in the early detection of infra-gravity waves and tsunami waves. Infra-gravity waves are the slow and long waves created by the combined energy of smaller waves. Tsunamis are massive and deadly waves that are mainly caused by underwater earthquakes. Systems such as seismometers and DART help with the early detection of these waves, but they are expensive and not widely available and, also as mentioned above, cannot detect failures, which is a significant weakness. A cheaper solution is to use fiber optic cables that detect underwater movement through the DAS process.

In the Alaska-Oregon region, the researchers collected data using this technique and other existing tools, such as DART surface buoys and tide gauges, from August to December 2021. These data showed the change in the speed of the tsunami waves. This change was dependent on the depth of the water and made it difficult to detect these waves. To ensure accuracy, they assumed a constant water depth for every part of the cable. As a result, reflections of long infra-gravity waves were seen bouncing across the ocean. The tsunami waves were detected only after the speed correction, but their timing aligned with that of the official instruments (DART surface buoys and tide gauges). They showed that the DAS technique serves as a proactive approach, detecting a coming tsunami early on through the speed correction method. However, the role of earthquakes in the formation of infra-gravity waves remains unanswered. Also, it is impractical to assume constant water depth since the depth of the real oceans vary, which affects the speed of the waves. In our opinion, further research is required to understand wave generation due to earthquakes and to find ways to adjust for varying water depths. Similarly, Chen *et al.* [2] focused on the use of signal processing techniques to analyze the data collected by fiber optic cables.

In their opinion, these cables can sense underwater vibrations that are detected by a light reflection technique, so they named it phase detection optical time-domain reflectometry (ϕ-OTDR). Ocean noise, which is generated from sources such as ocean currents and waves, interferes with these vibrations, which proves to be a challenge in the detection of seismic waves. The remedy they considered to remove this unwanted noise from the received signal is to use a high-pass filter. Furthermore, they added that machine learning algorithms can be used to remove any remaining noise. Hence, they detected seismic activity by monitoring underwater frequencies and wave patterns, which are usually obtained by using techniques like 2D fast Fourier transform (FFT) and short-time Fourier transform (STFT). Based on this information, they further estimated where the earthquake started (epicenter) and how strong it was (magnitude).

The study was carried out in the South China sea, where small earthquakes were clearly detected by processing the seismic signals. The epicenter and magnitude of the earthquake were also correctly estimated, which shows the reliability of the DAS technique. However, false alarms can be generated due to activities such as storms or ships moving in the ocean. In our opinion, training AI models with better data can help filter out these false alarms. In the next section, we will explain how AI models can help in the detection of earthquakes.

19.2 How AI can improve seismic monitoring?

In Lior *et al.* [4], the authors focused on seeing how much able DAS is in the underwater network. They used dataset from three underwater cables near large geological faults, two of the coast of Methoni, southwest Greece, and one cost of Toulon, south of France, as well as from seismometers, installed on land and underwater. These three underwater cables from Hellenic center for marine research (HCMR), neutrino extended submarine telescope with oceanographic research (NESTOR), and mediterranean eurocenter for underwater sciences and technologies (MEUST) were used. From the study, it was found that stronger earthquake signals were picked up where the seafloor was smooth and flat, and in rough areas, the signals were weaker. P waves were the fastest, but they created weak signals as they travel vertically, and DAS worked best for waves that travel horizontally. The main challenge of using this technique was the noise and unwanted signals in the DAS data which came from natural sources, from the fiber itself, and how the cables were connected to the floor of the sea. The depth of the water and the shape of the seafloor also affected the noise. To measure noise, they used a technique called power spectral density (PSD) to see the behavior of signals at different frequencies. This study did not provide information on how much coupling improves or degrades DAS signals. They mentioned that slow waves give strong signals and fast waves (P waves) give weak signals, but the reason was not explained.

Also, they did not give a solution to the problem of P waves losing energy in the DAS technique, nor did they say whether there is a way to overcome it and make it more reliable or if there is another technique where P waves can be used effectively. In our opinion, explaining this technique quantitatively can help in gaining more understanding. Likewise, Fernandez-Ruiz *et al.* [5] showed that the DAS transforms fiber optic cables into vibration sensors. These cables have a higher sensitivity over long distances compared to other cheaper traditional networks, making them more reliable. Light pulses were sent down the cable and the vibrations in the water made changes in the light that was received back. Different methods, such as using chirped pulses, helped make DAS more stable by improving its accuracy and reducing the noise in the received signals. Instead of using new cables, they tried to make DAS on the telecom fibers already used for phones and the internet, as they were more cost-effective. The efficiency of the DAS technique depends on the way well the cable is connected to the ocean floor. Even though DAS cannot be measured in 3D like the traditional tools, it still covers long distances and gives useful results. After the DAS data were recorded, they converted the strain into vibration data. Filters or

AI algorithms were used to reduce noise and then the data was compressed so that large files can be handled easily and effectively. In addition to earthquake and wave detection, DAS has a wide range of applications.

These include monitoring of small-scale seismic activity and mapping of underground surfaces on both land and sea. However, a DAS can produce large volumes of data. Fernandez-Ruiz *et al.* [5] did not say much about how they could handle all this data. In our opinion, the data that comes in can be instantly analyzed through real-time processing, and it can also be compressed using AI algorithms to reduce its size. Plevris [6] discussed how AI reduces the damage caused by earthquakes by improving current systems and tools. The need for improvement arises because old techniques are not good enough to predict and prepare people on time. AI computes a large amount of data faster, making quick predictions. In early warning systems, using AI effectively reduces the time taken to alert people, which is very useful in crowded places, and reduces false alarms by accurately processing data. AI-driven earthquake early warning systems (EEWS) can be further enhanced as AI continuously learns from previous data, hence continuously improving. Pairing this AI-driven system with good sensors and drones or data from satellites helps save more lives. In addition, AI-enabled structural health and risk assessment systems help to detect damaged points accurately, so repair can be done on time, which is much better than manual assessment. AI processes different massive datasets and tools such as accelerometers, strain gauges, drones, satellites, etc. are used.

In addition to checking structural health and risk assessment, AI can effectively help in making designs that are earthquake resistant by choosing the best materials for construction and making structures that are intelligent enough to adapt to seismic activity by using long short-term memory (LSTM) and fuzzy control. Those structural designs can first be simulated to check their performance [6]. A system for detecting multiple natural disasters is made by integrating different datasets for each of the disasters and then training the system on this dataset. One of the most important things is to make people comfortable with using this advanced technology and ensuring fair and morally correct use. In our opinion, the authors should have stressed the privacy and security regarding the datasets and compliance with local and international laws. They did not discuss how legal responsibility will be taken in case of false alarms. The widespread adoption and maintenance of the tools require more financial resources; they did not address the cost required for this in detail, especially in developing areas. In Chin *et al.* [7], the authors focused on improving the accuracy of earthquake detection using low-cost sensors and machine learning algorithms. For testing, they installed a network of these sensors called Palert. A specific dataset was chosen for analysis, which was then used to train machine learning models.

This dataset was constructed by filtering irrelevant signals. Only signals that qualified the minimum shake threshold were allowed to pass through, and their features were numerically summarized. The data were classified using three machine learning techniques. The first, K-nearest neighbors (KNN), analyzed the signal based on a k number of similar previous examples. The second, the classification tree, classified the data using a tree-like flowchart, where each branch of the tree asked for a yes or no call. The third, the support vector machine (SVM), used a mathematical boundary in space that separated the relevant and irrelevant signals. The detection quality and the time taken by each method were examined to ensure its effectiveness. Their study showed that machine learning methods provide us with more accurate and reliable results compared to other existing methods that rely on fixed thresholds. However, they did not explain how their system caters for sensor failures and network delays. In our opinion, multiple sensors can be installed in the same area to ensure backup in case of failure. In addition, we can avoid delays by processing the signals directly at the sensor or a nearby device, rather than relying on real-time transmission. Using their work as inspiration, we now divide this section into different AI models to explain which ones perform better in detecting earthquakes, taking into account all the challenges we discussed above.

19.2.1 Deep learning models for timely detection of earthquakes

[8] discussed several deep learning techniques in monitoring seismic activity, assessing damage caused by this activity, and risk assessment, basically, earthquake engineering. Using convolutional neural networks (CNN) and transfer learning (TL), vision-based damage assessment is made easy by extracting spatial features such as edges, textures, and shapes. For damage prediction and ground motion modeling, multi-layer perceptron (MLP) known as deep neural network (DNN) technique is used as it is easy to set up and train. The graph neural network (GNN) is also great for risk assessment. The recurrent neural network (RNN) is used for active control of signals for dampers and isolators, modeling seismic waveforms, and response prediction by simulating building motions. In generative adversial network (GAN), two networks are trained to produce realistic data, that is why it is a great technique for augmenting small damage image datasets. For detecting anomalies and filtering of noise, the autoencoder (AE) and the variational autoencoder (VAE) are used. Reinforcement learning (RL) and deep RL is extremely helpful in forming post-quake reconstruction strategies. Using these techniques together, increases the field of earthquake engineering greatly.

Vision-based techniques can quickly find the damage by analysis and using drones, satellites and cameras, etc., and raw data is used to make fast predictions about movement of buildings. It does not discuss how these techniques can be implemented in real-time monitoring systems, hardware requirements, and analysis of training time are not given. No new techniques are mentioned, like transformers, self-supervised learning, etc. The detection of seismic activity through manual networks can sometimes be slow and fallible. The authors in Zhuang *et al.* [9] replaced these methods which helped speed up the process. Deep learning models such as CNNs are trained to learn features such as the frequency and amplitude of seismic signals. CNNs outperform manual methods by using large data sets to study the irregular activity detected by DAS. They are used to process received signals, filter noise, and analyze waveforms to indicate the occurrence of earthquakes. Techniques like data augmentation, TL, and Bayesian optimization are used to identify and reform the best-performing parameters. This improves the reliability of the models over time as they continue to learn from the past data. Gradient-weighted class activation mapping (Grad-CAM) is also used for visual explanation to ensure better predictions.

Metrics such as the number of events observed, the accuracy of these observations, and the time taken by the models to signal an alert are evaluated and used in testing the performance of the deep learning models. Lightweight CNNs (like MobileNet) use image classification and pattern recognition to provide faster and more reliable results. However, they did not discuss the power efficiency of these systems if they are deployed in areas with low resources. In our opinion, further optimizing models for low-power and solar-powered devices can help conserve energy while maintaining efficiency. Similarly, in Rojas *et al.* [10], the authors discussed how earthquakes and volcanic eruptions can be monitored and studied using the end-to-end wireless sensor network (WSN). In addition to earthquakes, they also use them to find natural resources such as oil and gas buried deep underground. P waves (primary) and S waves (secondary) produced during seismic activity reflect (bounce back), refract (bend) and attenuate (lose energy) while moving through materials underground in Earth's layers. They placed sensor nodes evenly over a wide area, and these nodes detect the reflected waves that carry information about the underground structure. A graph or image is created using sensor's data. Currently, cable-based sensor networks are in use. Cables provide power to sensors and to transfer data, but can be expensive and time-consuming as they are difficult to set up and maintain.

A much easier, cheaper, and faster way is to use self-powered WSN. For this purpose, they identify solar power as a reliable energy source. One of the main topics they discuss in their work is the way to build a wireless geophone sensor network. In this network, each sensor node, including a signal conditioning module to clean the signal and a digital circuit module to process, is specifically designed to run on low power and be charged with solar power. WSN uses the mesh network topology, the coordinator (to manage the network), the router (to relay data), and the nodes to collect seismic data and enter sleep mode to save power. Their hardware structure comprised of a microcontroller (ESP32) to process data, a GPS (NEO-6M) that tracks node locations, solar-charged batteries and a Xbee-S2C module to manage wireless communication. Zigbee is employed as a communication protocol because of its low consumption of power and route optimization. In our opinion, they do not consider scalability as they test it only for six nodes. Hence, their work did not explain how the network behaves when the number of nodes is more than six. It also does not answer how the network performs in extreme environmental conditions. In addition, no details are given on the cost analysis. Given that the idea is employed to detect seismic activity, researchers in the underwater field can utilize this idea to examine such activities underwater.

19.2.2 Machine learning models for timely detection of earthquakes

Yang *et al.* [11] focus on ways to improve early warning systems using machine learning to predict the hypocenter (underground origin) of earthquakes fast and precisely. Identifying the origin of earthquakes helps us estimate the magnitude of the earthquake quickly and where the shaking can be felt the most. Classical methods such as the iterative method (using travel time of waves) or waveform method lack speed and accuracy. Machine learning improves the system by automating tasks like detecting and locating earthquakes and predicting their impacts. The exact moments when a seismic station detects seismic waves are called station trigger times, which tells us when the energy of the earthquake reached the station. Their study uses station trigger times instead of analyzing full seismic waveforms, which can be quite time-consuming. Earlier work used random forest (RF) with these trigger times, but did not distinguish between land and ocean quakes. Land and ocean quakes need to be treated differently as there are some stations in oceans and seismic waves travel differently from water, making detection harder. About 2.17 million seismic events were taken from the Japan meteorological agency to be used as data for this purpose. The stations were filtered to focus on dense coverage in two regions, the main islands of Japan and the northern areas of Japan, and then only 2066 stations remained.

Noise signals and earthquakes with magnitude less than or equal to 5 were removed from their testing. They tested three models which are RF, extreme gradient boost (XGBoost), and light gradient boosting machine (LightGBM) to see which one works the best with trigger times, built separate models for land and ocean quakes and added a classifier to determine if a quake is on land or in the ocean, and then tested the system's reliability. LightGBM outperformed the other two models due to boosting, even with 3 stations it predicted land quakes within approximately 2km error. This result enhances the entire early warning system, providing critical seconds for life-saving alerts. Their study does not tell whether the system gives alarms when there are no quakes or does not give alarms when there are quakes and if it does, what can be the solution to it. They discuss the hypocenter, latitudes, and longitudes of where the earthquake starts but not depth, which is significant for providing information on damage. Stations in ocean are fewer, they do not tell if AI can be trained equally on land and ocean data. One of the main questions also arises is how much we can rely on AI, what if it fails during a major earthquake. In Gomez and Kadri [12], the authors discuss that waves produced due to seismic activity underwater are called acoustic-gravity waves (AGW). These waves can travel at a long distance. They are detected by underwater microphones called hydrophones.

They also contain some important information that can be critical in predicting an earthquake or tsunami. By paying attention to what these waves convey, the slip type and the strength of the earthquake can be identified. The information carried by the waves can also help us assess the size and duration of the slip, along with the magnitude and location of the earthquake. They extracted this information from the 201 acoustic signals received from underwater activity. They did so using specific machine learning techniques, such as regression and classification. The type of slip of the earthquake was estimated by classification algorithms and its magnitude was estimated by regression algorithms. In this way, a correlation was defined between the signals generated by the seismic activity and the strength and magnitude of the earthquake. This helped them understand how serious and dangerous the earthquake actually was. In the slip-type classification, they identified the type of slip. For this, two types of classification approach were used. These were multi-class and binary. In binary classification, they study the motion of the earthquake. It can be horizontal, also known as the strike slip, or vertical, also known as the dip slip. Vertical motion is generally considered more dangerous and is often responsible for tsunamis.

Different methodologies were used to extract features from underwater signals. In addition, two more algorithms RF classifier and SVM were used and both of them showed similar accuracy. The examples of vertical motion showed better accuracy. In multi-class classification, these results were further studied. The data were classified into 3 parts: strike-slip (horizontal motion), thrust (upward motion), and normal (downward motion). Like in binary classification, this approach was accurate enough for strike-slip and thrust, but it was less accurate for the normal type earthquakes. The slip type classification showed that underwater acoustic signals can, in fact, provide clear information about the slip type. For the magnitude regression, their goal was to estimate the strength and magnitude of the earthquake with the help of underwater signals produced by seismic activity. In this process, two types of regression algorithms were used: support vector regression and random forest regression. With the help of these, four kinds of signal features were extracted and studied. Their results were generally accurate, proving that the technique is quite reliable. Although, interestingly, there were large errors for very high or very low magnitudes. They used five case studies to show the workings of both slip-type classification and magnitude regression. Their testing showed an accurate classification of the slip type.

The tested magnitudes also ended up being within 5 percent of the estimated ones. All predicted tsunamis ended up taking place, demonstrating the accuracy of machine learning models. They also tested twenty artificial earthquakes to test how well the machine learning models work for the vertical motion of these quakes. All of these samples were correctly classified. It was concluded that the wavelet feature of the signals and the SVM method gave the best results. In their work, the data that was studied were recorded by three hydrophone stations. These stations were located in the Indian and Pacific oceans. The four features of the signals discussed above were temporal, spectral, cepstral, and wavelet. These were the features studied for the testing and training of the machine learning models. They were tested using techniques such as 10-fold cross-validation and grid search. Once the type of slip and the magnitude of the earthquake were estimated, they were used to estimate the dimensions of the cause of the earthquake, known as the fault. This is done using a mathematical model called the inverse problem method. However, this model is only applicable to vertical motion. In our opinion, this issue is due to the limited dataset, containing only 201 signals. This can be addressed by expanding their dataset further. Another issue left unaddressed in their work is the accuracy of these techniques for deep-sea signals. They only performed tests for shallow earthquakes. To further evaluate the accuracy of the models, the effect of the depth of the signal should be studied.

19.3 Conclusion

In this chapter, we discuss different techniques that can be used for the detection of underwater earthquakes. We have studied different research from the literature to understand the monitoring of underwater activity and the analysis of the data extracted from underwater signals. Our main focus is the detection of acoustic signals and the advancements due to AI models. We have seen that these relevant data contain important information that can help us detect seismic activity. The technique that has been the primary focus is DAS because of its efficiency, accuracy, and smooth integration with machine learning models and signal processing models. This method requires the deployment of optical fiber cables in the ocean to detect waves and signals that could potentially lead to earthquakes and tsunamis. Not only is DAS considered to be more effective, it is also cheaper than other classical methods. We also have seen that we can improve the accuracy and reliability of DAS by real-time processing and filtering of the signals received from the ocean. Many machine learning models were trained and tested in the literature. These included deep learning, classification, and regression models. The chapter shows the workings of these techniques and demonstrates their accuracies under different conditions. DAS plays a major role in improving the reliability of detection, ensuring early warnings and correct interpretation of data. Generally, promising results have been achieved from the work done in the literature. However, some challenges still need to be addressed. In general, these works have shown significant potential for the detection of underwater earthquakes. These studies have a great contribution to geophysical research and early prediction of natural disasters. Further enhancement of these technologies is required for future progress.

References

[1] H. Xiao, Z. J. Spica, J. Li, and Z. Zhan, "Detection of earthquake infragravity and tsunami waves with underwater distributed acoustic sensing," *Geophysical Research Letters*, vol. 51, no. 2, p. e2023GL106767, 2024. e2023GL106767 2023GL106767. https://agupubs.onlinelibrary.wiley.com/doi/abs/10.1029/2023GL106767.

[2] S. Chen, J. Han, Q. Sui, K. Zhu, C. Lu, and Z. Li, "Advanced signal processing in distributed acoustic sensors based on submarine cables for seismology applications," *Journal of Lightwave Technology*, vol. 41, no. 13, pp. 4164–4175, 2023.

[3] B. G. Gorshkov, K. Yüksel, A. A. Fotiadi, *et al.*, "Scientific applications of distributed acoustic sensing: State-of-the-art review and perspective," *Sensors*, vol. 22, no. 3, pp. 1–41, 2022. https://www.mdpi.com/1424-8220/22/3/1033.

[4] I. Lior, A. Sladen, D. Rivet, *et al.*, "On the detection capabilities of underwater distributed acoustic sensing," *Journal of Geophysical Research: Solid Earth*, vol. 126, no. 3, p. e2020JB020925, 2021, e2020JB020925 2020JB020925. https://agupubs.onlinelibrary.wiley.com/doi/abs/10.1029/2020JB020925.

[5] M. R. Fernández-Ruiz, H. F. Martins, E. F. Williams, *et al.*, "Seismic monitoring with distributed acoustic sensing from the near-surface to the deep oceans," *Journal of Lightwave Technology*, vol. 40, no. 5, pp. 1453–1463, 2022.

[6] V. Plevris, "Ai-driven innovations in earthquake risk mitigation: A future-focused perspective," *Geosciences*, vol. 14, no. 9, pp. 1–28, 2024. https://www.mdpi.com/2076-3263/14/9/244.

[7] T.-L. Chin, C.-Y. Huang, S.-H. Shen, Y.-C. Tsai, Y. H. Hu, and Y.-M. Wu, "Learn to detect: Improving the accuracy of earthquake detection," *IEEE Transactions on Geoscience and Remote Sensing*, vol. 57, no. 11, pp. 8867–8878, 2019.

[8] Y. Xie, "Deep learning in earthquake engineering: A comprehensive review," 2024. *ASCE OPEN: Multidisciplinary Journal of Civil Engineering*, Vol. 3, no. 1, p. 03125001, 2025. https://doi.org/10.1061/AOMJAH.AOENG-0080.

[9] X. Zhuang, T. V. Tran, H. Nguyen-Xuan, and T. Rabczuk, "Deep learning-based post-earthquake structural damage level recognition," *Computers & Structures*, vol. 315, p. 107761, 2025. https://www.sciencedirect.com/science/article/pii/S0045794925001191.

[10] J. P. Rojas, R. Almazyad, A. Al Hayyah, *et al.*, "Self-powered end-to-end wireless sensor network for geophysical explorations," *IEEE Systems Journal*, vol. 19, no. 1, pp. 107–118, 2025.

[11] C. Yang, K. Zhang, G. Chen, Y. Pan, L. Zhang, and L. Qu, "Application of machine learning to determine earthquake hypocenter location in earthquake early warning," *IEEE Geoscience and Remote Sensing Letters*, vol. 21, pp. 1–5, 2024.

[12] B. Gomez, and U. Kadri, "Earthquake source characterization by machine learning algorithms applied to acoustic signals," *Scientific Reports*, vol. 11, no. 23062, pp. 1–15, 2021.

Chapter 20
Detection of tsunamis and earthquakes through underwater sensor networks

Sataish Elahi[1], Irfa Farooq[1], Hanzla Sajjad[1] and Sajjad Hussain[1]

Seismic activities underwater pose a great threat to tectonically active coastal communities. This review chapter will revolve around the discussion of modern technology that is being utilized or developed for the purpose of early detection and warning generation of these occurrences. This technology is inclusive of underwater sonar systems, detailed seafloor mapping techniques, and artificial intelligence (AI). Traditional monitoring systems, such as tide gauges and pressure sensors, fail to provide sufficient active readings for submarine landslides. In order to deal with their shortcomings, multi-beam echo sounder systems alongside remotely operated vehicles are used to construct an enhanced view of underwater structures. Additionally, a clearer identification of tsunami patterns is achieved using the digitization of tide data and the application of wavelet-based analysis techniques. AI, in conjunction with wireless sensor networks, is used to improve the efficiency of disaster detection. Such detection systems, when combined with edge computing and privacy-preserving methods such as federated learning, reduce the occurrence of false alarms. Underwater visible-light communication is also discussed as a high-speed alternative to acoustic systems and has sufficient potential when enhanced with intelligent beam alignment and adaptive modulation. In general, this study will circulate around the possibilities of integrating multiple methodologies to improve traditional methods and identify complex underwater geohazards before disasters occur.

20.1 Introduction

In modern ocean monitoring, underwater sonar systems are fundamental for detecting submarine earthquakes and tsunamis. In real-time, these sonar technologies identify seismic events, landslides, and seabed deformations by transmitting sound waves through water and meticulously measuring their reflections. These systems are indispensable because coastal or surface-based technologies often fail to detect underwater phenomena sufficiently quickly to issue effective warnings. Sonar technology has

[1]School of Electrical Engineering and Computer Science (SEECS), National University of Sciences and Technology (NUST), Pakistan

enabled scientists to monitor changes in the shape of the sea floor due to landslides, along with the detection of earthquakes, playing a critical role in early hazard detection. In addition to sonar systems, several other modern technologies have been developed for tsunami and earthquake detection. Tidal gauges, satellite altimetry, and bottom pressure recorders are commonly used to track sea level variations and seabed displacements [1]. These technologies work together with advanced data processing techniques such as least squares interpolation, bandpass filtering, wavelet analysis, and backward ray tracing (BRT). These approaches help locate earthquake epicenters, track tsunami propagation, and reconstruct event dynamics.

Bathymetric data, which refer to detailed measurements of the depth and topography of the seafloor, play a crucial role in detecting and modeling tsunamis. High-resolution bathymetry enables scientists to understand how underwater topography influences the travel and behavior of tsunami waves. Bathymetric maps can be obtained using sonar surveys and satellite altimetry techniques, and after significant seismic events, autonomous underwater vehicles (AUVs) can be deployed to update seabed data quickly. Digitization of tide records is another essential step, achieved by re-digitizing original analog recordings, fitting the data with least squares polynomial interpolation, and applying bandpass filters to ensure precision [2]. This allows researchers to identify tsunami signals above the noise level and determine sharp arrival times using methods such as wavelet spectral analysis. Despite significant advancements, current detection and warning systems still face major challenges. One key issue is the lack of high-resolution real-time seabed surveillance capable of detecting sudden slope failures or submarine landslides. Furthermore, the existing network of tide and pressure gauges remains sparse, especially near active tectonic boundaries.

Another limitation is that many tsunami warning models still rely on a single-source assumption, overlooking the potential complexity of dual-source events that involve both earthquakes and landslides. The advancement and implementation of underwater detection networks have become a global imperative due to the increasing threat that tsunamis pose to coastal populations, particularly within highly volatile tectonic regions. To improve tsunami detection and hazard assessment, it is essential to deploy denser networks of seabed sonar systems capable of detailed mapping and real-time seismic monitoring. Satellite-based altimetry can enhance early detection by capturing subtle changes in sea surface height in the open ocean. In addition, machine learning algorithms that analyze incoming waveform data can significantly improve the discrimination between purely earthquake-generated tsunamis and those triggered by secondary processes such as landslides. This chapter focuses on the technologies that can be used to help in the early detection of tsunamis and earthquakes. The rest of the chapter is arranged in the following manner: Section 20.2 discusses related work, Section 20.3 provides solutions with the aid of AI, Section 20.4 discusses visible-light communications, and Section 20.5 concludes the chapter.

20.2 Related work

The Southwest Iberian Margin constitutes a notably intricate domain, where substantial seismic hazards arise from the interaction between the Nubian and Eurasian tectonic plates. A comprehensive understanding of the mechanisms for the generation of tsunamis in this area is critical to mitigate risk and improve early warning effectiveness. The February 28, 1969 earthquake and tsunami on the Horseshoe Abyssal Plain, southwest of the Iberian Peninsula, serves as a crucial case study. A strong earthquake struck the region at 2:41 am UTC, with a moment magnitude of 8.0, making it the most powerful seismic event in continental Portugal of the twentieth century. The tremor was felt widely throughout Portugal, Morocco, France, and the Canary Islands, resulting in widespread disruption and casualties. However, despite the well-documented strength of the earthquake, the accompanying tsunami displayed anomalies that challenge conventional interpretations of only seismic events. Specifically, the Casablanca tide gauge recorded an earlier-than-expected tsunami arrival and inverse wave polarity, inconsistencies that cannot be explained by an earthquake source alone.

This observation leads to the central research question: could the 1969 tsunami be the result of both the earthquake and an underwater landslide triggered by the seismic event? Investigating this question reveals the limitations of traditional earthquake-only detection models and emphasizes the importance of dual-source modeling in tsunami hazard assessments. A study was conducted to observe the Atlantic Iberian earthquake and tsunami that occurred in 1969 [3], highlighting the use of updated tsunami records, wavelet analysis, and BRT. Twenty-one tide stations reported tsunami activity across Portugal, Spain, France, Morocco, and the United Kingdom, providing critical data for analysis. The tide stations closest to the epicenter were located along the west and south coasts of Portugal and Morocco, while distant stations in the Canary and Scilly Islands demonstrated the widespread impact of the tsunami. Tide records were extracted from original analog charts through a rigorous re-digitization process. Interpolated into polynomials using least-squares fitting and band-pass filtering, the cleaned data achieved an adjusted R-square greater than 0.995.

An adjusted R-square value close to one indicates that the polynomial model accurately fits the observed data, minimizing the discrepancy between the predicted and actual values. Wavelet spectral analysis, using Morlet's continuous wavelet approach, enabled the precise identification of tsunami arrival times and the detection of frequency shifts associated with first arrivals. Later, Guesmia *et al.* [4] performed a comprehensive reanalysis of the 1969 event. Previous research largely modeled the tsunami using single source earthquake fault parameters, typically based on Fukao [5] solution, successfully reproducing tsunami records in many locations but consistently failing to match the waveform observed at Casablanca. The major contribution of Baptista and Miranda [6] was the integration of a submarine landslide scenario west of Casablanca into their modeling. By expanding the digitized tide record dataset, applying advanced wavelet techniques, and simulating dual-source events, they showed that incorporating a moderate-volume landslide ($3\,km^3$) triggered shortly after the earthquake could explain both the anomalous early arrival and the inverted polarity recorded in Casablanca, while maintaining consistency with tsunami observations at other locations.

Although this work made a significant advancement, some uncertainties remain. Direct bathymetric evidence confirming the landslide is unavailable due to the absence of post-event seabed surveys. Instrumental errors, such as clock drifts from the tide gauge and limited near-source bathymetric resolution, introduce uncertainties. The exact timing and mechanisms that trigger the submarine landslide are mainly supported by indirect evidence. From an objective point of view, improving system performance requires a hybrid monitoring approach that combines AI-driven pattern recognition, real-time sonar mapping, and conventional seismology. Furthermore, multi-hazard simulations, incorporating both seismic and landslide sources, should become a standard element of tsunami risk assessments, particularly in tectonically active regions like the Atlantic Iberian Margin. The 1969 event serves as a critical reminder that natural disasters often involve complex processes that do not fit neatly into single-source models, and our detection systems must evolve accordingly. Some other underwater technologies used in the detection of seismic events are multibeam echo sounders (MBES) [7] and remotely operated vehicles (ROVs) for the detection of earthquakes and tsunamis using underwater technologies.

MBES systems fall into the category of sonar devices and are especially useful in mapping the sea floor. Their mode of operation revolves around the emission of sound waves from a ship and the notation of the time it takes for these waves to be reflected by the seafloor. They are also able to measure the intensity of the reflected waves, allowing the creation of highly detailed maps of the seafloor. It is worth noting that, in contrast to traditional single-beam echo sounders, MBES systems permit wider area coverage with improved accuracy due to the higher number of pings that can be sent at any given time. ROVs are compact unmanned underwater robots that are equipped with sensors and cameras, which allows the collection and inspection of visual data from the seafloor. Their role is critical in the modeling of underwater structures for analysis. Both technologies played a pivotal role in the examination of the root cause of the anomalous early arrival of the tsunami in Toyama Bay during the 2024 earthquake in the Noto Peninsula.

According to their study, MBES technology was used to obtain bathymetric surveys in January and February 2024. The data collected from these surveys was compared with previously conducted surveys in 2010. The discrepancy between the two allowed the researchers to identify negative depth changes in an approximate area of 3.5 km $\times$ 1 km, which revealed that submarine landslides had been triggered. The MBES bathymetric surveys provided comprehensive imagery of disturbed seabed areas which provided a quantitative perspective of seafloor deformation. Although the MBES bathymetric surveys provided an aerial view of the seafloor, an ROV named DepthUnit300 was used for direct visual confirmation of the underwater topography. The ROV captured extensive footage of disturbed sediments and fresh cliff formations, which provided visual data that verified the findings obtained through MBES. However, it is important to mention that because of the absence of an underwater positioning system in the ROV, its location had to be measured indirectly, which led to positional uncertainties in the visual data collected. The data collected using the MBES technology and the ROV was compared with the data obtained using a tsunami back-propagation model.

The researchers traced the arrival times of the tsunami to find the origin of the tsunami and discovered that this origin was the same area where submarine landslides were detected using MBES and ROV, confirming the role of underwater landslides in tsunami generation. This was a significant discovery because it provided direct evidence for previous research [8,9], which established the occurrence of submarine landslides due to earthquakes. This confirms the benefits of MBES and ROV technologies over earlier methods that relied on sparse soundings, but gaps remain in fully linking landslides to seismic events. To cover these gaps, the study should be expanded to other areas prone to hazards rather than focusing on a single canyon alongside the integration of real-time monitoring systems. Furthermore, more reliable navigation can be achieved by using ROVs with ultra-short baseline systems. Lastly, the combination of the bathymetric data obtained by the MBES systems with numerical tsunami models can aid in the quantification of risks. Therefore, we can conclude that to improve the overall detection of seismic events, the deployment of AUVs to survey larger areas that are frequently required, the integration of machine learning to identify landslides using bathymetric data and linking it to tsunami alert systems are necessary.

20.3 How AI can improve detection of tsunamis and earthquakes?

This section will focus on the use of AI-based detection and early warning systems to better deal with the increasing ferocity and number of natural disasters around the world. The necessity for this technology arises from the failure of conventional systems due to a lack of adaptability, delayed responses, and a high frequency of false positives. In Charan *et al.* [10], a multi-layered system was proposed that unified wireless sensor networks (WSNs) with artificial intelligence (AI) to create more effective real-time disaster early warning systems. A WSN is composed of a group of spatially distributed sensors that are used to monitor physical environments and transmit data to a core location. They are made up of nodes that are essentially individual sensors that transmit relevant information to the base station via radio waves. Therefore, WSNs play a crucial role in collecting environmental data and feeding it to AI algorithms that analyze this data to predict and detect natural disasters. Furthermore, WSNs alongside edge computing allow for localized processing, which with the integration of machine learning models allows the system to learn from past patterns and give more accurate predictions. The system presented by Charan *et al.* [10] were based on four layers: a sensor layer for data collection, a data processing layer, a communication layer, and an application layer.

The sensor layer was made up of fixed and mobile nodes that collect environmental data, which was then acted upon by the data processing layer that utilizes a variety of AI models, such as support vector machines, decision trees, and neural networks, to process sensor data. These models functioned under both supervised and unsupervised learning for anomaly detection, pattern detection, and predictive modeling. The communication layer used secure, redundant, and energy-efficient protocols such as Zigbee and LoRa for the transmission of data. The application layer was used to interact with the authorities through SMS alerts, mobile applications, and other various forms of communication. The entire system was also improved by using edge computing, cloud computing, and geographic information systems (GIS). Edge computing is a form of distributed information technology that processes data at the boundary of the network, as close as possible to the originating source of the data. This is especially useful for real-time processing at the node level in WSNs with minimal connectivity requirements. GIS, in conjunction with remote sensing, is also pivotal in the analysis of aerial visual data from aerial imaging for spatial mapping of data collection sites.

Through evaluation, it was determined that this four-tier system was highly accurate in determining disasters with minimal false alarms. The lifetime of sensor networks was also increased by the utilization of energy-efficient algorithms. Furthermore, edge processing allowed the system to function even in the absence of the internet, which is crucial for areas with weak infrastructure, as well as for disastrous natural events. Despite its efficiency, the system still has its drawbacks. These include inadequate exploration of the protocol challenges between various hardware-based sensors and network protocols, as well as the lack of discussion of the real-time working and constraints of this system. Another important issue that was not adequately addressed in their work is data privacy, considering that data breach issues might be inevitable for sensors placed in public zones. These issues can be overcome by using federated learning, which is a machine learning technique in which several users can jointly train a model without sharing raw data, thereby preserving privacy. Furthermore, autonomous drones can act as mobile sensor platforms for improved coverage and reliability in inaccessible disaster zones. In addition, it would be highly beneficial for further improvements to focus on the development of lightweight AI models that can reduce dependence on cloud computation.

Additionally, another technology that can be used for early warning systems is underwater gas monitoring, which can be implemented in conjunction with AI to facilitate the real-time analysis of anomalies within gas emissions, which could be acting as a possible precursor to seismic events. These systems, as discussed in the Bulgarian study by Pritchard *et al.* [11], are based on the hypothesis that sudden fluctuations in gas seepage levels underwater could be indicators of seismic activities that are yet to occur. The study proposed the installation of specialized gas capture devices directly above underwater gas seepage sites for the detection of changes within emission patterns. The data collected by these devices are transmitted to receivers on land for further processing, specifically for monitoring the correlation between anomalies in gas emission patterns and seismic events. When combined with modern sensors and AI, machine learning algorithms can be directly fed sensor data, allowing these gas-based detection systems to automatically measure risk levels, send alerts, and efficiently track real-time data with reduced human input. It is also to be noted that the AI enhancement of these systems allows for a reduction in false alarms and increased reliability in predicting earthquakes.

Future work in improving the early detection and warning systems for seismic activities should explore the integration of data from visual, acoustic, and gas detection technologies in conjunction with AI-based pattern detection. Visual and acoustic data can also be used for the detection of sudden gas discharge as per the Bulgarian study, which further emphasizes the importance of implementing sensor fusion to allow researchers to comprehensively understand seismic events triggered by anomalies, such as underwater gas emissions. Additionally, satellite-based remote sensing can also be utilized for surface-level detection of gas plumes. Therefore, integration of all these technologies allows for an efficient multi-layered system that can improve spatial mapping of vulnerable sites and improve the reliability of early warning systems. This approach can also be potentially implemented in the assessment of vibrogenesis effects in coastal areas that are susceptible to seismic events. Finally, extending these systems to areas lying on fault lines can potentially improve seismic monitoring globally.

20.4 Is underwater visible-light communication an alternative for the acoustic band?

This section will focus on an alternative to traditional underwater communication systems, underwater visible-light communication (UVLC). UVLC is a data transmission method that employs the use of the visible-light spectrum, 400–700 nm, for high-speed underwater data transmission that has fewer effects from interference compared to underwater acoustic communication systems. Traditional acoustic systems are capable of long-range communication, but they are prone to interference from noise signals and suffer from low bandwidth; on the other hand, RF waves also experience severe attenuation in water, which is why UVLC provides a reasonable trade-off between the two to enable communication with high bandwidth and security but with restrictions to the communication distance and line-of-sight (LOS) conditions. Having the core idea of UVLC, this section will now focus on the current state of UVLC technology, the challenges faced with its employment, and a general discussion of how different components affect the performance of a UVLC system. Light-emitting diodes (LEDs) are a cost-effective and energy-efficient light source that is frequently used in UVLC systems. They emit a broad beam, making them suitable for close-range communication.

Another frequently used light source is laser diodes (LDs). They provide narrow beams and are more powerful, which allows transmission over longer distances with higher data rates; however, they also require precise alignment to be of significant use. Common modulation techniques used in UVLC include on-off keying (OOK) and pulse position modulation (PPM). Advanced modulation techniques are based on orthogonal frequency division multiplexing (OFDM), while color shift keying (CSK) and quadrature amplitude modulation (QAM) are also in the experimental phase. OOK, as the name denotes, is implemented through the state of a carrier signal, where its presence denotes a 1 and its absence denotes a 0. It is a form of amplitude shift keying (ASK) used in the representation of digital data. With respect to UVLC, this form of modulation is achieved by turning on and off a light source for data transmission. Although OOK is easy to implement, it has a binary threshold-dependent nature, making it very energy-inefficient in murky water. PPM is a better alternative to OOK because of the way it functions. It transmits data by encoding the information in the time position of a pulse while keeping its amplitude constant; that is, the pulse is shifted with respect to a fixed time frame to represent a certain data. It is more resistant to noise, but requires precise timing synchronization, which makes it less robust, especially on channels with multi-path effects.

OFDM is a highly advanced and spectral technique that separates the signal into multiple carriers, each modulated at a low data rate. While OFDM provides a way to deal with the multi-path fading effect, it is still very complex computationally and has high power requirements, which makes it less suitable for many underwater applications. CSK and QAM are both experimental techniques that are under consideration for use in UVLC. CSK can increase data by multiplexing across different channels; however, it requires receivers capable of color discrimination, making it difficult to practically deploy this technology underwater, as maintaining color integrity in water is very hard. QAM, on the other hand, is more sensitive to channel impairments and requires very precise SNR control, which is hard to achieve underwater. In terms of the receiver side within the UVLC system, there are two commonly used components, photo-diodes (PDs) and avalanche photo-diodes (APDs). APDs are highly sensitive and suitable for long-range, low-light conditions because of a high internal gain, whereas PDs are more suitable for short-range UVLC because of their low energy and cost demand. In comparison, APDs are the better option in challenging environments, while PDs are more suitable for non-complex, low-power systems.

In a study conducted by Muthu and Cherubini [12], surveys were conducted for various UVLC systems, the evaluation being based on the data rate, transmission range, and bit error rate. These performance metrics were determined for various types of water according to the following categories: clear, coastal, and turbid. The survey revealed that while data rates up to several gigabits per second are achievable in clear water with short-range links, scattering and absorption in turbid waters lead to a visible deterioration in performance. While UVLC has several benefits, it also faces major challenges due to limited range in murky waters, the need for precision in laser-based systems, light interference that affects the PDs and beam divergence alongside tracking issues faced when using mobile nodes. These challenges could be overcome by integrating UVLC systems with AI, allowing real-time beam alignment, intelligent modulation selection based on channel estimation, and intelligent noise filtering. For example, using AI in conjunction with AUVs allows AUVs to intelligently adjust their positions to maintain optimal communication links. Furthermore, there is potential for research to investigate hybrid UVLC acoustic systems for long-range communication.

20.5 Conclusion

This chapter effectively discussed the role of integrating underwater sonar systems, modern sensing technologies, and AI in revolutionizing the detection of seismic disasters. Traditional methods, which include tide gauges and pressure sensors, although still valid, are being rapidly accompanied by high-resolution bathymetric surveys through MBES, ROVs, and AUVs. These advanced technologies allow for more efficient seafloor monitoring and better identification of complex seismic dual-source events. Furthermore, the application of AI in wireless sensing networks allows for the enhancement of early warning systems through faster and more precise disaster detection everywhere. Technologies such as underwater visible-light communication (UVLC) are also emerging as a replacement for traditional acoustic systems for high-speed data transmission underwater. In conclusion, the combination of all these analytical and data collection techniques highlights the importance of a technology-led approach to alleviating the risks in natural disasters in underwater environments.

References

[1] F. Imamura, S. P. Boret, A. Suppasri, and A. Muhari, "Recent occurrences of serious tsunami damage and the future challenges of tsunami disaster risk reduction," *Progress in Disaster Science*, vol. 1, p. 100009, 2019. https://www.sciencedirect.com/science/article/pii/S2590061719300092.

[2] M. A. Clare, G. Lintern, E. Pope, *et al.*, *Seismic and acoustic monitoring of submarine landslides*. American Geophysical Union (AGU), 2024, ch. 5, pp. 59–82. https://agupubs.onlinelibrary.wiley.com/doi/abs/10.1002/9781119750925.ch5.

[3] Baptista, M. A., Miranda, J. M., and Omira, R. "The 28th February 1969 earthquake and tsunami in the Atlantic Iberian margin," *Pure and Applied Geophysics*, vol. 182, pp. 1363–1380, 2025. https://doi.org/10.1007/s00024-025-03689-z

[4] Guesmia, M., Heinrich, P., and Mariotti, C. "Numerical simulation of the 1969 Portuguese tsunami by a finite element method," *Natural Hazards*, vol. 17, pp. 31–46, 1998.

[5] Fukao, Y. "Thrust faulting at a lithospheric plate boundary the Portugal earthquake of 1969," *Earth and Planetary Science Letters*, vol. 18, no. 2, pp. 205–216, 1973. https://www.sciencedirect.com/science/article/pii/0012821X73900587.

[6] Baptista, M. A. and Miranda, J. M. "Revision of the Portuguese catalog of tsunamis," *Natural Hazards and Earth System Science*, vol. 9, pp. 25–42, 2009.

[7] Minami, H., Umino, K., Tateishi, R. *et al.* "Detecting submarine landslides caused by the 2024 Noto Peninsula earthquake through repeat bathymetric surveys in Toyama Bay, Japan," *Landslides*, vol. 22, pp. 449–458, 2025.

[8] Hampton, M. A., Lee, H. J., and Locat, J. "Submarine landslides," *Reviews of Geophysics*, vol. 34, no. 1, pp. 33–59, 1996. https://agupubs.onlinelibrary.wiley.com/doi/abs/10.1029/95RG03287.

[9] Masson, D. G., Harbitz, C. B., Wynn, R. B., Pedersen, G., and Løvholt, F. "Submarine landslides: Processes, triggers and hazard prediction," *Philosophical Transactions of the Royal Society A: Mathematical, Physical and Engineering Sciences*, vol. 364, no. 1845, pp. 2009–2039, 2006.

[10] Charan, P., Siddiqui, M. M., Yadav, V., Pathak, C., Narayan, Y., and Khan, Z. H. "AI-enhanced early warning systems for natural disaster detection and mitigation using wireless sensor networks," in *2024 Second International Conference Computational and Characterization Techniques in Engineering & Sciences (IC3TES)*, 2024, pp. 1–6.

[11] Pritchard, M. E., Allen, R. M., Becker, T. W. *et al.*, "New opportunities to study earthquake precursors," *Seismological Research Letters*, vol. 91, no. 5, pp. 2444–2447, 2020.

[12] Muthu, B. A. and Cherubini, C. "Underwater digital twin sensor network-based maritime communication and monitoring using exponential hyperbolic crisp adaptive network-based fuzzy inference system," *Water*, vol. 17, no. 9, pp. 1–27, 2025. https://www.mdpi.com/2073-4441/17/9/1324.

Chapter 21

Underwater SONAR image classification and analysis with the aid of AI

Umar Bin Shakil[1]*, Rehan Ahmad Iftikhar*[1] *and Ayesha Kanwal*[1]

With an emphasis on improving the autonomy and operational capabilities of autonomous underwater systems, this chapter explores the crucial role that artificial intelligence (AI) plays in the classification and interpretation of underwater sound navigation and ranging (SONAR) pictures. Traditional optical sensors face difficulties in turbid underwater conditions with low light, making acoustic imaging technologies such as side-scan, forward-looking, and multibeam SONAR crucial for underwater perception. However, the complexity of interpreting noisy, low-resolution SONAR images has limited the effectiveness of autonomous underwater systems. AI, through deep learning (DL) and machine learning (ML) algorithms, has enabled automated feature extraction, real-time decision making, and improved generalization and classification of underwater objects in various marine environments. In this chapter, the use of several AI models in tasks such as object detection, segmentation, substrate categorization, and maritime debris monitoring is highlighted. It also evaluates specialized frameworks that address challenges such as limited data, intra-class variability, and semantic overlap in different settings. Although advances in AI have significantly improved SONAR-based perception, issues such as high computational demands, sensor noise, and dataset scarcity persist. The chapter concludes by discussing current limitations and proposing future directions that involve lightweight architectures, sensor fusion, and domain adaptation for scalable, robust, and energy-efficient underwater AI systems.

21.1 Introduction

SONAR is a sound wave-based technology that is used to detect and classify underwater objects. This is significant since traditionally employed vision sensors tend to perform poorly under deep underwater conditions due to turbidity, ambient lighting conditions, etc. [1]. SONAR data classification and analysis are very important in

[1]School of Electrical Engineering and Computer Science (SEECS), National University of Sciences and Technology (NUST), Pakistan

applications of marine exploration, search and rescue operations, military reconnaissance, and underwater archeology. Over time, many different schemes have been proposed to automate the SONAR data classification and interpretation process. Existing schemes for this task include classical methods that rely on manual feature extraction from SONAR images, using segmentation, geometrical analysis, or the time-frequency domain through techniques such as Mel-frequency cepstral coefficients and wavelet transform [1,2]. These existing classification models have been applied in some underwater environments, such as mine countermeasures, tracking marine mammals, detecting shipwrecks, and defense reconnaissance missions.

AI offers a solution by implementing robust and efficient machine learning (ML) and deep learning (DL) algorithms that learn directly from raw SONAR data to hierarchical representations while keeping the dependency on hand-tuned features at a minimum. Artificial intelligence (AI) can utilize data augmentation, synthetic data generation, and multisensor fusion to reduce data sparsity and improve detection robustness, making underwater detection systems more autonomous and robust. Traditional ML techniques, such as K-nearest neighbors (KNN), support vector machines (SVM), and hidden Markov model (HMM), provide improved robustness but still depend heavily on clean data. Modern DL architectures, such as convolutional neural networks (CNN), recurrent neural network (RNN), and transfer learning models (such as visual geometry group (VGG) and residual network (ResNet)), have enabled automatic feature extraction and better generalization [1]. Scaled-down residual networks (like ResNet18) and depth-wise separable CNNs are examples of lightweight designs used to manage the high computational cost of deep models.

In addition, time-varying acoustic signals, which are common in underwater environments, are better handled by sequence models, such as RNNs, long–short-term memory (LSTM) networks, gated recurrent units (GRUs), and attention mechanisms that dynamically focus on important features. These models are capable of generalizing to different conditions, improving noise robustness, and acquiring new targets with less retraining through transfer learning. These ML and DL approaches have greatly improved underwater SONAR systems. DL employs techniques such as data augmentation utilizing generative adversarial networks (GANs) and unsupervised learning (for example, deep belief networks (DBNs)) to address the primary issue of the lack of labeled datasets. Regularization strategies, dropout, and transfer learning from pretrained models are used to prevent overfitting that results from small datasets. In addition, conventional methods of image processing algorithms have been implemented for the detection of mine-like objects (MLO), while ML algorithms, including SVMs and KNNs, have been applied to automatically distinguish between ship noise and marine life noise [2].

However, conventional approaches do not scale, are susceptible to noise, and depend heavily on domain-specific feature engineering, as the seabed constantly changes. Moreover, the ultrasonic waves reflect off the floor, and the dummy causes a noise phenomenon. The actual conditions of the underwater environment, including low signal-to-noise ratios (SNR), changing record environments, and dense background interference, greatly affect performance. Signal distortion is often caused by noise and reverberations, which are minimized by extracting robust and noise-resistant characteristics. In order to facilitate real-time processing, pruning techniques and successful models such as MobileNet have been employed [1]. Finally, the limited diversity of current underwater datasets is addressed by synthetic data synthesis and multimodal learning, which brings together visual and auditory inputs. Combining ML and DL approaches improves the accuracy, robustness, and flexibility of underwater SONAR systems in the actual environment. To address the issues mentioned above, Nguyen *et al.* [3] proposed a computerized SONAR imaging, leveraging CNNs like AlexNet and GoogleNet for classification.

Conventional vision sensors, which are typically used for underwater exploration, rely on capturing clear optical images and are quiet effective in clear, well-lit water conditions. However, in environments where water turbidity is high and lighting is poor, their effectiveness drops significantly, making them inadequate for underwater imaging. This is where SONAR sensors come into action. SONAR sensors can capture images regardless of underwater conditions at the cost of lower image resolution and increased noise. Automatic detection using SONAR imaging plays a critical role in rescue and defense operations, reducing human error and reaction time by automating the detection of underwater objects. This chapter deals with the problem of SONAR image degradation due to underwater reflection noise and sensor angles. and discusses the possible opportunities to handle this issue. Incorporating AI into any environment benefits it and provides solutions to long-standing unsolved problems. The remainder of this chapter is arranged as follows. Related work in SONAR image classification and analysis is provided in Section 21.2, and possible solutions are discussed in Section 21.3, with Section 21.4 concluding the chapter.

21.2 Related work

Although earlier work had investigated SONAR imaging and hand classification, it had suffered from problems of noise and variable image quality. In Nguyen *et al.* [3], the authors tried to fill the gap with the application of CNNs to classify SONAR images even under given conditions, including noise due to background clutter, varied incidence angles, and polarization, more effectively. They contributed by verifying the capability of DL to improve SONAR-based detection. They did not, however, address problems such as low training data or sensor fusion, that is, combining outputs of different types of SONAR sensors. Their experiments showed conclusive proof that multibeam SONAR images have inferior resolution, which in turn makes image classification and detection inept. Moreover, they did not focus on complex architectures such as ResNet or transformer-based architectures, which can be considered in the future to provide further improvements in SONAR image classification. In addition, adding the dataset with an increased diversity of underwater environments, SONAR, and body positions can better generalize the model. Sensor fusion, fusion of SONAR with other inputs, such as partial vision data or acoustic Doppler sensors, can provide more information to improve detection performance in difficult instances.

Underwater optical cameras are typically not capable of shooting in turbid or low-light conditions, making acoustic imaging systems a more suitable alternative.

High-frequency SONAR systems, such as dual-frequency identification SONAR (DIDSON [4]), addressed this challenge by producing detailed acoustic images with near-optical resolution. DIDSON operated by transmitting sound waves and measuring their reflections, creating real-time visualizations of submerged objects. Unlike traditional SONARs that provide only range and intensity data, DIDSON delivered fine-grained, high-resolution imagery that enabled the use of image-processing algorithms that were originally designed for optical images. Autonomous vehicles, such as unmanned aerial vehicles (UAVs), unmanned ground vehicles (UGVs), and autonomous underwater vehicles (AUVs), require image processing, but vision systems face many difficulties in underwater environments. Due to their limited range, low visibility, and light distortion, optical cameras frequently perform poorly underwater. SONAR and other acoustic imaging offer an alternative, but their low resolution has historically limited their application in automatic object recognition and image processing.

To address these challenges, a lightweight neural model known as bidirectional associative memory (BAM) [5] was used to classify SONAR images, utilizing object shadows as key features. BAM is known for its speed and capacity to filter out noise and interpolate obscured images to recover full images. BAM is a type of neural network that stores and recalls input-output pairs. It uses a correlation matrix to link binary or bipolar pairs of vectors. When an element of the pair is presented as input, the network can recall the corresponding pair using a bidirectional iterative process. This allows BAM to retrieve complete outputs from noisy or incomplete inputs. Although computationally efficient, BAM is susceptible to local minima and spurious attractors, unintended stable states that degrade accuracy, especially with high-dimensional data or imbalanced binary patterns. A modified BAM with balanced output encoding and complementary dummy data improved classification robustness against up to 50% noise, but performance decreased with higher noise levels and similar object shapes. The use of confined dataset (four objects) and size-normalized images in their study limited its generalizability, highlighting the need for more scalable and noise-tolerant models. Particularly as the network dimensionality rises, its dependence on basic Hebbian learning frequently results in spurious attractors, which causes inaccurate pattern recalls.

Another issue with BAM is its limited storage capacity; unless data is carefully balanced (e.g., equal 1's and −1's distribution), storing multiple patterns increases the risk of interference and decreases reliability. Furthermore, misclassifications between cylinders and cubes demonstrate how performance suffers when exposed to geometrically similar objects or high noise levels. The design becomes more complex as a result of these limitations. To improve these limitations, the strategy used complementary dummy data to balance the distribution of 0's and 1's in output patterns in order to enhance the performance of BAM. In particular, in noisy or distorted environments, this enhanced recall accuracy, stabilized learning, and reduced interference. This improved the robustness and capacity of BAM to accurately classify underwater acoustic images, while balanced patterns helped to prevent recall errors that are frequently associated with unbalanced data. However, there are other alternatives that give the same or better performance and results for underwater image classification. For underwater image classification, several contemporary methods that offer greater scalability and robustness than BAM can be employed. CNNs are especially effective, as they can automatically extract spatial features from images and perform well in classification tasks even when there is noise or distortion.

Autoencoders, particularly denoising autoencoders, can be optimized for compressed representations of acoustic images and reconstruct them efficiently. When paired with custom features such as a histogram of oriented gradients (HOG) or Gabor filters, SVM offers a lightweight substitute that is especially helpful when data are scarce. Moving AUVs benefit from the ability of RNNs and LSTMs to capture temporal dependencies across SONAR image sequences. Similarly, *Wang et al.* [6] improved the accuracy of underwater SONAR image classification by addressing a core limitation of traditional CNNs, random weight initialization. They proposed an adaptive weights (AW)-CNN model, which integrates DBNs to pre-initialize CNN filter weights, enabling more accurate and stable feature learning in noisy underwater environments. In their study, they employed a CNN as a DL model that mimics the human visual system by using local receptive fields and shared weights to extract spatial features from SONAR images. Their experimental results demonstrated that AW-CNN significantly outperforms both standard CNN and DBN models in underwater SONAR image classification. Despite its improved performance, the AW-CNN approach presents several limitations. First, the SONAR dataset used is small, non-public, and limited to six object categories, restricting the generalizability of the model.

The fusion of a DBN with a CNN introduces architectural complexity, which may not be present in AUVs with limited computational and energy resources. Furthermore, the reliance of the model on extensive preprocessing raises concerns about robustness in real-world scenarios where such preprocessing may not be feasible or consistent. In addition, the DBN itself is an outdated architecture, less efficient, and harder to train than modern alternatives, and may not scale well with larger or more diverse SONAR datasets. To address these challenges, several improvements can be made. Replacing the DBN with more modern unsupervised or self-supervised learning methods, such as auto-encoders or contrastive learning (like simple contrastive learning of representations (SimCLR), momentum contrast (MoCo)), could enhance feature initialization while reducing model complexity. The AW-CNN architecture could be simplified by using lightweight CNNs such as MobileNet or EfficientNet to support real-time inference on AUVs. Incorporating domain adaptation techniques could help generalize the model to different SONAR environments, and using publicly available SONAR datasets can improve comparability and benchmarking. Lastly, integrating preprocessing directly into the neural network pipeline through a learnable augmentation or data normalization layer may reduce dependence on external processing steps and enhance the robustness of deployment.

21.3 How AI improves image classification and analysis?

AUVs are vital for oceanographic research, underwater surveillance, and seabed exploration. These self-guided systems operate without real-time human control and are designed to navigate complex, often GPS-denied environments. Unlike early

AUVs that relied on surface-level computation and manual oversight, modern versions are equipped with advanced on-board sensors, embedded processors, and AI-driven algorithms for real-time decision-making and data processing. Despite these advances, AUVs remain constrained by several operational challenges, particularly when deployed in dynamic and unpredictable underwater environments. AUVs are crucial for the collection and interaction of deep-sea data. Although AUVs still need human supervision, there is a growing movement toward complete autonomy [7]. However, it is still difficult due to factors such as poor visibility, limited underwater communication, and a lack of data, which make autonomous decision-making dangerous and uncertain. Due to the low quality of underwater optical imagery, AUVs are highly dependent on SONAR technologies such as side-scan SONAR (SSS), forward looking SONAR (FLS), and multibeam echo SONAR (MBES) [8]. Real-time SONAR data interpretation using DL models for classification, object detection, segmentation, and simultaneous localization and mapping (SLAM) improved the autonomy and capability of AUVs. Elmezain *et al.* [9] provided an extensive review of state-of-the-art DL models for underwater computer vision tasks such as image enhancement, object recognition, segmentation, and tracking.

They explored how DL architectures, ranging from CNNs to GANs and transformer-based models, addressed complex underwater challenges such as poor visibility, dynamic lighting, and color distortion. They highlighted specialized applications such as coral reef monitoring, marine life tracking, underwater infrastructure inspection, and marine debris detection. Various models such as dual attention mechanism deep neural network for underwater biological image classification (DAMNet) [10], marine litter detection (MLDet) [11], single object tracking in aqua-cultural environments for fish with coordinated attention mechanism (SiamFCA) [12], and dual segment anything model (SAM) Zhang *et al.* [13] were examined, showing high accuracy and robustness across different datasets. Their review also covered enhancement techniques using physics-based image formation models, attention mechanisms, and hybrid transformer-GAN networks. They showed that the benchmark results demonstrated exceptional performance; for example, the spectroformer [14] achieved the highest peak SNR and the structural similarity index measure (SSIM) scores, while models such as FocusDet [15], which focused on small object detection, are why named so, and FishTrack [16] excelled in detection and tracking accuracy under occlusions and scale changes. Furthermore, they included a detailed taxonomy and comparative analysis of models, evaluation metrics, and the most widely used datasets such as the underwater image enhancement benchmark (UIEB) [17], marine animal segmentation (MAS3K) [18], segmentation of underwater imagery (SUIM) [19], and the real-world underwater object detection (RUOD) [20].

They also emphasized the importance of preprocessing, multiscale features, and temporal data in enhancing underwater perception. This foundational work enables better system design for AUVs by identifying trends and capabilities in current DL research. Despite these advances, challenges persist. Chief among them are the high computational cost of deep models, the lack of standardized and publicly available underwater datasets, and limited model interpretability. These factors limit the deployment in real-time systems, particularly in power-constrained AUVs. Moreover, many models are heavily relying on synthetic or laboratory-generated data, which reduces their generalizability to real-world ocean environments. Occlusion handling, domain shifts, camouflaged objects, and low-resolution imagery remain unsolved problems for current DL models. To mitigate these issues, they recommended developing lightweight neural networks, domain adaptation techniques, and incorporating multimodal data fusion, leveraging SONAR, hyperspectral imaging, and stereo vision alongside traditional RGB data. The discussion above calls for future work to focus on interpretability, reduced data dependency through self-supervised learning, and hardware-efficient models for embedded underwater systems. It also encourages the standardization of benchmarking procedures to facilitate fair model comparisons and accelerate innovation. Ultimately, their work serves not only as a technical reference but also as a strategic guide for enhancing underwater autonomous perception through the lens of advanced DL technologies.

Another study in Xu *et al.* [21] introduced an ML approach for classifying substrate types, oyster density, and oyster sizes in bottom oyster farms using data collected from a low-cost single beam mechanical scanning SONAR. The objective was to enable precision aquaculture monitoring by replacing traditional, labor-intensive methods with a noninvasive, automated system. The experiments were carried out in a controlled laboratory tank simulating seabed conditions using real Chesapeake bay water and varied oyster configurations. Two models, a CNN-based modified LeNet-5 [22] and a dense neural network (DNN), were trained on SONAR data at five frequencies (500–900 kHz). Through extensive data augmentation techniques (image rotation, frequency overlays, vectorization), the models were evaluated on their ability to classify three parameters. CNN achieved up to 93.8% accuracy in size classification at 500 kHz, while DNN achieved more accuracy 94% in tasks using a tenfold increase in vectorized input data. To increase robustness, three-channel datasets that combine multiple frequencies were used, enhancing the depth of the characteristics. Field trials in an active oyster farm validated the real-world applicability of the models, achieving 86% accuracy in density classification and 90% size detection under intertidal conditions, indicating the potential of the proposed method for sustainable aquaculture operations. Despite its promising performance, the approach faces several challenges. The models showed reduced accuracy in field tests due to environmental noise, sensor motion, and varying substrate composition, particularly when classifying mixed shell and sand regions.

CNN's effectiveness was limited by data diversity and cropping errors that failed to capture representative features, while the DNN risked overfitting due to its dependency on synthetically expanded datasets. Furthermore, the system lacks support for fine-grained classification beyond basic categories, reducing scalability in commercial applications. To address these limitations, future work should include broader field data collection from diverse oyster habitats and the use of transfer learning via large-scale pre-trained networks to enhance generalization. Additionally, converting classification into a regression problem may yield more accurate continuous predictions for oyster size and density. The authors in Xu *et al.* [21] also suggested the deployment of the SONAR system on autonomous platforms such as AUVs or ROVs for automated seabed mapping and the integration of real-time denoising and navigation algorithms to counteract wave-induced distortions. Such advances could significantly improve the robustness and utility of ML-based monitoring tools for marine aquaculture. A learning paradigm for generalized zero-shot learning (GZSL) in SSS image classification is contrastive adaptation to domain augmentation (CADA) [23]. They addressed the issues of domain variability and a lack of labeled data in underwater environments. Due to irrational assumptions, underfitting, and overfitting to observed classes, traditional supervised and style transfer methods produced subpar results.

By combining SONAR-specific noise modeling and background fusion, CADA solved these problems and produced realistic pseudo-SSS samples of invisible classes. In order to improve class discrimination, they aligned the semantic and feature distributions of generated and original images using a contrastive Kullback-Leibler (KL) divergence loss. To address class imbalance and enable more reliable training in mixed-class situations, the framework added a benchmark dataset, GZSSS, and a logit adjustment mechanism. Compared to traditional and generative methods, the experiments showed that CADA improved harmonic mean accuracy by more than 20%. In classes with comparable grayscale patterns, the background fusion component is especially important to maintain important features and increase diversity. Noise injection enhances the robustness and generalization of the model. The use of shallow-angle photography, which closely resembles real-world SONAR shapes, greatly improves classification performance. However, CADA struggles with intra-class bias, where varied appearances within a class cause inconsistent recognition. In addition, it also faces semantic overlap, leading to misclassifications between visually similar classes like humans and airplanes. These issues limit generalization and reduce accuracy in complex real-world scenarios.

21.4 Conclusion

AI has significantly transformed the classification and analysis of underwater SONAR images, specifically enhancing the autonomy and perception capabilities of AUVs. Modern AUVs embedded with AI algorithms for the real-time interpretation of SONAR data enable robust object detection, segmentation, and navigation in visually degraded and GPS-deficient ocean environments. DL algorithms such as CNNs, GANs, and transformer-based models address critical underwater challenges, including image noise, low visibility, and acoustic distortions. These methods have shown success in diverse applications, from marine debris detection and coral monitoring to SONAR-based oyster farming, achieving classification accuracies (exceeding 90% in some experiments). However, limitations such as high computational demands, limited availability of labeled datasets, and poor generalizability in varying underwater conditions persist, causing problems. To address these problems, researchers have proposed domain adaptation strategies, regression-based modeling, and the deployment of lightweight networks such as MobileNet and EfficientNet for real-time embedded systems. Furthermore, synthetic data generation, multimodal data fusion, and self-supervised learning reduce data dependency and improve model robustness. Notably, advanced paradigms like CADA for GZSL offer viable solutions to classify foreign objects in SONAR data. Nevertheless, issues like intra-class variability, semantic overlap, and reliance on lab-generated datasets continue to hinder model performance in real-world environments. In summary, the integration of AI into SONAR systems marks a pivotal step toward fully autonomous, adaptive, and efficient AUVs. Henceforth, the operational capabilities and reliability of AUVs in complex and harsh marine environments are expanding. Future research could explore the integration of multimodal sensor fusion, trained on diverse and better datasets, such as GZSL, continual learning frameworks (like SimCLR, MoCo), and physics-informed neural networks to enable more adaptive, generalizable, and energy-efficient SONAR image interpretation in increasingly diverse and dynamic underwater environments.

References

[1] M. A. Aslam, L. Zhang, X. Liu, *et al.*, "Underwater sound classification using learning based methods: A review," *Expert Systems with Applications*, vol. 255, p. 124498, 2024. https://www.sciencedirect.com/science/article/pii/S0957417424013654.

[2] S. Hożyń, "A review of underwater mine detection and classification in SONAR imagery," *Electronics*, vol. 10, no. 23, pp. 1–22, 2021. https://www.mdpi.com/2079-9292/10/23/2943.

[3] H.-T. Nguyen, E.-H. Lee, and S. Lee, "Study on the classification performance of underwater SONAR image classification based on convolutional neural networks for detecting a submerged human body," *Sensors*, vol. 20, no. 1, pp. 1–13, 2020. https://www.mdpi.com/1424-8220/20/1/94.

[4] T. W. Kim, S.-C. Yu, and J. Yuh, "Neural network-based underwater image classification for autonomous underwater vehicles," *IFAC Proceedings Volumes*, vol. 41, no. 2, pp. 15991–15995, 2008, *17th IFAC World Congress*. https://www.sciencedirect.com/science/article/pii/S1474667016415673.

[5] Haines and Hecht-Nielsen, "A bam with increased information storage capacity," in *IEEE 1988 International Conference on Neural Networks*, vol. 1, pp. 181–190, 1988.

[6] X. Wang, J. Jiao, J. Yin, W. Zhao, X. Han, and B. Sun, "Underwater SONAR image classification using adaptive weights convolutional neural network," *Applied Acoustics*, vol. 146, pp. 145–154, 2019. https://www.sciencedirect.com/science/article/pii/S0003682X17311532.

[7] M. Aubard, A. Madureira, L. Teixeira, and J. Pinto, "SONAR-based deep learning in underwater robotics: Overview, robustness, and challenges," *IEEE Journal of Oceanic Engineering*, vol. 50, no. 3, pp. 1–19, 2025.

[8] Y. Steiniger, D. Kraus, and T. Meisen, "Survey on deep learning based computer vision for SONAR imagery," *Engineering Applications of Artificial Intelligence*, vol. 114, p. 105157, 2022. https://www.sciencedirect.com/science/article/pii/S0952197622002718.

[9] M. Elmezain, L. Saad Saoud, A. Sultan, M. Heshmat, L. Seneviratne, and I. Hussain, "Advancing underwater vision: A survey of deep learning models for underwater object recognition and tracking," *IEEE Access*, vol. 13, pp. 17830–17867, 2025.

[10] P. Qu, T. Li, L. Zhou, *et al.*, "DAMNet: Dual attention mechanism deep neural network for underwater biological image classification," *IEEE Access*, vol. 11, pp. 6000–6009, 2023.

[11] D. Ma, J. Wei, Y. Li, *et al.*, "MLDet: Towards efficient and accurate deep learning method for marine litter detection," *Ocean & Coastal Management*, vol. 243, p. 106765, 2023. https://www.sciencedirect.com/science/article/pii/S0964569123002909.

[12] Y. Mei, N. Yan, H. Qin, T. Yang, and Y. Chen, "SiamFCA: A new fish single object tracking method based on Siamese network with coordinate attention in aquaculture," *Computers and Electronics in Agriculture*, vol. 216, p. 108542, 2024. https://www.sciencedirect.com/science/article/pii/S0168169923009304.

[13] P. Zhang, T. Yan, Y. Liu, and H. Lu, "Fantastic animals and where to find them: Segment any marine animal with dual SAM," in *2024 IEEE/CVF Conference on Computer Vision and Pattern Recognition (CVPR)*, 2024, pp. 2578–2587.

[14] M. R. Khan, P. Mishra, N. Mehta, *et al.*, "Spectroformer: Multi-domain query cascaded transformer network for underwater image enhancement," in *2024 IEEE/CVF Winter Conference on Applications of Computer Vision (WACV)*, 2024, pp. 1443–1452.

[15] Y. Shi, Y. Jia, and X. Zhang, "FocusDet: An efficient object detector for small object," *Scientific Reports*, vol. 14, no. 10697, pp. 1–15, 2024.

[16] M. Dawkins, J. Prior, B. Lewis, *et al.*, "Fishtrack23: An ensemble underwater dataset for multi-object tracking," in *2024 IEEE/CVF Winter Conference on Applications of Computer Vision (WACV)*, 2024, pp. 7152–7161.

[17] C. Li, C. Guo, W. Ren, *et al.*, "An underwater image enhancement benchmark dataset and beyond," *IEEE Transactions on Image Processing*, vol. 29, pp. 4376–4389, 2020.

[18] L. Li, B. Dong, E. Rigall, T. Zhou, J. Dong, and G. Chen, "Marine animal segmentation," *IEEE Transactions on Circuits and Systems for Video Technology*, vol. 32, no. 4, pp. 2303–2314, 2022.

[19] S. Lian, H. Li, R. Cong, S. Li, W. Zhang, and S. Kwong, "Watermask: Instance segmentation for underwater imagery," in *2023 IEEE/CVF International Conference on Computer Vision (ICCV)*, 2023, pp. 1305–1315.

[20] C. Fu, R. Liu, X. Fan, *et al.*, "Rethinking general underwater object detection: Datasets, challenges, and solutions," *Neurocomputing*, vol. 517, pp. 243–256, 2023. https://www.sciencedirect.com/science/article/pii/S0925231222013169.

[21] M. Xu, N. Karapetyan, K. Rajasekaran, *et al.*, "Machine-learning based classification of on-bottom oyster farms using a mechanical scanning SONAR," *IEEE Sensors Journal*, vol. 25, no. 12, pp. 1–1, 2025.

[22] Y. Lecun, L. Bottou, Y. Bengio, and P. Haffner, "Gradient-based learning applied to document recognition," *Proceedings of the IEEE*, vol. 86, no. 11, pp. 2278–2324, 1998.

[23] Y. Jia, X. Ye, P. Li, and S. Guo, "Contrastive adaptation on domain augmentation for generalized zero-shot side-scan SONAR image classification," *IEEE Transactions on Instrumentation and Measurement*, vol. 74, pp. 1–13, 2025. https://doi.org/10.1109/TIM.2025.3551028

Chapter 22
Intelligent evaluation and categorization of SONAR images

Muhammad Talha[1], *Syed Muhammad Hamza Tauqeer*[2] *and Seung Yeob Nam*[3]

This chapter will discuss the integration of artificial intelligence (AI) into the analysis and classification of underwater sound navigation and ranging (SONAR) images, focusing on how AI enhances the efficiency and accuracy of interpreting complex underwater data. We start by examining the significance of underwater imaging and how SONAR technology contributes to advancements in underwater exploration. The chapter will delve into the challenges involved in interpreting SONAR images, such as noise, distortion, and the need for specialized human expertise. We will then introduce the emergence of AI in image analysis, highlighting how AI, particularly machine learning and deep learning, offers solutions to these challenges, automating and improving the classification process. Following this, we will examine the fundamentals of SONAR imaging, including sound propagation in water and the different SONAR systems used to capture underwater images. Traditional methods for SONAR image analysis, such as manual interpretation and statistical feature extraction, will be discussed, along with their limitations. The chapter will then cover AI-powered classification techniques, including machine learning models and deep learning architectures, such as convolutional neural networks (CNNs), which are revolutionizing the field. We will also explore real-world applications, such as mine detection, seabed mapping, and autonomous underwater vehicle (AUV) navigation, and conclude by addressing the challenges and future directions of AI in underwater SONAR image analysis.

22.1 Introduction

Underwater imaging is vital across numerous domains, including oceanographic research [1], marine archaeology [2,3], defense operations [4,5], and the upkeep

[1]School of Electrical Engineering and Computer Science (SEECS), National University of Sciences and Technology (NUST), Pakistan
[2]Department of Electrical Engineering, Military College of Signals (MCS), National University of Sciences and Technology (NUST), Pakistan
[3]Department of Information and Communication Engineering, Yeungnam University, South Korea

of underwater infrastructure [6]. When water turbidity and poor lighting hinder the effectiveness of optical imaging, acoustic methods such as SONAR become essential alternatives [5,7]. SONAR technology can generate two-dimensional visuals of submerged objects, even in murky waters where optical systems are ineffective [5]. This capability is crucial for long-range underwater surveillance and monitoring [5]. The development of more exposed farming sites and the increasing use of underwater vehicles necessitate effective underwater imaging for precision farming practices [6].

22.1.1 Overview of SONAR technology in underwater exploration

SONAR technology is widely utilized for underwater exploration due to its ability to provide acoustic images by emitting sound waves and analyzing the echoes [6,8]. Side-scan SONAR (SSS) is particularly valuable for seabed detection [9] and underwater target detection and recognition [10]. Active SONAR systems, such as forward-looking SONAR (FLS), are used to acquire underwater information by equipping remotely operated vehicles (ROVs) [11]. The data collected by SONAR systems can be used to construct near-photographic images of underwater areas, even in zero visibility conditions [8]. Dual-frequency identification SONAR (DIDSON) and acoustic Doppler current profile (ADCP) and imaging SONAR are utilized for detailed underwater monitoring. These systems enable the creation of detailed maps, identification of objects, and analysis of spatial and temporal dynamics in aquatic environments [12].

22.1.2 Challenges in underwater SONAR image interpretation

Despite its advantages, underwater SONAR imaging presents several challenges. SONAR images often suffer from low contrast, blurred edges, poor texture, and unsatisfactory quality [9]. The underwater environment introduces various noises, including speckle noise, sensor noise, and interference from other sources, which can degrade image quality [10]. Multipath effects caused by floating objects and particles further complicate acoustic wave transmission [13]. Moreover, obtaining reliable SONAR datasets for artificial intelligence (AI) model training is difficult because of restricted access to data, high costs, and privacy concerns [4]. Variations in the seawater environment lead to signal weakening and distortion, which hinders optimal performance in detecting objects using SONAR [13].

22.1.3 The emergence of artificial intelligence in image analysis

AI, especially through machine learning and deep learning, has transformed image classification by replicating human-like thinking and streamlining intricate decision-making tasks [14,15]. Deep learning techniques have shown potential in marine environment monitoring, such as underwater image classification [16]. CNNs are capable of autonomously identifying features in underwater SONAR images, leading to enhanced classification performance [1]. AI techniques are also applicable to image segmentation, a key process in identifying and recognizing underwater targets [17]. Explainable AI (XAI) is a growing field focused on making the workings of deep neural networks more transparent, thereby improving trust and dependability in sensitive areas like defense [14].

22.1.4 Chapter organization and scope

This chapter explores the intersection of underwater SONAR image classification and analysis with the aid of AI. It begins by outlining the importance of underwater imaging and providing an overview of SONAR technology. Subsequent sections delve into the challenges associated with interpreting underwater SONAR images and discuss the emergence of AI in addressing these challenges. The chapter will cover various AI techniques, including deep learning models and XAI, and their applications in enhancing the accuracy and reliability of underwater SONAR image analysis. Finally, the chapter will explore the use of simulated datasets and transfer learning to overcome the limitations of data availability in underwater SONAR imaging, aiming to advance the field and provide insights into future research directions.

22.2 Fundamentals of SONAR imaging

22.2.1 Sound propagation principles in water

Underwater acoustic imaging relies on the principles of sound propagation in water. Unlike electromagnetic waves, sound waves travel efficiently through water, making them ideal for underwater sensing and imaging [18]. However, sound propagation is affected by several factors, including [19],

- **Salinity**: Changes in salinity affect water density and thus sound speed.
- **Temperature**: Temperature gradients cause refraction of sound waves.
- **Pressure**: Increasing pressure with depth increases sound speed.

These factors lead to variations in the speed of sound, causing refraction, reflection, and scattering, all of which influence the quality and accuracy of SONAR images [20].

22.2.2 Types of SONAR systems

Various types of SONAR systems are used for underwater imaging, each with its own advantages and applications [20]. The comparative analysis of these systems is illustrated in Figure 22.1, which provides a comprehensive evaluation across multiple performance parameters.

22.2.2.1 Side-scan SONAR (SSS)

SSS is widely used for seabed mapping and object detection [21]. It operates by emitting acoustic waves to the side of the SONAR platform and recording the back scattered signals [20]. SSS provides high-resolution images of the seabed, which are useful for identifying objects, characterizing sediment types, and mapping underwater structures [22]. As demonstrated in Figure 22.1, SSS exhibits moderate performance across most parameters, with particular strengths in depth mapping

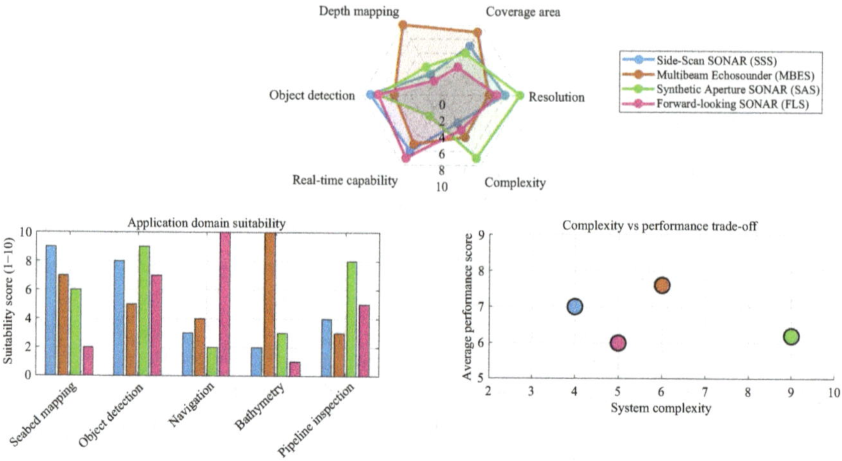

Figure 22.1 Comprehensive analysis of underwater SONAR systems: performance comparison across multiple parameters, including depth mapping, coverage area, object detection, resolution, real-time capability, complexity, application domain suitability, and complexity versus performance trade-offs.

capabilities (rated 7/10) and reasonable complexity levels (rated 6/10). SSS images are commonly used in underwater surveillance in both civilian and military sectors [4]. However, challenges include speckle noise and geometric distortions [21,23], which are reflected in its coverage area limitations shown in the radar chart analysis.

22.2.2.2 Multibeam echo sounders (MBES)

MBES emit multiple acoustic beams to create a swath of depth measurements, providing detailed bathymetric data. MBES are used for nautical charting, benthic habitat mapping, and seafloor geotechnical surveys. The combined acoustic response of the seabed can vary with the MBES operating frequency, making it valuable in seafloor characterization. Figure 22.1 reveals that MBES demonstrates superior performance in depth mapping (rated 9/10) and coverage area (rated 8/10), making them particularly suitable for large-scale bathymetric surveys. Multispectral MBES backscatter can improve seafloor characterization by providing different perspectives at different frequencies [24]. The application domain suitability analysis shows MBES excel in nautical charting and geological surveys, as indicated by their high performance ratings in the comparative bar chart.

22.2.2.3 Synthetic aperture SONAR (SAS)

SAS synthesizes a larger aperture by moving the SONAR transducer along a path, improving the cross-range resolution. SAS overcomes the physical limitations of SSS and provides superior resolution [25]. The comparative analysis in Figure 22.1 demonstrates that SAS achieves the highest resolution rating (10/10) and excellent object detection capabilities (9/10), justifying its use in high-precision applications. SAS systems are used in various applications, including pipeline inspection and object detection [20]. However, complexity versus performance trade-off analysis shows that SAS operates at higher complexity levels (rated 9/10), since SAS techniques generally demand dense sampling across both temporal and spatial dimensions [25].

22.2.2.4 Forward-looking SONAR (FLS)

FLS projects acoustic waves forward from the SONAR platform to detect objects in the vehicle's path [20,26]. FLS is commonly used for obstacle avoidance and navigation in underwater robotics [23,26]. According to the performance evaluation presented in Figure 22.1, FLS demonstrates exceptional real-time capability (rated 10/10) and strong object detection performance (rated 8/10), making it ideal for dynamic navigation scenarios. High-resolution and low-side lobe forward-look SONAR imaging can be achieved using deconvolution methods [26]. Multibeam FLS is also utilized to extract environmental features for simultaneous localization and mapping [21]. The application domain analysis indicates FLS's particular strength in AUV navigation and obstacle avoidance systems.

22.2.2.5 Performance rating justification

The quantitative ratings assigned to each SONAR system (scale 1–10) in Figure 22.1 are based on a comprehensive evaluation framework considering the following criteria:

Depth mapping capability: Ratings reflect the maximum operational depth and accuracy of depth measurements. MBES receives the highest rating (9/10) due to its specialized design for bathymetric mapping with multiple beam coverage, while SSS receives a moderate rating (7/10) as depth information is derived indirectly from shadow analysis.

Coverage area: Evaluated based on the spatial extent covered per survey pass. MBES achieves high ratings (8/10) due to its wide swath coverage, while FLS receives lower ratings (5/10) due to its narrow forward-focused beam pattern.

Object detection: Assessed based on the ability to identify and classify underwater objects. SAS achieves the highest rating (9/10) due to its superior resolution capabilities, while MBES receives moderate ratings (6/10) as they are primarily designed for bathymetric rather than object detection applications.

Resolution: Determined by the smallest distinguishable features in the acoustic imagery. SAS receives a maximum rating (10/10) due to its synthetic aperture processing that overcomes physical antenna limitations, while SSS receives a moderate rating (6/10) due to range-dependent resolution characteristics.

Real-time capability: Based on processing speed and data acquisition rates for immediate decision-making. FLS achieves a maximum rating (10/10) due to its design for real-time navigation applications, while SAS receives a lower rating (4/10) due to computationally intensive processing requirements.

System complexity: Evaluated considering hardware requirements, signal processing algorithms, and operational expertise needed. SAS receives the highest complexity rating (9/10) due to sophisticated motion compensation and focusing algorithms, while SSS receives a moderate rating (6/10) representing established, well-understood technology.

22.2.3 SONAR image formation and characteristics

SONAR image formation involves converting received acoustic signals into visual representations [20]. However, SONAR images are characterized by several unique features that present significant challenges for image processing and analysis, as demonstrated in Figure 22.2.

22.2.3.1 Speckle noise

Speckle noise is a granular noise that arises from coherent interference of acoustic waves [21,23]. As illustrated in Figure 22.2 (a), this multiplicative noise degradation

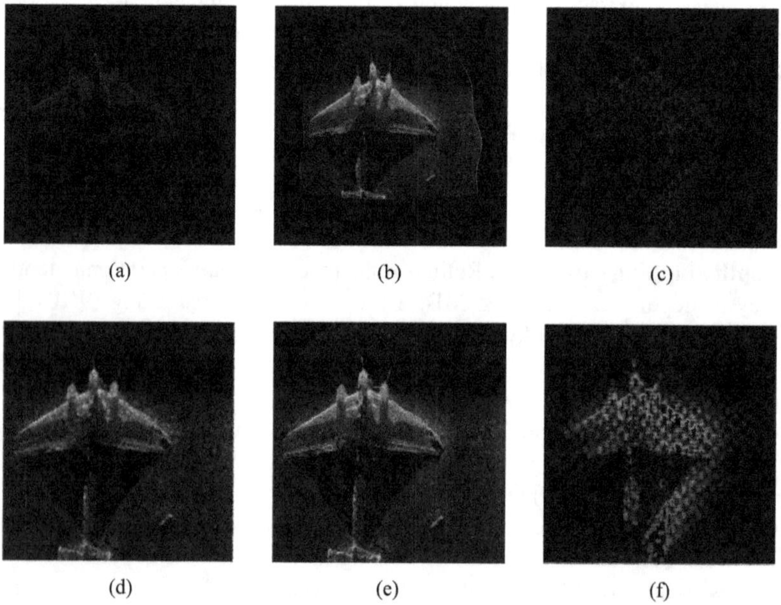

Figure 22.2 Comprehensive demonstration of SONAR image preprocessing techniques: (a) high-intensity speckle noise effect showing multiplicative noise degradation; (b) geometric distortion effect with extreme slant range compression and stretching; (c) intensity variations effect demonstrating angle-dependent backscatter, range attenuation, and acoustic shadows; (d) noise reduction result using Lee filter and median filtering to preserve edges while reducing speckle; (e) geometric correction result restoring proper spatial relationships; (f) radiometric correction result with time-varied gain compensation, shadow filling, angle correction, and contrast enhancement for uniform intensity distribution.

significantly affects image quality and makes target detection and recognition challenging [21]. Algorithms designed for optical images often degrade when applied to underwater SONAR scenes due to this characteristic noise pattern [27].

22.2.3.2 Geometric distortions

Geometric distortions occur due to the slant range geometry of SONAR systems and variations in sound speed. Figure 22.2 (b) demonstrates extreme slant range compression and stretching effects that can significantly affect measurement accuracy and spatial relationship interpretation. Real-time slant distance correction is essential for superior imaging of navigation strip data from SSS aboard AUVs [28].

22.2.3.3 Intensity variations

Intensity variations in SONAR images arise from changes in incidence angle, seabed composition, and object presence. Figure 22.2 (c) shows angle-dependent backscatter, range attenuation, and acoustic shadows that characterize these variations. Understanding these patterns is crucial for interpreting SONAR images and extracting meaningful information. Multispectral MBES backscatter serves as a tool for improved seafloor characterization [24].

22.2.4 Data acquisition and preprocessing for SONAR imagery

Data acquisition involves collecting SONAR data using appropriate SONAR systems and survey techniques. Preprocessing steps are essential to enhance image quality and correct for distortions [28]. The effectiveness of these preprocessing techniques is demonstrated in Figure 22.2 (d–f), which shows the results of applying various correction methods:

- **Noise reduction**: Figure 22.2 (d) demonstrates filtering techniques using the Lee filter and median filtering to remove speckle noise while preserving edges [27].
- **Geometric correction**: Figure 22.2 (e) shows correction for slant range distortion, restoring proper spatial relationships [28].
- **Radiometric correction**: Figure 22.2 (f) illustrates intensity value calibration with time-varied gain compensation, shadow filling, angle correction, and contrast enhancement for uniform intensity distribution [28].
- **Seafloor line detection**: Identifying lines or features on the seafloor from processed data [28].

Real-time processing sequences are crucial for on-the-fly image acquisition from SSS aboard AUVs. The comprehensive preprocessing demonstration in Figure 22.2 illustrates the transformation from degraded raw SONAR images to enhanced, analysis-ready data suitable for AI-based classification and interpretation.

22.3 Traditional methods for SONAR image analysis

Traditional methods for SONAR image analysis have played a crucial role in underwater target detection and recognition, seabed mapping, and various other marine applications [29,30]. These methods often rely on human expertise, statistical feature extraction, and model-based techniques. However, they also present certain limitations when dealing with complex underwater environments.

22.3.1 Manual interpretation and human expertise

In the early stages of SONAR image analysis, manual interpretation by trained experts was the primary method [20]. This involved visually examining the SONAR images and identifying objects or features of interest based on their shape, size, shadow, and context [31]. Manual interpretation relies heavily on the experience and knowledge of the interpreter, making it subjective and time-consuming. Despite advancements in automated methods, human expertise remains valuable for validating results and identifying subtle features that algorithms may miss [20,32]. The skill of the human interpreter often dictates the accuracy and efficiency of the analysis, particularly in complex or noisy environments. Conventional microscopic detection techniques often suffer from extended processing times, reduced accuracy at scale, and significant challenges in identifying rare microorganisms [20].

22.3.2 Statistical feature extraction

Statistical feature extraction involves quantifying the characteristics of SONAR images using mathematical and statistical measures [20]. These features can then be used for image segmentation, classification, and object recognition. Common statistical features include texture analysis, shape descriptors, and geometric properties [20,32,33].

22.3.2.1 Texture analysis

Texture analysis is a widely used technique in SONAR image analysis to characterize the spatial arrangement of pixel intensities [20,33]. Methods like the gray-level co-occurrence matrix (GLCM) and wavelet transforms are employed to capture texture information.

GLCM: GLCM determines how often specific pairs of pixel intensities appear in an image, offering insights into properties such as contrast, correlation, energy, and uniformity [34]. It is used to extract texture features from an image by analyzing the co-occurrence of gray levels at different spatial relationships [35].

Wavelets: Wavelet transforms break down an image into various frequency components, enabling analysis at multiple scales [17,36]. This is particularly useful for SONAR images, which often contain noise and artifacts at different scales. Wavelet-based techniques offer multiscale and multi-resolution capabilities, making them well-suited for detecting edges and extracting features from images [36].

Figure 22.3 presents an analysis of a SONAR image and its texture-based feature extractions. We can see from Figure 22.3(a), that the original SONAR image was used as the basis for further processing. Figure 22.3(b) illustrates the GLCM feature contrast, which captures the spatial relationships between pixel intensities and highlights texture variations. Figure 22.3(c) displays the results of a wavelet decomposition of the original image, separated into four frequency sub-bands: approximation (LL)

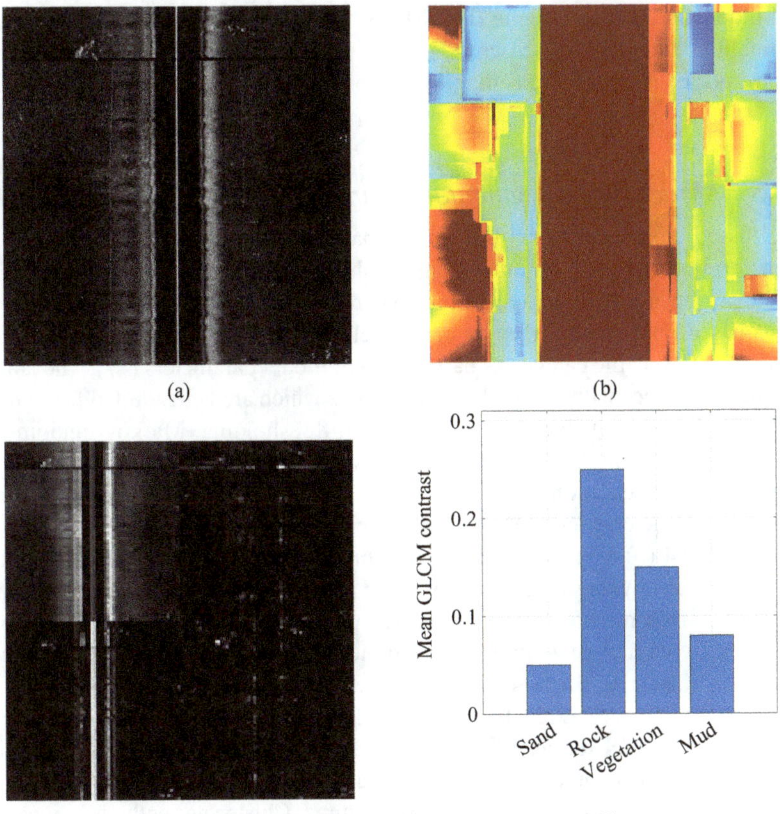

Figure 22.3 Comparison of the original SONAR image and its derived texture features: (a) original SONAR image; (b) GLCM feature contrast; (c) Wavelet decomposition results where top-L: approx (LL), top-R: horizontal (LH) bottom-L: vert (HL), bottom-R: diag (HH); (d) distribution of feature values across different textures.

in the top-left, horizontal details (LH) in the top-right, vertical details (HL) in the bottom-left, and diagonal details (HH) in the bottom-right. Finally, Figure 22.3(d) presents the distribution of feature values across different texture classes, providing insight into how these extracted features vary depending on the underlying texture patterns in the image. We can see in Figure 22.3 that the probability of the rock is very high and the feature value distribution across textures.

22.3.2.2 Shape descriptors

Shape descriptors are used to quantify the geometric properties of objects or regions in SONAR images [20,32]. These descriptors can include area, perimeter, circularity, and other geometric measures that help to distinguish different shapes [33]. Shape features describe the geometric shape of objects [20]. Precise detection of the fetal facial ultrasound standard plane (FFUSP) is crucial for reliable screening of facial abnormalities and related diseases [37]. However, the traditional method of obtaining standard planes is manual screening by doctors leading to large errors in the results [37].

22.3.3 Model-based segmentation and classification

Model-based approaches involve using mathematical models to represent the expected properties of different regions or objects in a SONAR image [29]. These models can be used for segmentation, where the image is divided into different regions based on how well they fit the model, and for classification, where objects are assigned to specific categories based on their model parameters [17]. The authors advocate techniques with detailed characteristics which are better suited for SONAR image segmentation, including edge detection, thresholding, Markov random field (MRF), and clustering [17]. Figure 22.4 shows the application of several common model-based segmentation techniques on a representative SONAR image. Figure 22.4 (a) shows the original SONAR image, serving as the baseline for comparison.

Figure 22.4 (b) presents the results of thresholding-based segmentation, which separates regions by applying intensity value cutoffs to distinguish objects from the background. This method is simple and computationally efficient but may struggle with noisy data or low-contrast boundaries. Figure 22.4 (c) demonstrates edge detection, highlighting the boundaries between different regions by identifying discontinuities in intensity. Edge detection can effectively delineate object contours but often requires careful tuning to avoid false edges. Finally, Figure 22.4 (d) shows clustering-based segmentation, where pixels are grouped into clusters based on similarity metrics, such as intensity or texture features. Clustering methods can capture complex region characteristics and handle variability in SONAR images but may be computationally intensive. Together, these examples underscore the strengths and limitations of various model-based approaches for segmenting SONAR imagery, highlighting the need to select appropriate methods based on image properties and application goals.

22.3.4 Limitations of Traditional Approaches

Despite their utility, traditional methods for SONAR image analysis have several limitations [38]:

Sensitivity to noise: SONAR images are often corrupted by speckle noise, sensor noise, and other forms of interference [30,39]. Traditional methods can be highly sensitive to this noise, leading to inaccurate results. In SONAR images, the speckle noise significantly hinders both automatic image recognition and human interpretation [39]. In Figure 22.5, we can easily see the effects of how the noise can affect future data processing on the SONAR images.

Dependence on image quality: The performance of traditional methods is heavily dependent on the quality of the SONAR images. Factors such as low resolution, uneven intensity distribution, and geometric distortions can significantly degrade their performance [40].

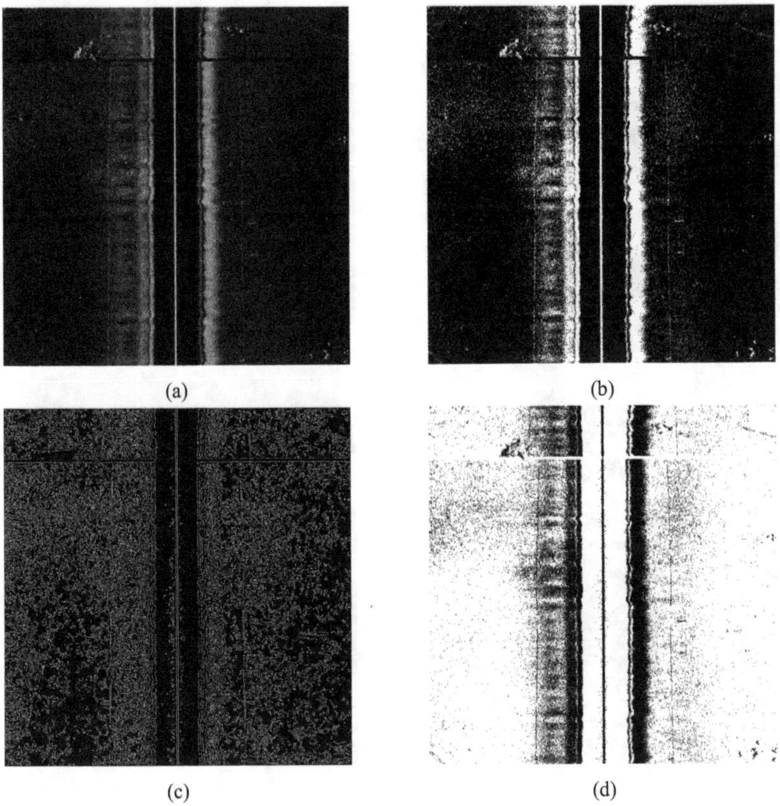

Figure 22.4 Comparison of different segmentation techniques applied to a SONAR image. (a) Original SONAR image. (b) Segmentation using thresholding. (c) Edge detection results highlighting boundaries. (d) Segmentation based on clustering methods.

Limited feature representation: Statistical feature extraction relies on hand-crafted features that may not capture the full complexity of the information present in SONAR images [32].

Lack of generalization: Model-based approaches often require specific models for different types of objects or environments. This limits their ability to generalize to new or unseen situations [41].

Computational cost: Some traditional methods, such as those involving complex statistical calculations or model fitting, can be computationally expensive, making them unsuitable for real-time applications [42]. For underwater target recognition, enhancing accuracy and real-time performance is critical as it depends on SONAR systems to filter out noise and interpret the data effectively [42]. Figure 22.6 shows a comparison of how much computational cost is required for different SONAR data processing methods, that is, GLCM, complex model fitting, and simple thresholding.

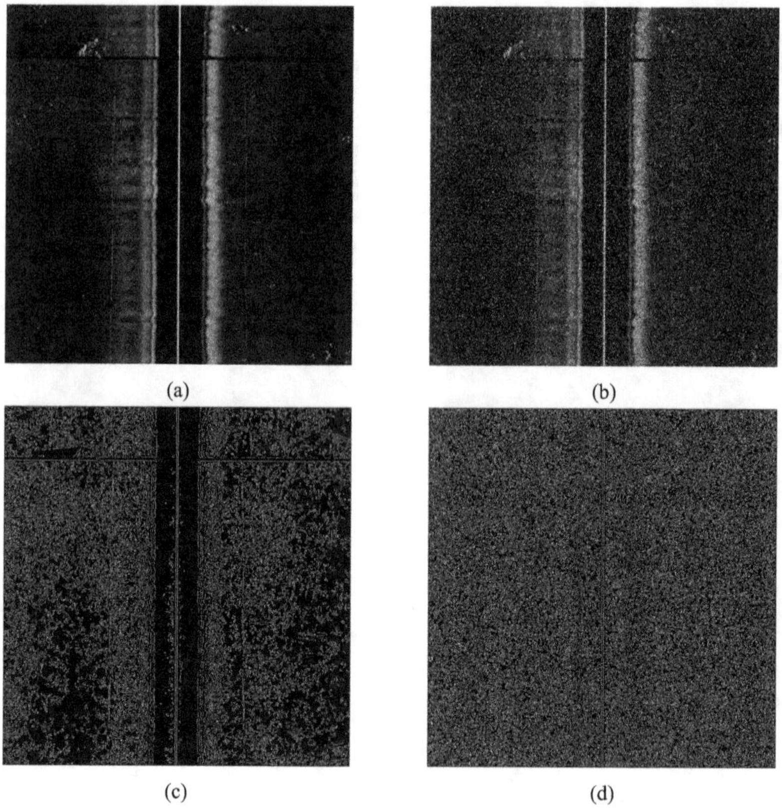

Figure 22.5 Effect of noise and image quality degradation on SONAR image processing, illustrating how low resolution and intensity distortions can hinder effective analysis: (a) Clean SONAR image, (b) Pepper and salt noise image (0.2), (c) Edges of clean image, and (d) Edges of noised image

22.4 The role of artificial intelligence in SONAR image analysis

The role of AI in SONAR image analysis is increasingly significant, driven by the need to enhance detection accuracy and automate complex decision-making processes in underwater environments [43,44]. SONAR imaging, which relies on the transmission and reception of acoustic pulse waves, is crucial for underwater archaeology, energy exploration, and oceanographic surveying [45]. However, SONAR images face several inherent challenges that complicate automated analysis, as demonstrated in Figure 22.7. These images often suffer from low signal-to-noise ratio (see Figure 22.7 (b)), where reduced signal strength combines with background clutter to obscure target features. Additionally, noise interference from multiple sources including speckle and interference patterns (see Figure 22.7 (c)) further degrades image quality. The relatively low resolution compared to optical images, with quantization effects limiting detail preservation (Figure 22.7 (d)), makes object detection

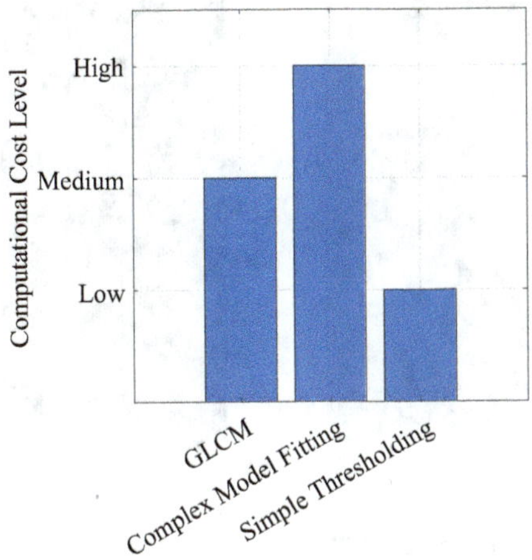

Figure 22.6 Computation cost comparison among different methods.

particularly challenging when compared to the baseline reference image (Figure 22.7 (a)) [45,46].

AI techniques, particularly deep learning, offer comprehensive solutions to address these degradation challenges illustrated in Figure 22.7. These solutions include image enhancement algorithms to improve signal-to-noise ratio and reduce noise interference, advanced feature extraction methods that can operate effectively despite resolution limitations, and robust object classification systems that maintain accuracy even under adverse imaging conditions [47]. The integration of AI methodologies enables automated processing of degraded SONAR imagery, transforming the challenging conditions shown in Figure 22.7 into actionable intelligence for underwater applications.

22.4.1 Overview of AI/machine learning concepts

Underwater SONAR image classification and analysis have been significantly advanced by the application of machine learning and AI [43]. SONAR imaging is crucial for underwater surveillance, object detection, and environmental monitoring, especially where optical imaging is limited by turbidity and low light conditions [5,47]. The integration of deep learning techniques has revolutionized this field by automating complex decision-making processes and mimicking human cognition [43].

22.4.1.1 Supervised, unsupervised, and reinforcement learning

Underwater SONAR image classification has greatly benefited from AI through supervised, unsupervised, and reinforcement learning techniques as illustrated in Figure 22.8. This framework demonstrates how different AI approaches address underwater SONAR image analysis challenges. The supervised learning branch faces challenges due to scarce labeled datasets in underwater environments [48].

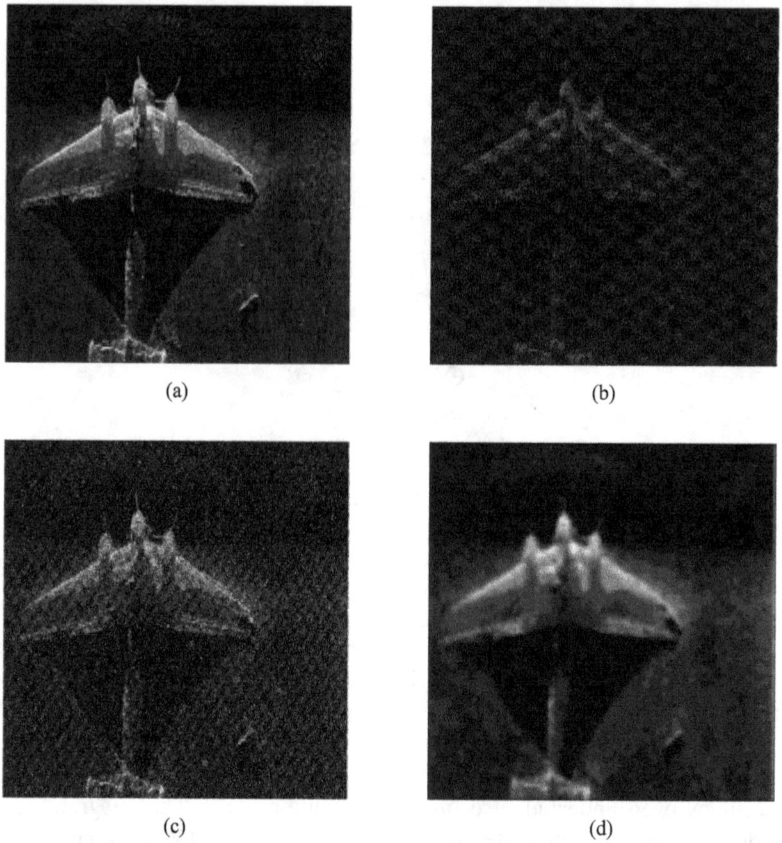

Figure 22.7 Common degradation effects in SONAR imaging systems: (a) baseline reference image, (b) reduced signal strength with background clutter, (c) multiple noise sources including speckle and interference patterns, and (d) resolution limitations with quantization effects.

As shown in Figure 22.8, the framework addresses these limitations through transfer learning and few-shot learning techniques, which leverage pre-trained models to bridge the gap between limited underwater training data and robust classification performance [5,49]. The unsupervised learning pathway utilizes self-supervised learning methodologies that enable learning meaningful image representations without extensive labeled datasets [48]. This approach extracts valuable features from unlabeled SONAR images by learning inherent data patterns, making it particularly suitable for underwater applications where ground truth data is scarce. Reinforcement learning, while less common in SONAR image classification, offers advantages for image enhancement processes. Figure 22.8 shows how reinforcement learning models enhancement as a Markov decision process, enabling systematic image quality improvement through iterative learning [50]. The integration of these three paradigms creates enhanced accuracy and efficiency in underwater SONAR image classification.

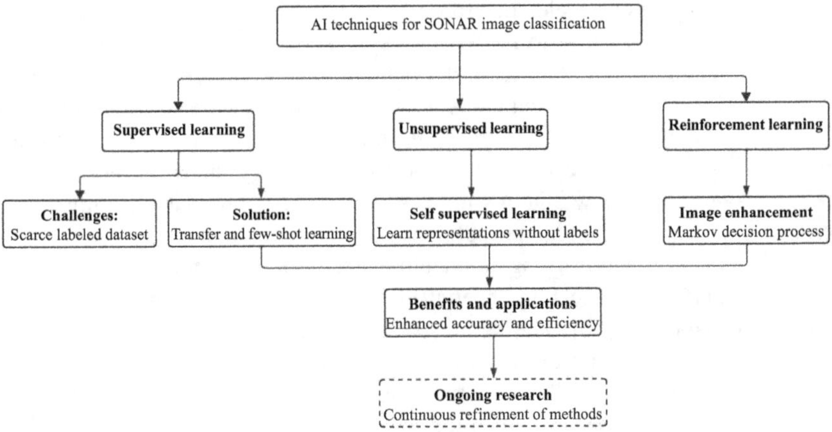

Figure 22.8 Artificial intelligence techniques for SONAR image analysis.

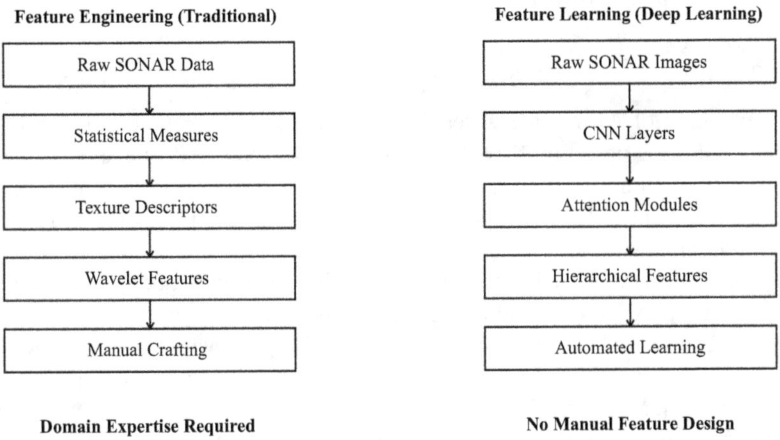

Figure 22.9 Comparison of feature learning versus feature engineering.

Ongoing research continues to explore and refine these AI-driven methods, developing hybrid models that leverage each learning paradigm's strengths while mitigating individual limitations [51].

22.4.1.2 Feature learning versus feature engineering

In underwater SONAR image classification, feature learning and feature engineering represent two distinct approaches to extracting relevant information from raw data, as illustrated in Figure 22.9. This comparison highlights the fundamental differences between traditional manual approaches and modern automated learning methods. The traditional feature engineering pathway, shown on the left side of Figure 22.9, involves manual crafting of features based on domain expertise and prior knowledge [52]. This approach processes raw SONAR data through statistical measures, texture descriptors, and wavelet transforms to create handcrafted features that require significant domain expertise [53]. Traditional signal processing techniques extract informative features from SONAR return data, though this process can be challenging in complex acoustic environments. Conversely, the feature learning approach, depicted on the right side of Figure 22.9, utilizes deep learning models, particularly CNNs, to automate feature extraction [43,54,55]. As shown in the flowchart, this pathway processes raw SONAR images through CNN layers and attention modules to extract hierarchical features automatically, eliminating manual feature design requirements [47]. Deep learning architectures incorporate attention mechanisms to focus on important channels and spatial locations, significantly enhancing feature extraction capabilities [56]. The key distinction highlighted in Figure 22.9 is that feature engineering requires extensive domain expertise while feature learning eliminates manual feature design. This automated approach is particularly advantageous in underwater SONAR image classification, where underlying physical factors affecting image formation are complex and difficult to model explicitly [43]. The choice between these approaches depends on specific application requirements, available data, and domain expertise accessibility.

22.4.2 Why AI for SONAR images? Addressing the challenges

The application of AI in underwater SONAR image classification is driven by the unique challenges posed by SONAR imagery [5,47]. SONAR images often suffer from low resolution, poor contrast, and significant noise due to the complex underwater environment [57]. Traditional image processing techniques may struggle to effectively address these issues [53]. AI, particularly deep learning, offers powerful tools to overcome these limitations [47,55]. Deep learning models are capable of capturing complex patterns and dependencies in data, allowing them to efficiently improve, denoise, and categorize SONAR images [47]. Transfer learning allows pre-trained models to be adapted to small SONAR datasets, enhancing their classification performance [5]. Moreover, XAI methods can help to interpret the decisions made by deep learning models, providing insights into the features that are most relevant for SONAR image classification [43]. Techniques like local interpretable model-agnostic explanations (LIME) can explain the predictions of any machine learning model [58]. This explainability is crucial for deploying AI systems in high-security domains such as defense, where trust and transparency are paramount [43].

22.4.3 Data requirements and augmentation for artificial intelligence training

AI models, especially deep learning ones, generally need vast datasets to train effectively [59]. However, acquiring sufficient underwater SONAR image data can be challenging due to the time-consuming and cost-intensive nature of SONAR data acquisition [60]. Data augmentation techniques play a critical role in addressing this data scarcity problem by artificially expanding the training dataset through systematic transformations [59]. Figure 22.10 demonstrates that data augmentation generates additional training samples by applying different transformations to the original

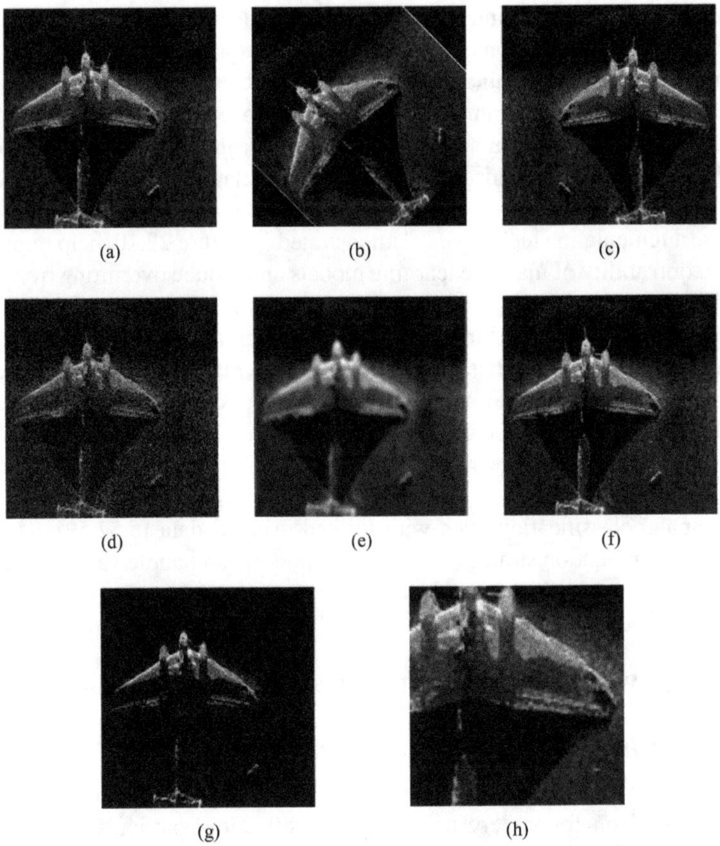

Figure 22.10 Comprehensive data augmentation techniques for SONAR image enhancement: (a) original SONAR image serving as baseline reference; (b) 45-degree rotation augmentation for rotational invariance training; (c) horizontal flip augmentation to increase dataset diversity and mirror symmetry learning; (d) additive Gaussian noise injection to simulate real-world acoustic interference and improve noise robustness; (e) Gaussian blur filter application to simulate depth-of-field effects and enhance edge detection training; (f) brightness enhancement to simulate varying illumination conditions and acoustic intensity variations; (g) contrast adjustment using histogram stretching to improve feature visibility and dynamic range optimization; (h) center crop with scaling to focus on region of interest and simulate different SONAR range settings.

images. The comprehensive augmentation strategy shown includes geometric transformations such as rotation (see Figure 22.10 (b)) and horizontal flipping (see Figure 22.10 (c)), which enhance rotational and reflectional invariance of the model. Photometric augmentations include Gaussian noise injection (see Figure 22.10 (d)) to simulate real-world acoustic interference, Gaussian blur application (see Figure 22.10 (e)) to enhance edge detection capabilities, brightness adjustment (see Figure 22.10 (f)) to simulate varying illumination conditions, and contrast enhancement (see Figure 22.10 (g)) using histogram stretching for improved feature visibility [52,56]. Additionally, spatial transformations such as center cropping and zooming (see Figure 22.10 (h)) focus on regions of interest while simulating different SONAR range settings.

These augmentation techniques, demonstrated in Figure 22.10, help improve the generalization ability of machine learning models and reduce overfitting by exposing the model to diverse variations of the original data [61]. Transfer learning and few-shot learning techniques can further mitigate data scarcity by leveraging knowledge gained from pre-trained models on related tasks [5]. Furthermore, generative models like generative adversarial networks (GANs) can be used to generate synthetic SONAR images, further expanding the training dataset. The combination of systematic data augmentation approaches shown in Figure 22.10, along with transfer learning and generative models, enables the training of robust AI models for underwater SONAR image classification even with limited original data [5,52,59]. This comprehensive augmentation strategy ensures that models can handle various real-world conditions and acoustic variations encountered in underwater environments.

22.5 AI-powered SONAR image classification techniques

22.5.1 Machine learning approaches

Machine learning approaches have significantly enhanced AI-powered SONAR image classification for underwater image classification, particularly since 2022. These methods offer automated and efficient solutions to overcome traditional challenges [43,51]. SONAR, valuable for ocean exploration, benefits immensely from deep learning, which has infused new vitality into underwater SONAR image classification [51]. Underwater SONAR image classification refers to the use of AI to recognize and classify underwater objects from SONAR images. It holds significant military and civilian value. Support vector machine (SVM), random forest, and K-nearest neighbors (KNN) are commonly used machine learning algorithms, each offering unique advantages in underwater SONAR image classification, as discussed below:

22.5.1.1 Support vector machines (SVMs)

SVM is a powerful classification algorithm that classifies data by finding the optimal hyperplane in a high-dimensional space [62,63]. In underwater SONAR image classification, SVM maps image features into a high-dimensional space to better distinguish between different underwater objects [64,65]. For instance, Zhu *et al.* [64] proposed a method that integrates KNN, SVM, and an improved CNN for underwater target classification. Their approach involves preprocessing and segmenting the acoustic image, computing Zernike moments as feature vectors from the binary image, and finally classifying using KNN [64]. Yin *et al.* [66] used an SVM model combined with positron emission tomography/computed tomography

(PET/CT) imaging to improve diagnosis of lymph node metastasis in non-small cell lung cancer. Alimjan *et al.* [67] introduced a hybrid SVM-KNN method for remote sensing image classification based on SVM class separability and spatial/spectral characteristics of remote sensing data. The advantage of SVM lies in its strong generalization ability, effective performance in high-dimensional spaces, and robustness in small sample scenarios [63]. However, its disadvantages include high computational complexity and sensitivity to parameter tuning. To improve SVM performance, researchers often incorporate feature selection and parameter optimization techniques [65,68]. For example, Shakil *et al.* [68] studied the impact of feature selection techniques on dragon fruit disease recognition and found that different feature selection methods affect the accuracy of SVM classification.

22.5.1.2 Random forests

Random forest is an ensemble learning algorithm that classifies by building multiple decision trees and integrating their results [62,63]. Each decision tree is trained using a random subset of features, which reduces overfitting and improves generalization [62,63,69]. In underwater SONAR image classification, random forest handles high-dimensional features effectively and is robust to noise and outliers [69–71]. For example, Zhang *et al.* [69] used SVM, KNN, random forest, decision tree, and multilayer perceptron (MLP) to extract forest land information from sentinel-2 images. The benefits of random forest include fast training speed, ability to handle large feature sets, and built-in feature importance evaluation [69,71,72]. However, its downside is poor interpretability, making individual decision processes of trees difficult to understand. To enhance its performance, researchers use feature engineering and model optimization techniques [68,73]. For instance, Radwan *et al.* [73] used optimized machine learning models and feature selection techniques to classify potato leaf diseases.

22.5.1.3 K-nearest neighbors (KNN)

KNN is a simple and intuitive algorithm that classifies a sample by finding its K nearest neighbors and voting based on their classes [62,63,74]. In underwater SONAR image classification, KNN can directly use raw image features without complex feature extraction or model training [75]. For example, Rana *et al.* [75] used methods like maximum distance, KNN, SVM, random forest, K-dimensional (KD)-KNN, and classification and regression tree (CART) to map flood risks caused by extreme rainfall. KNN's strengths are its simplicity, ease of implementation, and minimal assumptions about data distribution [74,76]. However, it suffers from high computational cost, large memory requirements, and sensitivity to the choice of K. To enhance KNN performance, researchers often use dimensionality reduction and distance metric learning [62,77]. For example, Kouadio *et al.* [77] proposed a K-means featurizer method to improve performance on complex datasets.

22.5.2 Deep learning for SONAR images

22.5.2.1 Convolutional neural networks (CNNs) architectures

CNNs are widely used in image classification because they can automatically learn hierarchical features directly from raw pixel inputs [78]. CNNs, such as AlexNet,

VGG, and ResNet, are high-computing power deep neural networks [79]. These architectures include convolutional layers, pooling layers, and fully-connected layers [80]. In SONAR image processing, CNNs are pivotal for automatic target recognition and classification, enhancing accuracy and efficiency [81].

- **LeNet:** LeNet-5, one of the pioneering CNN architectures developed in 1998, introduced fundamental concepts of CNNs, such as convolutional, pooling, and fully-connected layers [80].
- **AlexNet:** AlexNet made a significant impact in 2012 by demonstrating the power of deep CNNs for large-scale image classification. It utilizes techniques like ReLU activation functions and dropout for better performance [80].
- **VGG:** VGG architectures (VGG-16 and VGG-19) are known for their simplicity and the use of 3×3 convolutional filters, proving that deeper networks can achieve better image-related task performance [80].
- **ResNet:** Introduced in 2015, ResNet solves the vanishing gradient problem in deep neural networks by using residual connections, which allow for the effective training of very deep models [80].
- **U-Net:** U-Net is commonly used for image segmentation, particularly in medical imaging, because of its symmetric encoder-decoder design that includes skip connections [82].

For underwater SONAR images, a novel CNN model incorporating a deep separable residual module has been proposed for multiscale feature extraction and noise suppression [83]. The model enhances the segmentation accuracy of target regions by fusing multichannel information and suppressing noise [83]. MobileNet and EfficientNet, which are also CNN architectures, have been used in underwater item detection due to their efficiency [84]. These models are compared to assess their performance in marine surveys and underwater object classification [84].

22.5.2.2 Transfer learning in SONAR image classification

Transfer learning utilizes pre-trained models to address the issue of scarce data in SONAR image classification [85]. Pre-trained models on large datasets can be adapted to specific SONAR image tasks, minimizing the requirement for large amounts of training data and computing power [80]. For instance, the classification of non-small cell lung cancers has benefited from pre-trained deep CNNs via transfer learning [85]. In the context of underwater SONAR images, transfer learning can help in the classification of maxillary sinus pathologies [86]. By using pre-trained models, the amount of required labeled data can be reduced [80]. GANs can be combined with CNNs to further enhance the classification performance [86].

22.5.2.3 Recurrent neural networks (RNNs) for sequential data

RNNs are built to process sequential data, making them ideal for analyzing video sequences that rely on temporal relationships [87,88]. RNNs, including long short-term memory (LSTM) and gated recurrent units (GRU) [87], can capture the temporal dynamics of SONAR images in video format [89]. Hybrid models such as CNN-RNN combine the spatial feature extraction of CNNs with the temporal modeling capabilities of RNNs. In environmental sound classification, CNN-RNN architectures, along with data augmentation techniques using deep convolutional (DC)-GAN, have been used to improve classification accuracy [90]. While the literature do not focus specifically on underwater SONAR image video sequences, the successful application of CNN-RNN in similar contexts suggests their potential utility in analyzing temporal patterns in SONAR videos [90].

22.5.2.4 Data augmentation and synthesis with generative adversarial networks (GANs)

GANs include two components: a generator that creates new data similar to the training set, and a discriminator that aims to differentiate between real and generated data [91–93]. GANs can address the data scarcity problem, which is common in SONAR image classification due to the high cost of data sampling [94]. By generating synthetic SONAR images, GANs can augment the training dataset and improve the performance of deep learning models [91]. A new network (ResNet-asymmetric convolution and weight reuse (ACW)), which builds on residual networks and incorporates a few-shot learning strategy inspired by GANs, improves recognition accuracy when trained on small datasets [94]. Also, an attention-enhanced GAN can remove noise from photo-acoustic images [95]. In the context of medical imaging, GANs have been used to generate images for Maxillary sinus classification [86]. Although not explicitly discussed for SONAR images, the application of GANs in similar domains suggests their potential for data augmentation and improved classification accuracy in underwater SONAR image processing [86].

22.5.3 Object detection and segmentation in SONAR images with AI

The application of AI in underwater SONAR image analysis has seen remarkable progress, particularly in object detection and segmentation. Traditional SONAR image analysis faces challenges like low resolution, noise, and blurred object boundaries, which are being addressed using advanced deep-learning techniques [96,97]. Region-Based CNNs, single-shot detectors, and methods for semantic and instance segmentation are pivotal in this evolution.

22.5.3.1 Region-based CNNs (R-CNN, fast R-CNN, faster R-CNN)

R-CNNs represent a foundational approach in object detection. These networks operate in two stages: first, they generate region proposals, and second, they classify these proposals and refine their boundaries [98,99]. R-CNNs, fast R-CNNs, and faster R-CNNs have been instrumental in various image analysis tasks, including underwater object detection [100,101]. For instance, an improved dung beetle optimization (IDBO) algorithm with R-CNN was proposed for underwater object image detection and classification [100]. While traditionally used in optical images, adaptations of R-CNNs are now being explored for SONAR images [102,103]. However, the computational cost and complexity associated with R-CNN can be significant limitations, especially for real-time applications [104].

22.5.3.2 Single-shot detectors (SSD, YOLO)

Single-shot detectors (SSDs) offer a more streamlined approach by performing object detection in a single pass. Unlike R-CNN that require multiple stages, SSDs predict bounding boxes and class probabilities directly from feature maps, making them faster and more efficient [98,99]. YOLO (you only look once), a prominent SSD, has gained traction in underwater object detection due to its speed and reasonable accuracy [98]. For example, rotated convolutional center-aware detector (CCAD)-YOLOv3 has been used for oversized ore detection, demonstrating the adaptability of YOLO-based methods to specific industrial needs [105]. Lightweight versions of YOLO, such as PP-YOLO and YOLOX-Nano, are designed for edge devices, indicating a trend toward real-time object detection in resource-constrained environments [98]. The use of SSDs addresses some limitations of R-CNN by enhancing speed and efficiency [99].

22.5.3.3 Semantic and instance segmentation

Semantic segmentation involves classifying each pixel in an image, while instance segmentation goes a step further by differentiating individual objects of the same class [20,102]. Mask R-CNN, an extension of faster R-CNN, is a popular choice for instance segmentation [98]. It combines object detection with pixel-level segmentation, providing detailed masks for each object instance [102]. In SONAR imagery, mask R-CNN has been used for tasks such as detecting the eye of tropical cyclones from satellite images [106]. Furthermore, research is exploring lightweight semantic segmentation methods based on mask R-CNN to balance accuracy and computational efficiency [102]. The effective strategy for SONAR images instance segmentation (ESSIIS) introduces ASception, a new module for SONAR image instance segmentation [20]. Multibeam SONAR target segmentation algorithms based on bottleneck supervised (BS)-Unet have also been developed to improve target segmentation accuracy in low-visibility conditions [107]. Instance segmentation provides granular insights critical for complex underwater tasks. CNN-transformer hybrid architectures are also being employed for salient object detection (SOD) of FLS images [96].

22.6 Applications and case studies

Recent advancements in AI have significantly enhanced underwater SONAR image classification capabilities across various marine applications. AUVs equipped with AI-driven SONAR systems now perform complex underwater tasks with unprecedented accuracy and efficiency [108,109]. The integration follows a systematic five-stage workflow as illustrated in Figure 22.11, encompassing data acquisition through various SONAR modalities, signal processing, AI analysis, decision-making, and final action implementation.

22.6.1 Mine countermeasures and unexploded ordnance detection

22.6.2 Seabed mapping and habitat classification

AI plays a crucial role in processing SONAR data for comprehensive seabed mapping and marine habitat classification. The workflow's data acquisition stage utilizes

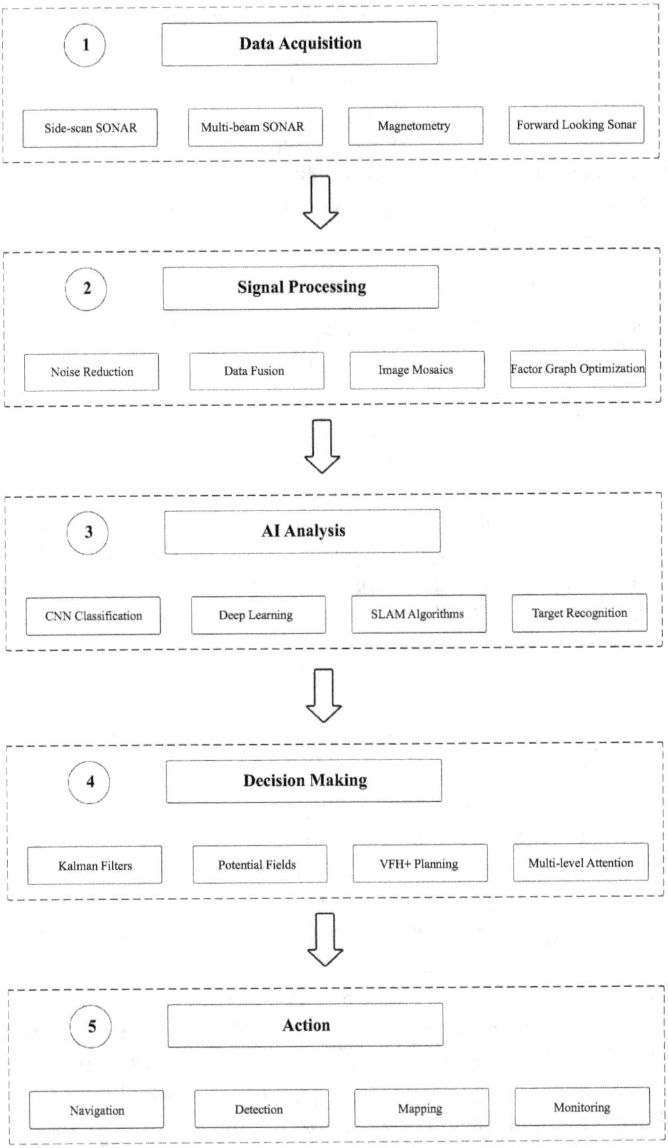

Figure 22.11 Artificial intelligence underwater SONAR technology integration workflow.

SSS systems mounted on AUVs to capture detailed seafloor imagery. The signal processing stage (stage 2) employs factor graph optimization techniques to enhance navigation precision, which is critical for creating high-quality SONAR image mosaics that accurately delineate seafloor topography and feature locations [110]. During the AI analysis stage, deep learning frameworks process the SONAR data through multi-level attention adaptive segmentation algorithms, enabling fine-grained recognition of seabed targets [111]. This AI-driven approach facilitates comprehensive habitat monitoring, including fish identification and counting, along with algae, sea grass, and coral recognition, which is essential for biodiversity assessment [112]. The system's decision-making capabilities enable the construction of detailed three-dimensional underwater models and bathymetric maps, supporting underwater construction, navigation, and environmental studies [112].

22.6.3 Underwater object recognition

AI-driven underwater object recognition encompasses the identification of various submerged structures including pipelines, shipwrecks, and marine life, significantly improving underwater inspection capabilities and marine conservation efforts. The workflow's AI analysis stage implements deep CNNs to classify SSS images, though research has identified vulnerabilities to adversarial attacks that require ongoing attention [113]. Advanced deformable deep aggregation network models address the challenging detection of underwater cultural objects, effectively handling poor imaging conditions and variations in object shapes [114]. The system also incorporates automatic target recognition capabilities for geolocating natural gas seeps, which is environmentally critical due to their negative impact on marine habitats and contribution to greenhouse gas emissions [115].

22.6.4 Autonomous underwater vehicle navigation and obstacle avoidance

The integration of AI algorithms significantly enhances AUV navigation and obstacle avoidance capabilities, enabling safer and more efficient underwater missions. The workflow's data acquisition stage utilizes forward-looking high-frequency active SONAR systems that feed into three-dimensional collision avoidance controllers. These systems employ clustering algorithms to differentiate objects during the signal processing stage [116]. The AI analysis stage implements simultaneous localization and mapping algorithms based on image SONAR salient target detection, optimizing inertial navigation system errors for long-range AUV operations [2]. Multibeam FLS enables real-time underwater simultaneous localization and mapping implementation [21]. The decision-making stage utilizes modified potential field algorithms and vector field histogram plus planning to determine appropriate maneuvers [116,117]. Advanced implementations include efficient reactive obstacle avoidance systems that balance computational resources, precision, and safety using 2.5-dimensional FLS [118]. Bio-inspired cooperative control schemes facilitate obstacle avoidance for unmanned underwater vehicle swarms, effectively addressing communication constraints in multivehicle operations [119].

22.6.5 Environmental monitoring and oceanography

AI facilitates comprehensive environmental monitoring and oceanographic research by enabling AUVs to autonomously collect and analyze marine data. The workflow supports multisensor data fusion approaches that integrate data from closed-circuit television cameras, gyroscopes, accelerometers, and rotating SONAR devices for comprehensive environmental assessment [120]. The internet of underwater things (IoUT) framework enhances underwater applications through wireless communication networks, with software-defined IoUT implementations utilizing software-defined networking and AI for load balancing and quality-of-service provisioning [121,122]. AUVs and ROVs monitor and upscale marine ecosystem restoration efforts in deep-sea environments, addressing critical habitat loss concerns [109]. Multi-frequency noise reduction methods implemented during the signal processing stage mitigate the impact of AUV vibration on shear probe performance, significantly enhancing the accuracy of oceanographic measurements [123]. This integrated approach provides essential data for climate research and ecosystem management, demonstrating the comprehensive capabilities of AI-enhanced underwater SONAR systems.

22.7 Challenges and future directions

SONAR image analysis, particularly when combined with AI and deep learning, holds immense potential for underwater object detection and classification [124]. However, several challenges hinder its widespread adoption and effectiveness [9]. Addressing these challenges will pave the way for future advancements in autonomous underwater operations.

22.7.1 Data scarcity and annotation difficulties

One of the primary challenges in applying deep learning to SONAR imagery is the scarcity of labeled data [9,125]. Deep learning models, especially CNNs, require substantial amounts of training data to achieve high accuracy [18]. In underwater environments, acquiring high-quality SONAR datasets is difficult due to financial constraints, data confidentiality, and the time-consuming nature of data collection [17,125]. Furthermore, annotating SONAR images is a labor-intensive process that requires expert knowledge to accurately identify and label objects of interest [126]. The characteristics of underwater SONAR imaging, such as low contrast, blurred edges, poor texture, and unsatisfactory quality, exacerbate these difficulties [127]. Following are the potential solutions:

- **Data augmentation:** Employing techniques to artificially expand the size of the existing dataset by creating modified versions of existing images [17].
- **Synthetic data generation:** Using simulators to generate synthetic SONAR images [17,125]. For example, the S3 simulator dataset utilizes advanced simulation to create a benchmark dataset for SSS images [125]. GANs can also be employed to generate realistic SONAR images, which can be used to train object detection models [17].
- **Transfer learning:** Utilizing models pre-trained on large datasets like ImageNet and adapting them to smaller SONAR datasets enables the transfer of learned knowledge from other domains to underwater applications [128].

- **Semi-supervised learning:** Combining labeled and unlabeled data to train models [1].

22.7.2 Generalization across different SONAR systems and environments

SONAR systems vary in frequency, resolution, and operating principles [129]. Furthermore, underwater environments differ significantly in terms of water turbidity, salinity, and bottom topography [124]. These variations pose challenges for the generalization of AI models trained on one specific SONAR system or environment to others [14]. A model trained on data from clear water might perform poorly in turbid conditions. Following are the potential solutions:

- **Domain adaptation techniques:** Employing techniques that adapt models trained on one domain (e.g., a specific SONAR system or environment) to perform well on a different domain [128].
- **Multi-domain training:** Training models on data from multiple SONAR systems and environments to improve their generalization capability [128].
- **Physics-based modeling:** Incorporating physics-based models of SONAR signal propagation and underwater acoustics into deep learning architectures to improve their robustness to environmental variations.

22.7.3 Real-time processing and computational efficiency

Many underwater applications, such as autonomous navigation and mine detection, require real-time processing of SONAR data [9]. Deep learning models, particularly complex CNNs, can be computationally expensive, making them challenging to deploy on embedded systems with limited processing power [9]. Following are the potential solutions:

- **Model compression techniques:** Applying techniques such as pruning, quantization, and knowledge distillation to reduce the size and computational complexity of deep learning models without significantly compromising their accuracy [130].
- **Hardware acceleration:** Utilizing specialized hardware, such as GPUs and FPGAs, to accelerate the computation of deep learning models.
- **Efficient network architectures:** Designing lightweight and efficient network architectures that can achieve high accuracy with fewer parameters and computations [130].

22.7.4 Interpretability and explainable AI (XAI) for critical applications

Deep learning models are often considered “black boxes” due to their complex internal workings [5]. This lack of interpretability poses challenges for critical applications, such as defense and structural health monitoring, where it is essential to understand the reasoning behind a model’s decisions [5,131]. XAI aims to develop techniques that can provide insights into the decision-making processes of AI models [5]. Following are the potential solutions:

- **Local interpretable model-agnostic explanations (LIME):** Using LIME to explain the predictions of underwater SONAR image classification models [5].
- **Attention mechanisms:** Incorporating attention mechanisms into deep learning architectures to highlight the regions of the input image that are most relevant to the model's prediction.
- **Rule extraction:** Developing methods to extract explicit rules from trained deep learning models.

22.7.5 Fusion of SONAR data with other sensor modalities (e.g., optical)

SONAR sensors offer advantages in turbid water conditions, while optical sensors provide high-resolution images in clear water [128,131]. Fusing data from multiple sensor modalities can provide a more comprehensive and robust perception of the underwater environment [132]. Following are the potential solutions:

- **Early fusion:** Concatenating features extracted from different sensor modalities before feeding them into a deep learning model.
- **Late fusion:** Training separate deep learning models for each sensor modality and combining their predictions using a fusion rule.
- **Attention-based fusion:** Using attention mechanisms to dynamically weigh the contributions of different sensor modalities based on their relevance to the current task.

22.7.6 Emergent AI techniques (e.g., federated learning, self-supervised learning)

Emerging AI techniques offer promising solutions to address the challenges in SONAR image analysis [133,134].

- **Federated learning:** Enables collaborative training of AI models across multiple devices or organizations without sharing the raw data [134]. This approach can address data scarcity and privacy concerns in underwater SONAR image analysis.
- **Self-supervised learning:** Leverages unlabeled data to learn useful representations [134]. This can be particularly beneficial in underwater environments where unlabeled SONAR data is abundant.

22.7.7 Ethical considerations in autonomous underwater operations

As AUVs become more prevalent, it is crucial to address the ethical implications of their deployment [9]. These include issues such as:

- **Bias in AI models:** Ensuring that AI models are not biased against certain objects or environments [135].

- **Accountability:** Determining who is responsible for the actions of autonomous underwater systems.
- **Transparency:** Ensuring that the decision-making processes of autonomous systems are transparent and understandable.

Tackling these ethical issues is essential for the responsible design and implementation of AI-driven underwater systems.

22.8 Conclusion

22.8.1 Summary of key findings and progress

This chapter explored the integration of AI into underwater SONAR image classification and analysis, emphasizing its role in addressing traditional challenges such as noise, low contrast, and data scarcity. Key findings highlight the superiority of deep learning architectures, particularly CNNs, in automating feature extraction and improving classification accuracy compared to manual interpretation and statistical methods. Techniques like transfer learning and GANs effectively mitigate limited data availability by leveraging pre-trained models and synthetic data generation. Case studies demonstrated AI's practical utility in mine detection, seabed mapping, and AUV navigation, showcasing significant improvements in efficiency and precision. However, challenges persist, including annotation complexity, environmental variability, and computational constraints in real-time applications.

22.8.2 The transformative impact of AI on SONAR analysis

AI has revolutionized SONAR analysis by enabling scalable, automated solutions that transcend human expertise and traditional model-based approaches. Deep learning models, such as mask R-CNN and YOLO, have enhanced object detection and segmentation accuracy in low-resolution SONAR imagery, while attention mechanisms and XAI frameworks improve interpretability for critical military and environmental applications. The adoption of GANs and physics-informed simulations has expanded training datasets, fostering robustness in diverse underwater conditions. These advancements not only reduce operational costs but also empower autonomous systems like AUVs to perform complex tasks, such as obstacle avoidance and habitat monitoring, with unprecedented reliability.

22.8.3 Outlook for the field

Future advancements in underwater SONAR analysis will hinge on addressing data scarcity through federated learning and synthetic data pipelines, while improving model generalization across heterogeneous environments through domain adaptation. Lightweight architectures and hardware acceleration will bridge the gap between computational demands and real-time processing needs. The fusion of SONAR with optical, LiDAR, and IoT sensors promises holistic underwater perception, enhancing applications in climate research and infrastructure inspection. Emerging techniques like self-supervised learning and bio-inspired algorithms offer untapped potential, but ethical frameworks must evolve to ensure transparency and accountability in autonomous systems. Collaborative efforts across academia, industry, and defense sectors will drive innovation, cementing AI as a cornerstone of next-generation underwater exploration.

References

[1] Wang X, Jiao J, Yin J, *et al.* Underwater SONAR image classification using adaptive weights convolutional neural network. *Applied Acoustics*, 2019;146:145–154. https://www.sciencedirect.com/science/article/pii/S0003682X17311532.

[2] Liu J, Guan W, Liu J, *et al.* Underwater Slam algorithm based on image SONAR salient target detection. *SSRN Electronic Journal*, 2022;25(11):1–52.

[3] Özdaş KNH. Archeological assessment of two naval battlefields in Çeşme and Sinop, Turkey. *Journal of Marine Architecture*, 2022;15:365–392.

[4] Kamal Basha S and Nambiar A. S3Simulator: a benchmarking side scan SONAR simulator dataset for underwater image analysis. In: *Pattern Recognition: 27th International Conference, ICPR 2024, Kolkata, India*, December 1–5, 2024, Proceedings, Part XVI. Berlin, Heidelberg: Springer; 2024. pp. 219–235. https://doi.org/10.1007/978-3-031-78444-6_15.

[5] Chungath TT, Nambiar AM, and Mittal A. Transfer learning and few-shot learning based deep neural network models for underwater SONAR image classification with a few samples. *IEEE Journal of Oceanic Engineering*, 2024;49(1):294–310.

[6] Zhang Q, Bloecher N, Evjemo LD, *et al.* Farmed Atlantic salmon (*Salmo salar* L.) avoid intrusive objects in cages: the influence of object shape, size and colour, and fish length. *Aquaculture*, 2024;581:740429. https://www.sciencedirect.com/science/article/pii/S0044848623012036.

[7] Nguyen HT, Lee EH, and Lee S. Study on the classification performance of underwater SONAR image classification based on convolutional neural networks for detecting a submerged human body. *Sensors*, 2020;20(1):1–13. https://www.mdpi.com/1424-8220/20/1/94.

[8] Javidan R. *Underwater Acoustic Image Processing.* Koln, DEU: LAP Lambert Academic Publishing; 2012.

[9] Ruan F, Dang L, Ge Q, *et al.* Dual-path residual "Shrinkage" network for side-scan SONAR image classification. *Computational Intelligence and Neuroscience*, 2022;2022(1):6962838. https://onlinelibrary.wiley.com/doi/abs/10.1155/2022/6962838.

[10] Ge LHMYea Q. Shuffle-RDSNet: a method for side-scan SONAR image classification with residual dual-path shrinkage network. *Journal of Supercomputing* 2024;80:19947–19975.

[11] Tang Z, Lu J, Wang Z, *et al.* Three dimensional height information reconstruction based on mobile active SONAR detection. *Applied Acoustics*,

2020;169:107459. https://www.sciencedirect.com/science/article/pii/S0003682X20305636.

[12] Schmidt SMMB. Hydroakustik zur Analyse von Fischbeständen und Fischverhalten – Fallstudien aus Österreich. *Österr Wasser- und Abfallw*, 2020;72:213–222.

[13] Qin KS, Liu D, Wang F, *et al.* Improved YOLOv7 model for underwater SONAR image object detection. *Journal of Visual Communication and Image Representation*, 2024;100:104124. https://www.sciencedirect.com/science/article/pii/S1047320324000798.

[14] Natarajan P and Nambiar A. *Underwater SONAR Image Classification and Analysis using LIME-based Explainable Artificial Intelligence*, 2024. https://arxiv.org/abs/2408.12837.

[15] Anastasia A and Theodosiou RCR. Artificial intelligence, machine learning and deep learning: potential resources for the infection clinician. *Journal of Infection*, 2024;87(4):287–294.

[16] Zhu K, Zhao Y, Zheng L, *et al.* Underwater acoustic image classification based on image denoising and transfer learning. In: *2023 China Automation Congress (CAC)*; 2023. pp. 7605–7610.

[17] Cai W, Zhu J, and Zhang M. From classical approach to deep-learning: a review on underwater target segmentation with SONAR image. *Neurocomputing*, 2025;637:130087. https://linkinghub.elsevier.com/retrieve/pii/S0925231225007593.

[18] Ge L, Singh P, and Sadhu A. Advanced deep learning framework for underwater object detection with multibeam forward-looking SONAR. *Structural Health Monitoring*, 2024;24(4):1991–2007. https://doi.org/10.1177/14759217241235637.

[19] Moghimi MK and Mohanna F. Real-time underwater image enhancement: a systematic review. *Journal of Real-Time Image Processing*, 2021;18:1509–1525.

[20] Shi B, Cao Q, Ge YL, and Wang Z. SONAR image intelligent processing in seabed pipeline detection: review and application. *Measurement Science and Technology* 2024;35(4):045405.

[21] Cheng C, Wang C, Yang D, *et al.* Underwater localization and mapping based on multi-beam forward looking SONAR. *Frontiers in Neurorobotics*, 2022;15:801956. https://www.frontiersin.org/journals/neurorobotics/articles/10.3389/fnbot.2021.801956.

[22] Wang X, Wang L, Li G, *et al.* A robust and fast method for sidescan SONAR image segmentation based on region growing. *Sensors*, 2021;21(21):6960. https://www.mdpi.com/1424-8220/21/21/6960.

[23] Shin YS, Cho YG, Choi TG and Kim A. Comparative study of SONAR image processing for underwater navigation. *Journal of Ocean Engineering and Technology*, 2016;30(3):214–220.

[24] Brown CJ, Beaudoin J, Brissette M, *et al.* Multispectral multibeam echo sounder backscatter as a tool for improved seafloor characterization. *Geosciences*, 2019;9(3):126. https://www.mdpi.com/2076-3263/9/3/126.

[25] Choi HM, Yang HS, and Seong WJ. Compressive underwater SONAR imaging with synthetic aperture processing. *Remote Sensing*, 2021;13(10):1924. https://www.mdpi.com/2072-4292/13/10/1924.

[26] Liu X, Fan J, Sun C, *et al.* High-resolution and low-sidelobe forward-look SONAR imaging using deconvolution. *Applied Acoustics*, 2021;178: 107986. https://www.sciencedirect.com/science/article/pii/S0003682X21000797.

[27] Zhao B, Zhou Q, Huang L, *et al.* Minutia reconstruction in SONAR images with diffusion probabilistic models. *Engineering Applications of Artificial Intelligence*, 2024;135:108850. https://www.sciencedirect.com/science/article/pii/S095219762401008X.

[28] Tang Y, Wang J, Jin S, *et al.* Real-time processing and high-quality imaging of navigation strip data using SSS based on AUVs. *Journal of Marine Science and Engineering*, 2023;11(9):1–24. https://www.mdpi.com/2077-1312/11/9/1769.

[29] Russell GT, Bell JM, Holt PO, *et al.* SONAR image interpretation and modelling. In: *Proceedings of Symposium on Autonomous Underwater Vehicle Technology*; 1996. pp. 317–324.

[30] Yang D, Wang C, Cheng C, *et al.* Semantic segmentation of side-scan SONAR images with few samples. *Electronics*, 2022;11(19):1–14. https://www.mdpi.com/2079-9292/11/19/3002.

[31] Wang Z, Wang B, Guo J, and Zhang S. SONAR objective detection based on dilated separable densely connected CNNs and quantum-behaved PSO algorithm. *Computational Intelligence and Neuroscience*, 2021;18.

[32] Jiao S, Zhao C, and Xin Y. Research on convolutional neural network model for SONAR image segmentation. In: *MATEC Web of Conferences*, Vol. 220. EDP Sciences, 2018. p. 10004.

[33] He J, Chen J, Xu H, *et al.* SONARNet: hybrid CNN-transformer-HOG framework and multifeature fusion mechanism for forward-looking SONAR image segmentation. *IEEE Transactions on Geoscience and Remote Sensing*, 2024;62:1–17.

[34] Juneja M, Minhas JS, Singla N, *et al.* Fused framework for glaucoma diagnosis using optical coherence tomography (OCT) images. *Expert Systems with Applications*, 2022;201:117202. https://www.sciencedirect.com/science/article/pii/S0957417422005887.

[35] Chiesa-Estomba CE, Echaniz O, Larruscain E, *et al.* Radiomics and texture analysis in laryngeal cancer. Looking for new frontiers in precision medicine through imaging analysis. *Cancers (Basel)*. 2019;11(10):1409.

[36] Gan HS, Ramlee MH, Wahab AA, *et al.* From classical to deep learning: review on cartilage and bone segmentation techniques in knee osteoarthritis research. *Artificial Intelligence Review* 2021;54:2445–2494.

[37] Wang X, Liu Z, Du Y, *et al.* Recognition of fetal facial ultrasound standard plane based on texture feature fusion. *Computational and Mathematical Methods in Medicine*, 2021;2021(1):6656942. https://onlinelibrary.wiley.com/doi/abs/10.1155/2021/6656942.

[38] Ma P, Li C, Rahaman MM, *et al.* A state-of-the-art survey of object detection techniques in microorganism image analysis: from classical methods to deep learning approaches. *Artificial Intelligence Review*, 2023;56:1627–1698.

[39] Chen H, Li X, Li C, *et al.* What can machine vision do for lymphatic histopathology image analysis: a comprehensive review. *Artificial Intelligence Review*, 2024;57(2):71.

[40] Tian Y, Lan L, and Sun L. A review of SONAR image segmentation for underwater small targets. In: *Proceedings of the 2020 International Conference on Pattern Recognition and Intelligent Systems. PRIS'20*. New York, NY: Association for Computing Machinery; 2020. https://doi.org/10.1145/3415048.3416098.

[41] Li C, Chen H, Li X, *et al.* A review for cervical histopathology image analysis using machine vision approaches. *Artificial Intelligence Review*, 2020;53:4821–4862.

[42] Ha OK, Lee K, Kim WJ, *et al.* Effective parallelization method for object recognition in 2D SONAR images based on task partitioning. *Scientific Programming*, 2019;2019(1):8908950. https://onlinelibrary.wiley.com/doi/abs/10.1155/2019/8908950.

[43] Natarajan P and Nambiar A. Underwater SONAR image classification and analysis using LIME-based explainable artificial intelligence; 2024. https://arxiv.org/abs/2408.12837.

[44] Liu X. Review of object detection algorithms for SONAR images based on deep learning. *Recent Patents on Engineering*, 2025;19(3):43–53. https://eurekaselect.com/article/135020.

[45] Shi P, He Q, Zhu S, *et al.* Multi-scale fusion and efficient feature extraction for enhanced SONAR image object detection. *Expert Systems with Applications*, 2024;256:124958. https://www.sciencedirect.com/science/article/pii/S0957417424018256.

[46] Gu X, Tang S, Cao Y, Yu C. Underwater object detection in SONAR imagery with detection transformer and zero-shot neural architecture search; 2025. https://arxiv.org/abs/2505.06694.

[47] Chai Y, Yu H, Xu L, *et al.* Deep learning algorithms for SONAR imagery analysis and its application in aquaculture: a review. *IEEE Sensors Journal*, 2023;23(23):28549–28563.

[48] Preciado-Grijalva A, Wehbe B, Firvida MB, Valdenegro-Toro M. Self-supervised learning for SONAR image classification. In: *2022 IEEE/CVF Conference on Computer Vision and Pattern Recognition Workshops (CVPRW)*; IEEE Computer Society, CA, USA. 2022:1498–1507. https://doi.ieeecomputersociety.org/10.1109/CVPRW56347.2022.00156

[49] Ochal M, Vazquez J, Petillot Y, *et al.* A comparison of few-shot learning methods for underwater optical and SONAR image classification. In: *Global Oceans 2020: Singapore – U.S. Gulf Coast*; 2020. pp. 1–10.

[50] Sun S, Wang H, Zhang H, *et al.* Underwater image enhancement with reinforcement learning. *IEEE Journal of Oceanic Engineering*, 2024;49(1):249–261.

[51] Peng Y, Li H, Zhang W, *et al.* Underwater SONAR image classification with image disentanglement reconstruction and zero-shot learning. *Remote Sensing*, 2025;17(1):1–22. https://www.mdpi.com/2072-4292/17/1/134.
[52] Chen W, Wang X, Li M, *et al.* An underwater SONAR small sample image transformation for feature domain optimization. In *2023 IEEE 6th International Conference on Electronic Information and Communication Technology (ICEICT)*; 2023. pp. 1377–1386.
[53] Christensen A, Sen Gupta A, Kirsteins I. Underwater small target classification using sparse multi-view discriminant analysis and the invariant scattering transform. *Journal of Marine Science and Engineering*. 2024;12(10):1–32. https://www.mdpi.com/2077-1312/12/10/1886.
[54] Xu S, Chen S, Xu R, *et al.* Local feature matching using deep learning: a survey. *Information Fusion*, 2024;107:102344. https://www.sciencedirect.com/science/article/pii/S1566253524001222.
[55] Mittal S, Srivastava S, and Jayanth JP. A survey of deep learning techniques for underwater image classification. *IEEE Transactions on Neural Networks and Learning Systems*, 2023;34(10):6968–6982.
[56] Wu G, Jin E, Sun Y, *et al.* Deep attention fusion hashing (DAFH) model for medical image retrieval. *Bioengineering*, 2024;11(7):1–15. https://www.mdpi.com/2306-5354/11/7/673.
[57] Gong T, Zhang M, Zhou Y, *et al.* Underwater image enhancement based on color feature fusion. *Electronics*, 2023;12(24):1–17. https://www.mdpi.com/2079-9292/12/24/4999.
[58] Dwivedi R, Dave D, Naik H, *et al.* Explainable AI (XAI): core ideas, techniques, and solutions. *ACM Computing Surveys*, 2023;55(9):1–33. https://doi.org/10.1145/3561048.
[59] Steiniger Y, Bueno A, Kraus D, *et al.* Tackling data scarcity in SONAR image classification with hybrid scattering neural networks. In: *OCEANS 2023 – Limerick*; 2023. pp. 1–7.
[60] Kim WK, Bae HS, Son SU, *et al.* Neural network-based underwater object detection off the coast of the Korean Peninsula. *Journal of Marine Science and Engineering*, 2022;10(10):1–16. https://www.mdpi.com/2077-1312/10/10/1436.
[61] Anitha E and Aravindhar DJ. Efficient retinal detachment classification using hybrid machine learning with levy flight-based optimization. *Expert Systems with Applications*, 2024;239:122311. https://www.sciencedirect.com/science/article/pii/S0957417423028130.
[62] García-Pedrajas NE, Cuevas-Muñoz JM, Cerruela-García G, *et al.* A thorough experimental comparison of multilabel methods for classification performance. *Pattern Recognition*, 2024;151:110342. https://www.sciencedirect.com/science/article/pii/S0031320324000931.
[63] Boateng OJE and Abaye D. Basic tenets of classification algorithms K-nearest-neighbor, support vector machine, random forest and neural network: a review. *Journal of Data Analysis and Information Processing*, 2020;8:341–357.

[64] Zhu K, Tian J, and Huang H. Fusion of underwater object classification methods in SONAR images. In: Li C, Yu H, Pan Z, *et al.*, editors. *Tenth International Conference on Graphics and Image Processing (ICGIP 2018). vol. 11069. International Society for Optics and Photonics*. SPIE; 2019. p. 110691X. https://doi.org/10.1117/12.2524270.

[65] Pathak DK and Kalita SK. Spectral spatial feature based classification of hyperspectral image using support vector machine. In: *2019 6th International Conference on Signal Processing and Integrated Networks (SPIN)*; 2019. pp. 430–435.

[66] Yin G, Song Y, Li X, *et al.* Prediction of mediastinal lymph node metastasis based on 18F-FDG PET/CT imaging using support vector machine in non-small cell lung cancer. *European Radiology* 2021;31(6):3983–3992.

[67] Alimjan G, Sun T, Liang Y, *et al.* A new technique for remote sensing image classification based on combinatorial algorithm of SVM and KNN. *International Journal of Pattern Recognition and Artificial Intelligence*, 2018;32(7):1859012.

[68] Shakil R, Islam S, Shohan YA, *et al.* Addressing agricultural challenges: an identification of best feature selection technique for dragon fruit disease recognition. *Array*, 2023;20:100326. https://www.sciencedirect.com/science/article/pii/S2590005623000516.

[69] Zhang C, Liu Y, and Tie N. Forest land resource information acquisition with Sentinel-2 image utilizing support vector machine, K-nearest neighbor, random forest, decision trees and multi-layer perceptron. *Forests*, 2023;14(2):1–18. https://www.mdpi.com/1999-4907/14/2/254.

[70] Kumar S, Janet B, and Neelakantan S. IMCNN: Intelligent malware classification using deep convolution neural networks as transfer learning and ensemble learning in honeypot enabled organizational network. *Computer Communications*, 2024;216:16–33. https://www.sciencedirect.com/science/article/pii/S0140366423004772.

[71] Mukherjee B, Kar, and Sain K. Machine learning assisted state-of-the-art-of petrographic classification from geophysical logs. *Pure and Applied Geophysics*, 2024;181:2839–2871.

[72] Mouafik M, Chakhchar A, Fouad M, *et al.* Remote sensing technologies for monitoring Argane forest stands: a comprehensive review. *Geographies*, 2024;4(3):441–461. https://www.mdpi.com/2673-7086/4/3/24.

[73] Radwan M, Alhussan AA, Ibrahim A, *et al.* Potato leaf disease classification using optimized machine learning models and feature selection techniques. *Potato Research*, 2024;67:63.

[74] Lakshmi DGV. A review on K nearest neighbour classification technique in machine learning. *International Journal of Scientific Research in Science, Engineering and Technology* 2025;12(1):257–260.

[75] Rana VK, Linh NTT, Ditthakit P, *et al.* Mapping and analysing framework for extreme precipitation-induced flooding. *Earth Science Informatics*, 2023(16):4213–4234.

[76] Umar R, Riadi I, and Faroek DA. A komparasi image matching menggunakan metode K-nearest neightbor (KNN) dan support vector machine (SVM). *Journal of Applied Informatics and Computing*, 2020;4(2):124–131.

[77] Kouadio KL, Liu J, Liu R, *et al.* K-means featurizer: a booster for intricate datasets. *Earth Science Informatics* 2024;17:1203–1228.

[78] Liu C. Research on image classification leveraging deep convolutional neural networks and visual cognition. *Applied and Computational Engineering*, 2024;32:200–209.

[79] Kolobaev M, Prytyka D, Mikhailova D, *et al.* Modern architectures of convolutional neural networks used for image classification tasks. *Soft Measurements and Computing*, 2023 01;8:69–78.

[80] Heenaye-Mamode Khan, M., Reesaul, P., Auzine, M.M. *et al.* Detection of Alzheimer's disease using pre-trained deep learning models through transfer learning: a review. *Artificial Intelligence Review*, 2024;57:275.

[81] Du X, Sun Y, Song Y, *et al.* A comparative study of different CNN models and transfer learning effect for underwater object classification in side-scan SONAR images. *Remote Sensing*, 2023;15(3):1–13. https://www.mdpi.com/2072-4292/15/3/593.

[82] Ahmed FA. Deep learning for surgical instrument recognition and segmentation in robotic-assisted surgeries: a systematic review. *Artificial Intelligence Review* 2025;58(1):1–31.

[83] Wang Z, Guo J, Huang W, *et al.* Side-scan SONAR image segmentation based on multi-channel fusion convolution neural networks. *IEEE Sensors Journal*, 2022;22(6):5911–5928.

[84] Arjun PA, Suryanarayan S, Viswamanav RS, *et al.* Unveiling underwater structures: MobileNet vs. efficientNet in SONAR image detection. *Procedia Computer Science*, 2024;233:518–527.

[85] Atiya SU, Ramesh NVK, and Reddy BNK. Classification of non-small cell lung cancers using deep convolutional neural networks. *Multimedia Tools and Applications*, 2024;83:13261–13290.

[86] Alhumaid M and Fayoumi AG. Transfer learning-based classification of maxillary sinus using generative adversarial networks. *Applied Sciences*, 2024;14(7):1–28. https://www.mdpi.com/2076-3417/14/7/3083.

[87] Liu P, Qian W, Zhang H, *et al.* Automatic sleep stage classification using deep learning: signals, data representation, and neural networks. *Artificial Intelligence Review*, 2024;57(301):1–68.

[88] Wang K, Fu Y, Duan X, *et al.* Abnormal traffic detection system in SDN based on deep learning hybrid models. *Computer Communications*, 2024;216:183–194. https://www.sciencedirect.com/science/article/pii/S0140366423004484X.

[89] Islam MS, Kabir MN, Ghani NA, *et al.* Challenges and future in deep learning for sentiment analysis: a comprehensive review and a proposed novel hybrid approach. *Artificial Intelligence Review*, 2024;57(62):1–79. https://doi.org/10.1007/s10462-023-10651-9.

[90] Bahmei B, Birmingham E, and Arzanpour S. CNN-RNN and data augmentation using deep convolutional generative adversarial network for environmental sound classification. *IEEE Signal Processing Letters*, 2022;29: 682–686.

[91] Bansal MA, Sharma DR, and Kathuria DM. A systematic review on data scarcity problem in deep learning: solution and applications. *ACM Computing Surveys*, 2022;54(10s):1–29. https://doi.org/10.1145/3502287.

[92] Hanassab S, Abbara A, Yeung AC, *et al.* The prospect of artificial intelligence to personalize assisted reproductive technology. *NPJ Digital Medicine*, 2024;7(55):1–19.

[93] Redhu A, Choudhary P, Srinivasan K, *et al.* Deep learning-powered malware detection in cyberspace: a contemporary review. *Frontiers in Physics*, 2024;12:1–29. https://www.frontiersin.org/journals/physics/articles/10.3389/fphy.2024.1349463.

[94] Dai Z, Liang H, and Duan T. Small-sample SONAR image classification based on deep learning. *Journal of Marine Science and Engineering*, 2022;10(12):1–22. https://www.mdpi.com/2077-1312/10/12/1820.

[95] Wei X, Feng T, Huang Q, *et al.* Deep learning-powered biomedical photoacoustic imaging. *Neurocomputing*, 2024;573:127207. https://www.sciencedirect.com/science/article/pii/S0925231223013309.

[96] Lei J, Wang H, Lei Z, *et al.* CNN-transformer hybrid architecture for underwater SONAR image segmentation. *Remote Sensing*, 2025;17(4):1–32. https://www.mdpi.com/2072-4292/17/4/707.

[97] Li D, Qu D, Li X, *et al.* Lightweight global adaptive feature enhancement network for underwater object detection with SONAR image. *Journal of Physics: Conference Series*, 2024;2914(1):012023. https://dx.doi.org/10.1088/1742-6596/2914/1/012023.

[98] Mittal P. A comprehensive survey of deep learning-based lightweight object detection models for edge devices. *Artificial Intelligence Review*, 2024;57(242):1–49.

[99] Epifani L and Caruso A. A survey on deep learning in UAV imagery for precision agriculture and wild flora monitoring: datasets, models and challenges. *Smart Agricultural Technology*, 2024;9:100625. https://www.sciencedirect.com/science/article/pii/S2772375524002302.

[100] Jasmine JJ, Raja SE, Muniraj R, *et al.* An improved dung beetle optimization with recurrent convolutional neural networks for efficient detection and classification of undersea water object images. *Earth Science Informatics* 2024;17:3651–3671.

[101] Yadav A and Kumar E. Object detection on real-time video with FPN and modified mask RCNN based on inception-ResNetV2. *Wireless Personal Communications* 2024;138:2065–2090.

[102] Liu M. Light-weight semantic segmentation based on mask RCNN. *Transactions on Computer Science and Intelligent Systems Research*, 2024;5: 385–389.

[103] Fontana G, Calabrese M, Agnusdei L, *et al.* SolDef_AI: an open source PCB dataset for Mask R-CNN defect detection in soldering processes of electronic components. *Journal of Manufacturing and Materials Processing*, 2024;8(3):1–26. https://www.mdpi.com/2504-4494/8/3/117.

[104] Islam MS. Deep learning-based SONAR image object detection system. *International Journal of Information Engineering and Applications*, 2024;6(2):186–199.

[105] Li Q, Wang W, Xie S, *et al.* Oversized ore detection using rotated CCAD-YOLOv3 and ellipse fitting. *Expert Systems with Applications*, 2024;255:124545. https://www.sciencedirect.com/science/article/pii/S095741742401412X.

[106] Pal SK, Biswas S, and Dutta D. Granulated mask RCNN and eye detection index (EDI) for detection and localization of eye of tropical cyclone from satellite imagery. *Journal of Data, Information and Management*, 2024;6:255–275.

[107] Zhang W, Zhang X, Zhang Y, *et al.* Multi-Beam SONAR target segmentation algorithm based on BS-Unet. *Electronics*, 2024;13(14):1–15. https://www.mdpi.com/2079-9292/13/14/2841.

[108] Mao Y. Research and applications of autonomous underwater vehicles. *Science and Technology of Engineering, Chemistry and Environmental Protection*, 2024;1(5):1–7.

[109] Aguzzi J, Thomsen L, Flögel S, *et al.* New technologies for monitoring and upscaling marine ecosystem restoration in deep-sea environments. *Engineering*, 2024;34:195–211. https://www.sciencedirect.com/science/article/pii/S2095809924000286.

[110] Zhang L, Guan L, Zeng J, *et al.* Autonomous underwater vehicle navigation enhancement by optimized side-scan SONAR registration and improved post-processing model based on factor graph optimization. *Journal of Marine Science and Engineering*, 2024;12(10):1–23. https://www.mdpi.com/2077-1312/12/10/1769.

[111] Wang Q, Zhang Y, and He B. Intelligent marine survey: lightweight multi-scale attention adaptive segmentation framework for underwater target detection of AUV. *IEEE Transactions on Automation Science and Engineering*, 2025;22:1913–1927.

[112] González-Sabbagh SP, Robles-Kelly A. A Survey on Underwater Computer Vision. ACM Comput Surv. 2023;55(13s):1–39. https://doi.org/10.1145/3578516.

[113] Ma Q, Jiang L, Yu W. Lambertian-based adversarial attacks on deep-learning-based underwater side-scan SONAR image classification. *Pattern Recognition*, 2023;138:109363. https://www.sciencedirect.com/science/article/pii/S003132032300064X.

[114] Yang Y, Liang W, Zhou D, *et al.* Object detection for underwater cultural artifacts based on deep aggregation network with deformation convolution. *Journal of Marine Science and Engineering*, 2023;11(12):1–17. https://www.mdpi.com/2077-1312/11/12/2228.

[115] Bresciani M, Zacchini L, Topini A, *et al.* Automatic target recognition and geolocalisation of natural gas seeps using an autonomous underwater vehicle. *Control Engineering Practice*, 2024;145:105864. https://www.sciencedirect.com/science/article/pii/S0967066124000248.

[116] Williams GN, Lagace GE, and Woodfin A. A collision avoidance controller for autonomous underwater vehicles. In: *Symposium on Autonomous Underwater Vehicle Technology*; 1990. pp. 206–212.

[117] Ma C, Zou H, and An X. A complete coverage path planning approach for an autonomous underwater helicopter in unknown environment based on VFH+ algorithm. *Journal of Marine Science and Engineering*, 2024;12(3):1–18. https://www.mdpi.com/2077-1312/12/3/412.

[118] Mane P, George AJ, Makam R, *et al.* EROAS: 3D efficient reactive obstacle avoidance system for autonomous underwater vehicles using 2.5D forward-looking SONAR; 2024. https://arxiv.org/abs/2411.05516.

[119] Wang Z, Wang H, Yuan J, *et al.* Bio-inspired cooperative control scheme of obstacle avoidance for UUV swarm. *Journal of Marine Science and Engineering*, 2024;12(3):1–19. https://www.mdpi.com/2077-1312/12/3/489.

[120] Li C, Chen K, Li H, *et al.* Multisensor data fusion approach for sediment assessment of sewers in operation. *Engineering Applications of Artificial Intelligence*, 2024;132:107965. https://www.sciencedirect.com/science/article/pii/S0952197624001234.

[121] Ali ES, Saeed RA, Eltahir IK, *et al.* A systematic review on energy efficiency in the internet of underwater things (IoUT): Recent approaches and research gaps. *Journal of Network and Computer Applications*, 2023;213:103594. https://www.sciencedirect.com/science/article/pii/S1084804523000139.

[122] Shi Y, Yang Q, Huang X, *et al.* An SDN-enabled framework for a load-balanced and QoS-aware internet of underwater things. *IEEE Internet of Things Journal*, 2023;10(9):7824–7834.

[123] Mao B, Yang H, Li W, *et al.* Multi-frequency noise reduction method for underwater radiated noise of autonomous underwater vehicles. *Journal of Marine Science and Engineering*, 2024;12(5):1–14. https://www.mdpi.com/2077-1312/12/5/705.

[124] Aromiwura AA, Settle T, Umer M, *et al.* Artificial intelligence in cardiac computed tomography. *Progress in Cardiovascular Diseases*, 2023;81: 54–77.

[125] Xie K, Yang J, and Qiu K. A dataset with multibeam forward-looking SONAR for underwater object detection. *Scientific Data*, 2022;9(739):1–8.

[126] Yow AP, Wong D, Zhang Y, *et al.* Artificial intelligence in optical lens design. *Artificial Intelligence Review* 2024;57(193):1–21.

[127] Lee S, Park B, Kim A. Deep learning from shallow dives: SONAR image generation and training for underwater object detection; 2018. https://arxiv.org/abs/1810.07990.

[128] Sen Gupta A, Kirsteins IP, Christensen AJ, *et al.* Discovery and learning of feature geometry in underwater acoustic sensing: case studies and challenges in active SONAR applications. *The Journal of the Acoustical Society of America*, 2024;155(3):A47–A47.

[129] Yang D, Wang C, Cheng C, *et al.* Data generation with GAN networks for sidescan SONAR in semantic segmentation applications. *Journal of Marine Science and Engineering*, 2023;11(9):1–16. https://www.mdpi.com/2077-1312/11/9/1792.

[130] Islam H, and Guedes Soares C. Assessment of uncertainty in the CFD simulation of the wave-induced loads on a vertical cylinder. *Marine Structures*, 2021;80:103088. https://www.sciencedirect.com/science/article/pii/S0951833921001416.

[131] Wang H, Wang W, Xiao S, *et al.* Improving artificial Bee colony algorithm using a new neighborhood selection mechanism. *Information Sciences*, 2020;527:227–240. https://www.sciencedirect.com/science/article/pii/S0020025520302449.

[132] Kaur M and Vijay S. Underwater images quality improvement techniques for feature extraction based on comparative analysis for species classification. *Multimedia Tools and Applications*, 2022;81:19445–19461.

[133] Sampaio AZ. Project management in office: BIM implementation. *Procedia Computer Science*, 2022;196:840–847. International Conference on ENTERprise Information Systems/ProjMAN – International Conference on Project MANagement/HCist – International Conference on Health and Social Care Information Systems and Technologies 2021. https://www.sciencedirect.com/science/article/pii/S1877050921023061.

[134] Zhang C, Hallbeck MS, Salehinejad H, *et al.* The integration of artificial intelligence in robotic surgery: a narrative review. *Surgery*, 2024;176(3):552–557. https://www.sciencedirect.com/science/article/pii/S0039606024000680.

[135] Neupane D and Seok J. A review on deep learning-based approaches for automatic SONAR target recognition. *Electronics*, 2020;9(11):1–30. https://www.mdpi.com/2079-9292/9/11/1972.

Chapter 23
Real-time modeling and verification of SONAR systems

Hajrah Ibrahim Virk[1], *Hamza Ahmad*[1], *Sidra Sultana*[1] *and Nazia Perwaiz*[1]

This chapter presents a modeling and verification scenario activity focused on simplified sound navigation and ranging (SONAR) based subsystems, developed, and verified using the UPPAAL model checker. The primary objective is to explore timing behavior, synchronization mechanisms, and verify correctness properties. Each of the three small-scale models that have been developed represents an abstraction of the underwater environment based on acoustic communication with different levels of functionality and interactivity. These models aim to capture crucial operational aspects such as signal strength, signal reception, signal time emission, reflection, detection, and time synchronization to preserve the reliability of the system. Moreover, formal methods are applied to check the correctness of the defined temporal properties, which include safety and deadlock-freeness.

23.1 Introduction

SONAR systems are powerful technologies for emitting acoustic signals into water to detect, identify, and track objects by analyzing the echoes that are returned. SONAR is one of the most effective methods used for underwater sensing and communication since sound waves travel easily, compared to electromagnetic waves, which get absorbed in water. SONAR systems have numerous applications, including marine biology, where they are used to track sea life, oceanography for seabed mapping, and offshore industries to locate underwater pipelines and geological formations. In the defense sector, SONAR systems are essential for detecting submarines, identifying underwater mines, and providing navigation assistance in complicated underwater landscapes. These systems are also mounted on autonomous underwater vehicles (AUVs) which allows the vehicles to autonomously navigate and physically avoid obstacles.

[1]School of Electrical Engineering and Computer Science (SEECS), National University of Sciences and Technology (NUST), Pakistan

One of the main challenges with SONAR systems is the modeling and handling of time-critical interactions between the emitters/transmitter nodes, receiver nodes, and target nodes. The problems are worse with multipath propagation, Doppler shifts, and the presence of surrounding noise, all of which can alter and lag behind the reception of signals. SONAR systems have an operational range in the frequencies between some kilohertz (kHz) and several hundred kHz; lower frequencies are used for long-distance detection, while high frequencies are for high-resolution imaging. These systems also form the basis for communication networks that operate underwater, where reliable data transfer between various sensor nodes is of the utmost importance. To model such networks, formal tools like UPPAAL can be employed to ensure strict synchronization and correctness of system reliability with respect to real-time constraints.

This chapter describes three communication and detection scenario models verified in UPPAAL. The first simulates a passive signal reception system which listens for acoustic signals but does not send any out. This model is an illustration of silent receivers in many sensor-based communication systems that are used to secure underwater networks. The second model describes a synchronized full-duplex exchange between two sensor nodes, utilizing a token-based protocol that guarantees one-way transmission and prevents signal interference or overlapping. However, the third focuses on time-based object detection, where the delay between a signal's emission and its echo is used to classify the proximity of an external entity which reflected the signal as either dangerous or safe. Each model is built using automated time and emphasized timing constraints and state transitions relevant to real-world scenarios. Through model checking, key properties such as deadlock freedom and safety are verified, ensuring that each system behaves predictably under defined timing thresholds. These verifications strengthen the reliability of such time-critical communication and detection mechanisms.

23.2 Related work

Recent work on real-time underwater systems, including SONAR applications, has focused mainly on simulation-based testing and signal processing, with relatively limited attention to formal verification of system-level timing properties. Cerqueira *et al.* [1] developed a graphics processing unit (GPU)-accelerated SONAR simulator combining OpenSceneGraph, Gazebo, and the robot construction kit (Rock) framework for real-time rendering of mechanical scanning and forward-looking SONAR. While valuable for visualizing SONAR behavior and prototyping, this framework lacks formal guarantees on system correctness across all possible execution paths. Formal verification approaches address this limitation by exhaustively analyzing system behaviors. Kumar *et al.* [2] applied formal verification techniques to underwater sensor networks by modeling the time-delayed allocation medium access control (TDA-MAC) protocol using process meta language (PROMELA) and a simple PROMELA interpreter (SPIN). Their work revealed concurrency and reachability issues that are overlooked by conventional simulations. However, SPIN is limited to discrete event models and lacks native support for continuous-time semantics, which are essential for real-time SONAR systems. Addressing this gap, Larsen *et al.* [3] introduced UPPAAL, a model checker for real-time systems based on timed automata. The implementation of difference bound matrices (DBMs) and state-space reduction methods in UPPAAL allows for the efficient evaluation of dense real-time systems, thus supporting its use in applications with rigid timing constraints.

Quijano *et al.* [4] presented the AUV UPPAAL and verified the safety properties of AUV using timed computation tree logic (TCTL), showcasing the practical value of UPPAAL. Their work validated that fundamental mission behaviors can be encoded using timed automata, avoiding the need for complicated hybrid models. Complementing these efforts, Wang *et al.* [5] proposed signal processing methods to mitigate reverberation and interference from active SONAR systems through beam space data normalization. Although not involving formal verification, this highlights the operational challenges and stringent timing demands faced in real-world SONAR systems. Ferrère *et al.* [6] extended formal verification capabilities by translating metric interval temporal logic (MITL) into timed automata via temporal testers, enabling formal specification of precise timing requirements. Using UPPAAL, these automata can be verified, allowing for in-depth examination of time-sensitive attributes such as guaranteeing that SONAR pings get valid echoes within preset time limits. Although tools like Gazebo and SPIN excel in specific domains, they fail to handle the dense real-time constraints and clock interactions inherent to SONAR systems. UPPAAL offers a comprehensive framework for modeling and verifying time-critical underwater systems, combining expressive timed automata, efficient verification algorithms, and demonstrated success in underwater applications. This makes UPPAAL the preferred choice for our work in real-time modeling and formal verification of SONAR systems.

23.3 System model

In this section, we define the basic components that form our model scenarios. These components increase in complexity, from basic communication elements to advanced systems that mimic real-world SONAR systems. Before going into detail on each component of our model, we first explain why we chose UPPAAL and what its key features are.

23.3.1 Why UPPAAL?

For this study, the tool choice has been UPPAAL due to its unique support for describing embedded systems in real-time using timed automata. Its integration into SONAR systems is where its real-valued clocks become useful, as they enable the simulation of "bounded" time transitions, which are required for precise timing reproduction of the physical signals. UPPAAL's support for real-valued clocks makes it possible to simulate time-bounded transitions, which is crucial for reproducing the timing physical realities of sent and received signals. These capabilities enable designers to ensure that emissions, echoes, and responses from SONAR components function within the tight constraints defined by the temporal boundaries of the system. Such precision is needed because performance and safety in SONAR systems are critical to time. This renders dependable timing essential for analyzing the worst-case scenario delays and deadlines.

The implementation of synchronization channels such as *ping*! and *echo*? is the other most important feature of UPPAAL. It enables close coordination and communication among the constituents. This capability is very helpful for modeling the behavior of SONAR system emitters and receivers involving controlled propagation and reflection of acoustic signals. The provided channels allow for explicit simulation of events such as signal sending and receiving as well as acknowledgment, so the behavior of each constituent is driven by events. In addition, UPPAAL allows the configuration of urgent and committed transitions, which prevent time from advancing during certain tasks, capturing precise or almost precise electronic signal events. Accomplishing precision is hard with general-purpose simulation tools because they lack adequate frameworks for real-time synchronization and discrete event control.

The formal critical system properties that UPPAAL can check are already verified beyond simulation with model evaluation of echo timings and dead systems. Liveness guarantees paired with safety are also enforced and include things like no-death states. Prompt detection and reporting of states is a guarantee. Since UPPAAL supports systems with all state-dependency exploration, it is certain to prove escalation possibilities like deadlocks, race conditions, or unbounded delays become impossible for a defined system model. This allows edge case behavior validation ensuring the dependability of SONAR systems that are not feasible through brute-force simulation testing. Unlike informal evaluation or nonspecific function testing with simulators, reliance on UPPAAL ensures mandate checking accuracy, zero redundancy, responsive behavior, and system-defect checks. It is unparalleled when precision SONAR systems and time-critical redundancy-level safety measures are integrated.

23.3.2 *Key UPPAAL modeling concepts*

In UPPAAL, timed automata are the primary modeling framework which allows system components such as the emitter, hydrophone, and reflecting object to be modeled as entities represented by state machines. These automata are composed of discrete locations (or states) and actions (or events) that advance the automata, referred to as transitions, determined by a combination of events and the passage of time. Each component has its own behavior, is modeled independently, and interacts with other components using synchronization channels for communication. The time in the system can be manipulated with local clock variables, such as t, which keep track of the time spent in a state, allowing for controlled conditions. Controlled transitions can be set with guards like $t \geq 5$ to send signals. Invariants such as $t \leq 10$ can restrict how long a system can remain in a state. The combination of clocks, guards, and invariants in UPPAAL allows us to simulate real-time embedded systems and monitor their behavior under different constraints. In UPPAAL, the synchronization is done with the help of synchronization channels.

Consider a channel named "ping" to simulate a real ping in a SONAR system. So, a *ping*? action in one automaton can only occur when another *ping*! action occurs in another automaton, which ensures synchronization between different automata or within an automaton. These channels can simulate the SONAR signals that are sent by the transmitter nodes and received by the receiver nodes. Guards and invariants are also important modeling constructs in UPPAAL. Guards are conditions that must be satisfied for an automaton to transition from one state to another. For example, if $t \geq 2$ is a guard on a transition where t is a clock variable, it means that the transition can only occur if the clock value at that instant is greater than 2. Invariants, on the other hand, are used to restrict the time it takes for an automaton to stay in a particular state. Thus, an invariant $t \leq 10$ in a particular state where t is a clock variable ensures that the system leaves that state before t exceeds 10-time units. Using guards, invariants, and channels, we can ensure that a model behaves as expected in terms of timing, synchronization, and order of events.

23.3.3 Transmitter in SONAR system

Just like terrestrial networks, underwater networks also have transmitters and receivers that emit and receive sound signals, respectively. Active transmitters are present in underwater SONAR systems. Active transmitters emit sound waves that reflect back as echoes by bouncing off underwater objects. In UPPAAL, the transmitter is a simple automaton that periodically simulates a *ping* using the *ping*! synchronization channel. For a receiver to receive a signal, we need an automaton to generate that signal, and this component achieves that goal. Whenever a ping is simulated by the transmitter, it can be received by the receiver, and the timing and repetition of this cycle can be analyzed using UPPAAL.

23.3.4 Receiver in SONAR system

The receiver model represents a passive SONAR system. Passive SONAR detects and analyzes sounds emitted by a transmitter without sending any signals of its own. It is stealthier than active SONAR but typically provides only directional information about the detected source. The receiver model is designed to continuously monitor for pings emitted by the transmitting component. By modeling the emission and reception of discrete communication signals using UPPAAL channels and the *ping*! and *ping*? actions, it demonstrates basic synchronization processes. This simple setup acts as a reference point for assessing how accurate real-time reception behaviors are. It also makes it possible to formally verify important system characteristics under minimal operating assumptions, such as reachability and deadlock freedom. Before adding more complexity in later system improvements, the model offers a controlled setting to lay the foundation for guarantees.

23.3.5 Half-duplex SONAR communication

This model represents a basic active SONAR sequence. Active SONAR transmits sound pulses into water and listens for returned echoes to detect, locate, and classify objects. It tells how far away and in what direction the objects are, but it can also give away the transmitter's location. In the half-duplex SONAR simulation, a transmitter emits a ping; this signal travels through some medium, typically water. A target (which may move due to ocean currents from its current position) reflects it after a delay, and a receiver listens to the echo (the reflected signal). This setup verifies if the echo is received within a specified time window (e.g., within 60 time units); it models the core behavior of range-limited detection systems. This configuration is critical to understanding the operational constraints of active SONAR systems, especially in scenarios that involve object detection, localization, and distance estimation in underwater environments.

23.3.6 Full-duplex SONAR communication

In this model, we consider two underwater nodes that can transmit and receive signals simultaneously. Each node is equipped with its own transmitter and receiver (which shows full duplex communications). They exchange pings bidirectionally, simulating underwater communication where both sides initiate and respond to signals. One node sends out a signal that travels through the medium of transmission and is received by the other node, and vice versa. This configuration introduces mutual synchronization, timing dependencies, propagation delays, and more intricate state interactions that have to be accounted for at both ends. This scenario models a simplified but meaningful version of underwater communication. Such simulations, which we will discuss in the next section, are essential to evaluate latency, signal integrity, and protocol robustness in environments where traditional electromagnetic communication is impractical.

23.3.7 Object detection through SONAR systems

This model simulates a real-time object detection scenario using a SONAR system. To detect a potential surrounding object, the SONAR node emits an acoustic ping and listens for an echo. The round-trip time of the signal determines the distance of the object: the sooner the echo is received, the closer the object is assumed to be. If the echo returns within a predefined threshold (for example, 60 time units), the object is classified as dangerously close, triggering an immediate alert. If the echo arrives after the threshold, the object is considered safe distance away, and the system returns to its idle monitoring state. During this interaction, a separate object model simulates the environment by receiving the ping and sending back an echo. This abstraction does not rely on explicit spatial coordinates, but infers distance through timing alone. The model showcases a time-critical detection protocol and uses formal verification to ensure that safety alerts are only triggered under correct conditions.

23.4 Design and objectives of system model

The primary objective of all models is to accurately reproduce signal timing, with a special emphasis on the intervals at which SONAR components send, receive, or react to signals. The key focus of these models is to simulate the timing constraints that are present in a real-world system. For example, consider that a model sends out a ping and waits for an echo to be received. If this echo is not received within a specific time, it might imply that either the ping was lost or it was not received by the model even after being reflected. Thus, it is important for these models to verify these timing constraints to detect anomalies and ensure system safety. Together, these measures provide strong assurance that the system behaves as intended in all relevant scenarios. Modeling constructs such as synchronization channels, guards, and invariants in UPPAAL help to simulate how interactions take place in a real-time system. These constructs are used to make sure that communication between elements is synchronized and the system transitions to expected states under different timing constraints. With the help of the UPPAAL model verifier, the expected behavior of the models can also be verified. Conditions such as deadlock-freeness, liveness, safety, mutual exclusion, correctness, and system-specific constraints can all be verified with the help of this verifier. If a certain condition is not met, it means that the model needs some modifications to work correctly.

23.4.1 Model structure overview

Each SONAR model is built from templates that represent logical roles in the system. There are three main components, the emitter/transmitter automaton, the environment/target automaton, and the receiver/hydrophone automaton. The transmitter automaton periodically sends a signal (e.g. *ping*!) when a clock threshold (e.g. 10 time units) is reached. The target automaton waits for the signal (e.g. *ping*?), introduces a delay, and emits a response (e.g., *echo*!). However, the receiver automaton waits in a listening state to receive the response (*echo*?) and, depending on timing constraints, either continues operation or triggers a fatal alert. All components synchronize through channels and use local clocks to simulate deterministic time-driven behavior consistent with real-world SONAR systems. The final object detection model also introduces state-based classification logic (safe versus fatal), demonstrating the power of UPPAAL to handle hybrid logically timed interactions.

23.4.2 First scenario: watch silently – detecting SONAR pings

In this scenario, we define a model consisting of a hydrophone that passively monitors acoustic activity. The main focus of this model is to capture the repetition and timing of the signals that are received by the hydrophone. This is an abstract model, but it can provide a basis for real-world acoustic systems. We consider that this hydrophone operates in an underwater environment and passively listens to any nearby acoustic signals without revealing its position. This is called passive SONAR because the hydrophone is not actively emitting SONAR pings but is listening for them. This model does not simulate mobility or any decision-making, but demonstrates how a passive SONAR system behaves.

23.4.2.1 Model architecture and design overview

As mentioned above, this scenario models a passive SONAR system that receives signals that might have been emitted by any external source. Figure 23.1 shows the

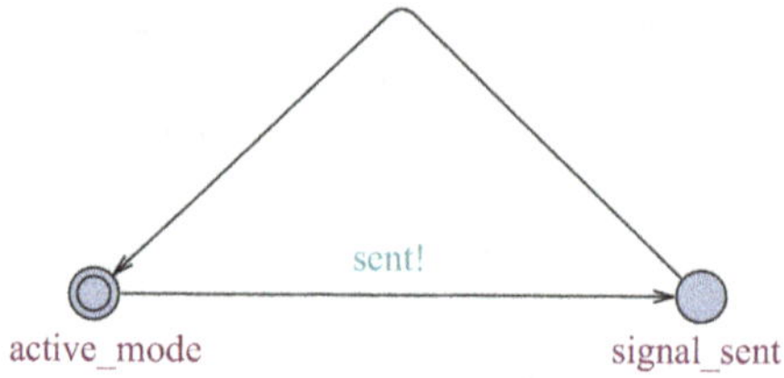

Figure 23.1 The emitter model

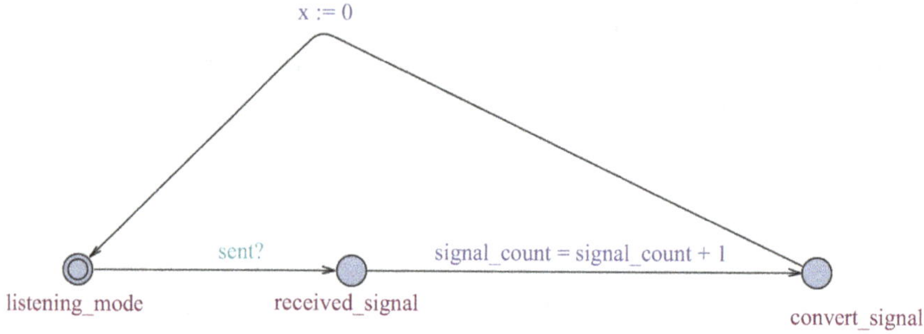

Figure 23.2 The hydrophone model

emitter automaton, which has two states: *active_modeandsignal_sent*. This emitter is capable of generating SONAR pings periodically. It starts in its initial state *active_mode* and simulates a *ping* through the *sent*! synchronization channel and transitions to the *signal_sent* state. After this, it transitions back to the *active_mode* state and continues to repeat the same cycle of generating pings.

Figure 23.2 shows the hydrophone automaton, which passively listens to any acoustic signals. It starts in *listening_modestate*, and waits for the emitter automaton to trigger the *sent*? synchronization channel to simulate a *ping*. When this happens, the hydrophone transitions to the *received_signal* state. After that, the hydrophone transitions to the state *convert_signal*, and during this transition, it increments the variable *signal_count* to keep track of all the signals that have been received to identify how frequent or repetitive these signals are. Finally, the hydrophone transitions to *listening_mode* and sets the clock value to 0 to mark the arrival time of the next signal and continues in *listening_mode*.

This whole simulation is illustrated through the UPPAAL symbolic simulator, which provides us with a trace shown in Figure 23.3 and the full model in Figure 23.4. The trace shows one cycle of successful signal generation and reception: the emitter triggers the *sent*! synchronization channel to simulate a *ping*. The hydrophone receives and processes it and listens for the next signal. To ensure that the model behaves as expected, several verification properties were defined and evaluated using the UPPAAL model checker. These properties were designed to confirm fundamental aspects of the system's operation, including its ability to avoid deadlock, ensure the reception of signals, and successfully transition through critical processing stages.

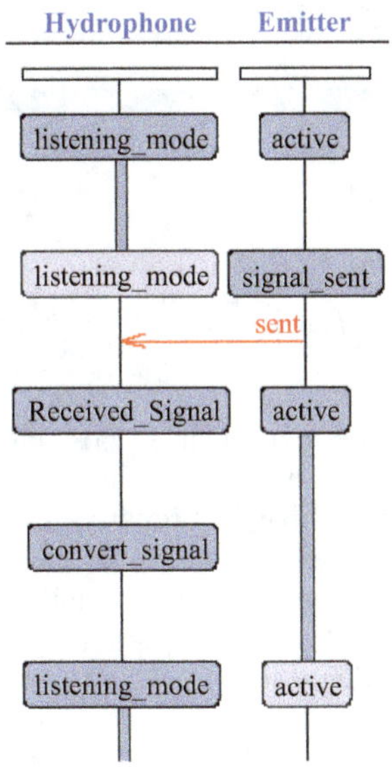

Figure 23.3 The simulation trace

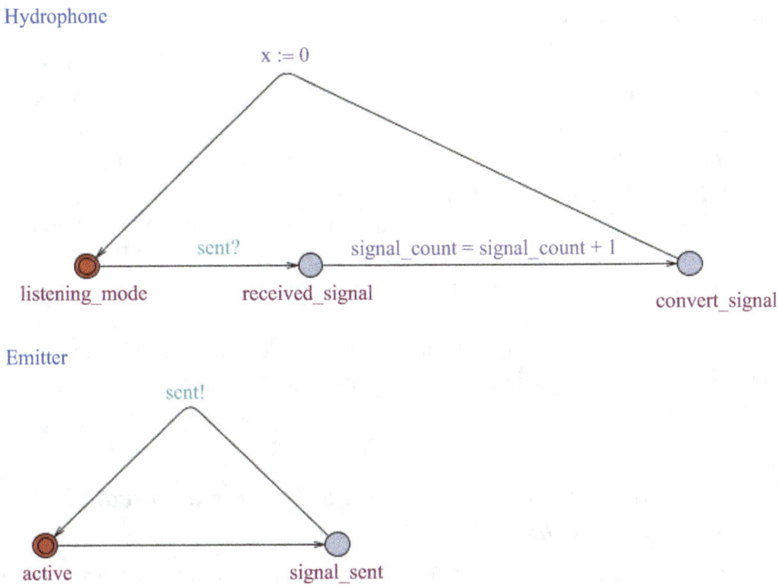

Figure 23.4 The model structure

Together, these checks establish confidence in the correctness and reliability of the hydrophone's behavior under all modeled conditions.

The first verification property (23.1) ensures that the system is deadlock-free, so that at no point does the system enter a combination of states where progress

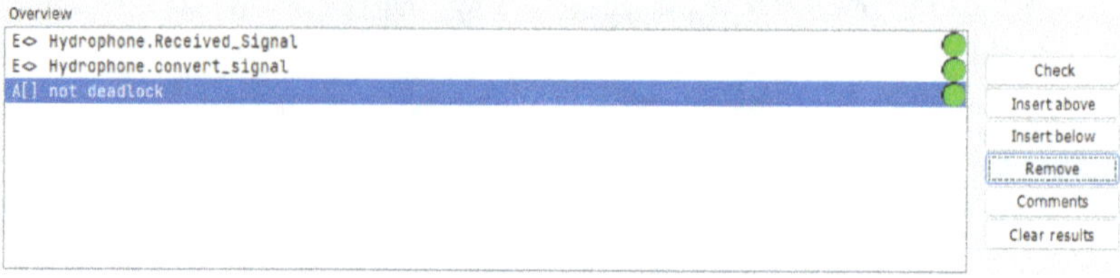

Figure 23.5 Verification properties of the first scenario

is not possible. This makes sure that under all possible paths, the hydrophone and the emitter automaton can continue communicating indefinitely without any states where no further transition is possible. In property (23.2), the liveness of the model is validated. This property ensures that whenever the emitter automaton simulates a *ping* using the *sent*! channel, that *ping* is eventually received by the hydrophone automaton. The third property (23.3) confirms that whenever the system receives a signal, it reaches the *convert_signal* state to ensure that *received_signal* is processed to complete the communication cycle.

$$\texttt{A[] not deadlock} \tag{23.1}$$

$$\texttt{E<> Hydrophone.Received\_Signal} \tag{23.2}$$

$$\texttt{E<> Hydrophone.convert\_signal} \tag{23.3}$$

All of the properties defined above help ensure that the model defined to simulate this scenario behaves correctly in all possible states. In this model, the most important aspects that need to be checked are deadlock-freeness, reception of a generated signal, and conversion of that received signal. All properties defined for this model were verified using the UPPAAL verifier, as indicated by the green circle against each property in Figure 23.5. With these verification properties, the model defined in this scenario provides a basic abstraction of real-world passive SONAR systems. This scenario shows how a SONAR system might work in a situation where it is important to conceal its presence, listen for signals in its surroundings, and analyze them to extract critical information. It also demonstrates how formal methods can be used to analyze the functioning of different components in underwater acoustic systems and ensure their correctness.

23.4.3 Second scenario: full-duplex acoustic communication between stealth units

In this scenario, a full-duplex acoustic communication system has been modeled between two underwater nodes. A token-based protocol is used for these nodes to communicate, ensuring that the nodes can only send signals to each other when they hold the token. This ensures that only one node sends the signal, while the other node listens for it. Through this protocol, any overlapping transmission is avoided and turn-based communication is achieved. This scenario reflects how underwater acoustic systems communicate by exchanging low-energy SONAR pings to interact reliably.

23.4.3.1 Model architecture and design overview

In this model, two identical hydrophone automatons communicate with each other with the help of a token generator that releases the first token to initialize communication. The hydrophone automaton is shown in Figure 23.6. Both hydrophone automatons have their *active_mode* state as their initial state and wait for a *token*? signal or a *sent*? signal. Once the hydrophone automaton receives a *token*? signal, it passes to the *generate_signal* state and simulates a SONAR *ping* by triggering the *sent*! synchronization channel. Then, it passes through the *transmitted_signal* state and releases the token by triggering the *token*! synchronization channel, allowing the other hydrophone to generate its signal. This ensures that only one hydrophone can send a signal at any given instant. The other hydrophone that does not possess the token at first acts as a listener and waits for the *sent*? synchronization channel to be invoked, and when it does, it transitions to the *receive_signal* state and then finally transitions to *convert_signal* state, and goes back to *active_mode* state and waits for the token to send its own signal.

The token generator automaton, shown in Figure 23.7, supports communication between the two hydrophones. Only communication between two hydrophones is initialized by releasing the token for the first time when the simulation begins. This token is passed to only one of the hydrophones and thus only that hydrophone is capable of sending the signal, while the other hydrophone can only listen. This enforces mutual exclusion, which allows for this back-and-forth communication to take place. If both hydrophones send signals at the same time or listen for them at the same time, this will cause a deadlock, which must be avoided. So, the token generator automaton, although simple, allows one-time token generation to initialize the communication. Figure 23.8 shows the output trace generated when simulations are run for this model

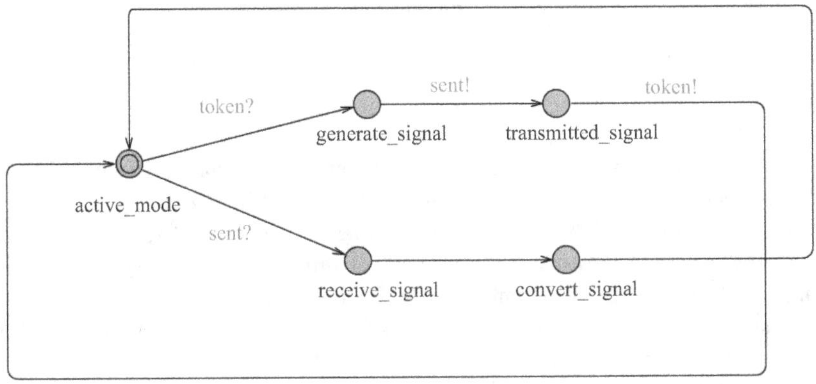

Figure 23.6 The hydrophone template

Figure 23.7 Token generator

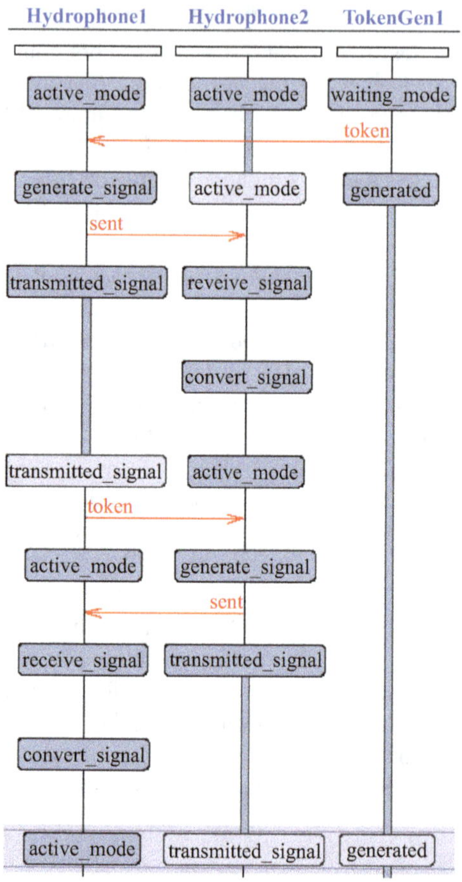

Figure 23.8 The communication trace of the hydrophone

in UPPAAL. First, *TokenGen*1 releases the token to *Hydrophone*1, which has the ability to send the signal while *Hydrophone*2 listens for it. *Hydrophone*1 simulates a *ping* through the *sent*! synchronization channel, *Hydrophone*2 receives and converts it, and then *Hydrophone*2 receives the token through the *token*? synchronization channel and initiates its own transmission. In this way, there is no overlap in the transmission, and the system remains deadlock-free. The complete model with both hydrophones and the token generator is shown in Figure 23.9.

The token-based protocol enables us to use the same model capable of transmitting and receiving pings for both hydrophones, while ensuring that no simultaneous transmissions take place. With UPPAAL, we can verify the correct behavior of our model by proving properties such as deadlock-freeness, communication liveness, and mutual exclusivity of sender/receiver actions. These properties have been verified by the UPPAAL verifier for this model as shown in Figure 23.10, highlighted by green signals.

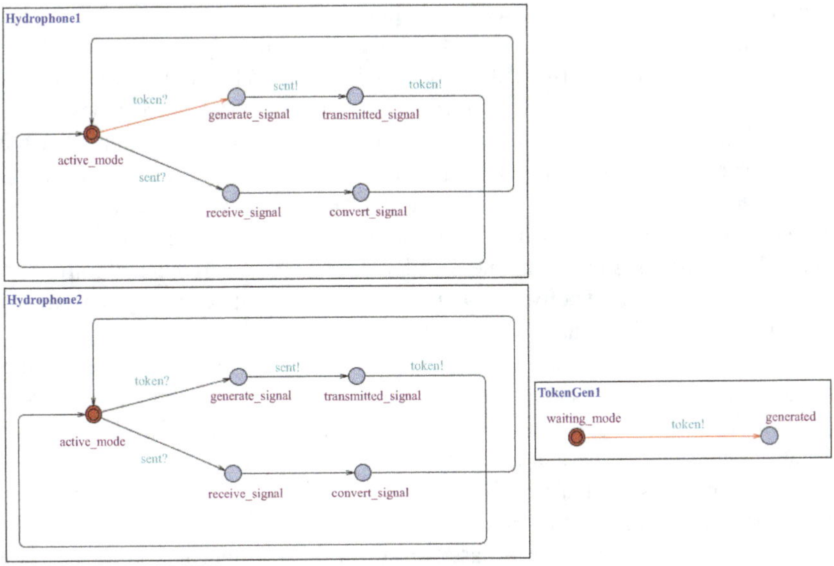

Figure 23.9 The model structure of second scenario.

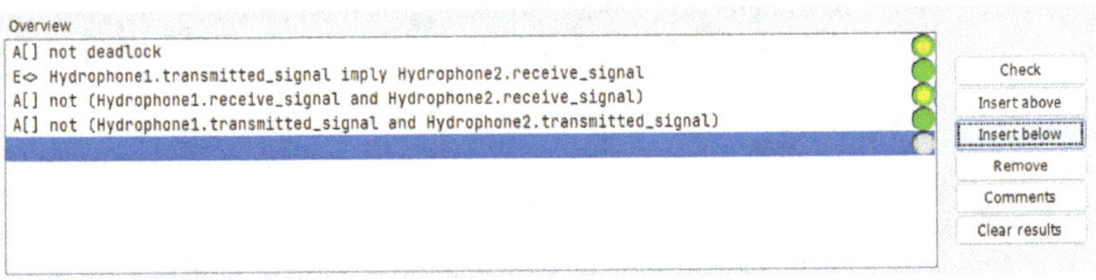

Figure 23.10 Verification properties of the second scenario

$$\texttt{A[] not deadlock} \tag{23.4}$$

$$\texttt{E<> Hydrophone1.transmitted\_signal imply Hydrophone2.receive\_signal} \tag{23.5}$$

$$\texttt{A[] not (Hydrophone1.receive\_signal and Hydrophone2.receive\_signal)} \tag{23.6}$$

$$\texttt{A[] not (Hydrophone1.transmitted\_signal and Hydrophone2.transmitted\_signal)} \tag{23.7}$$

23.4.4 Third scenario: real-time object detection using acoustic pings

In this scenario, the model simulates an object detection system that uses active SONAR to assess the proximity of underwater obstacles or any enemy nodes. A hydrophone periodically emits SONAR pings and listens for the echo reflected from nearby objects. The core behavior revolves around the time it takes for the echo to return: this timing determines whether an object is considered critically close or safely distant. If the echo is received within 60 time units, the system classifies the object as a potential hazard and immediately triggers an alert to prompt evasive or defensive action. In contrast, if the echo takes longer than 60 time units, the object is deemed to be at a safe distance, and the system resumes its scanning cycle uninterrupted.

This model captures a critical decision-making process found in real-world naval defense systems, such as those deployed on AUVs or fixed surveillance nodes. Timely classification of objects based on echo return time is essential in environments where seconds can separate routine monitoring from collision or compromise. In the model, the surrounding object is represented as a passive entity that reflects any received ping without any additional behavior or response logic. This interaction forms the basis for evaluating system correctness under strict time constraints and mirrors the fundamental mechanics of active SONAR-based proximity detection systems used in both military and civilian underwater operations.

This scenario is an enhanced extension of the half-duplex SONAR simulation. The base model consists of a transmitter that emits a signal to simulate a SONAR ping or communication burst. The transmitter is shown in Figure 23.11. The signal is received and reflected by a target node, which acts as a dummy object. It does not perform any processing, but simply returns the signal using a synchronized channel. The target is illustrated in Figure 23.12. The reflected signal is captured by a receiver automaton that transforms based on synchronized input and time-based constraints. The receiver's responsibility is to detect the presence of an echo and respond accordingly. The receiver is shown in Figure 23.13.

The real-time object detection model builds on this foundation by incorporating timing-based decision logic. Rather than simply detecting the presence of an echo,

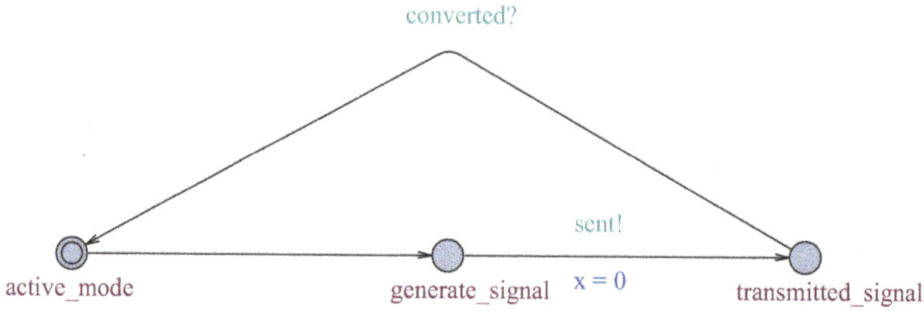

Figure 23.11 The transmitter template

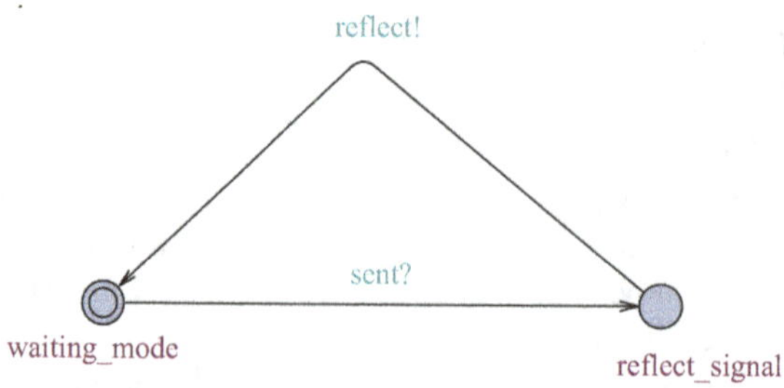

Figure 23.12 The target template

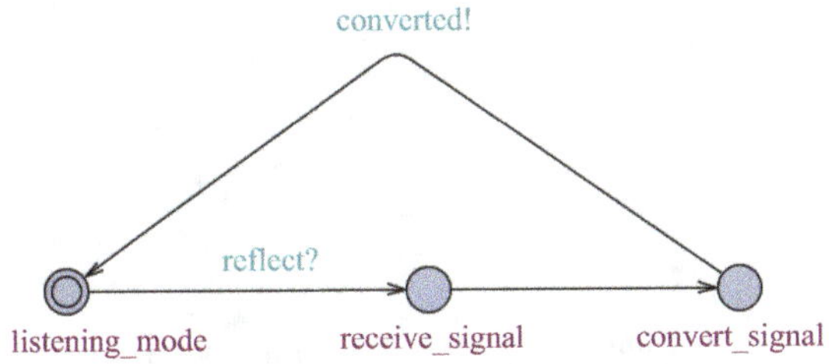

Figure 23.13 The receiver template

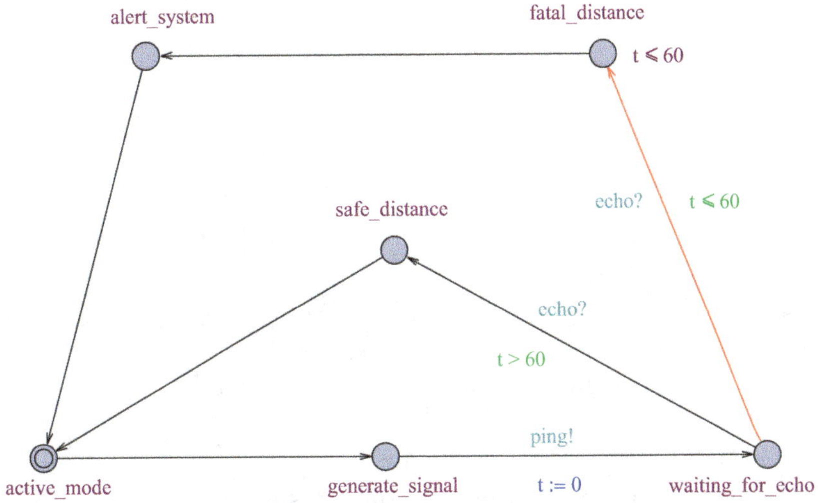

Figure 23.14 The hydrophone template

the system actively evaluates the echo return time to classify the object's proximity. This enhancement allows the model to differentiate between safe and dangerous situations, forming a more realistic and reactive system useful for modeling critical safety behaviors in SONAR-based environments.

23.4.4.1 Model architecture and design overview

The hydrophone automaton (see Figure 23.14) models the core behavior of an active SONAR detection system. It begins in the *active_mode* state and transitions to the *generate_signal*, where it emits a *ping*! signal and resets the clock with $t := 0$. After that, it moves into the *waiting_for_echo* state, where it waits for an *echo*? signal from a nearby object. If the echo is received within 60 time units ($t \leq 60$), the hydrophone changes to *fatal_distance*, indicating a dangerously close object. This immediately leads to the *alert_system* state raising an alarm. If the echo is received after 60 time units ($t > 60$), the hydrophone instead switches to *safe_distance*, signaling that no immediate threat is present. After either path, the automaton returns to *active_mode* to restart the scanning cycle.

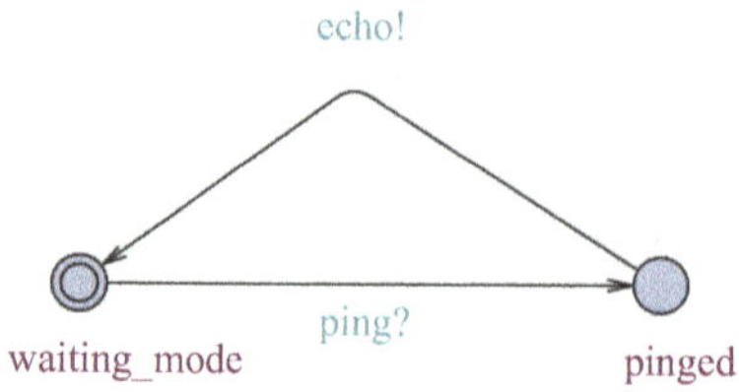

Figure 23.15 The object template

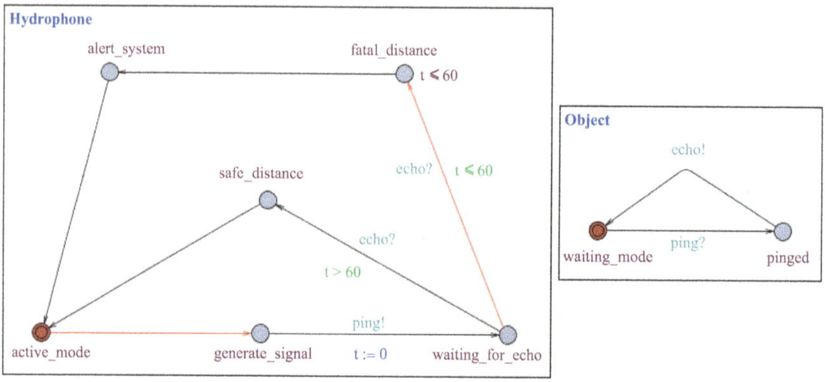

Figure 23.16 The model structure of the third scenario

The object automaton (see Figure 23.15) functions as a passive reflector. It begins in the *waiting_mode* state and then switches to pinged upon receiving a *ping*? signal. It immediately responds with an *echo*! signal and returns to *waiting_mode*, ready to reflect the next incoming ping. This automaton abstracts real-world underwater objects, such as mines or terrain features, that do not emit signals themselves but can reflect SONAR pulses.

The system shown in Figure 23.16 consists of the hydrophone and an object automata interacting through synchronized channels *ping* and *echo*. This setup enables SONAR-based proximity evaluation using timing constraints. Two representative traces demonstrate the behavior of the system: in the first trace shown in Figure 23.17, the echo returns after more than 60 time units, and the hydrophone classifies the object as safe. In the second trace shown in Figure 23.18, the echo is received in 60 time units, and the system raises an alert in response to the detected proximity threat.

23.4.4.2 Distance calculation

For distance calculation, we need to measure the time it takes for a SONAR ping to travel to the object and for the echo to return to the hydrophone. Since UPPAAL does not allow us to directly store the value of a clock in a variable at a specific instant, we rely on execution traces to observe the clock value after a transition occurs, from

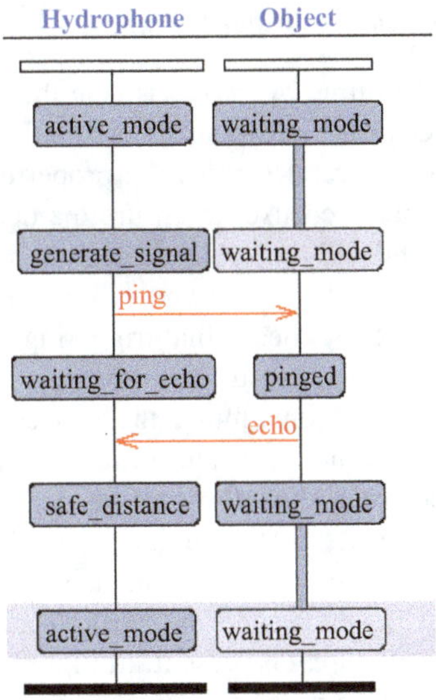

Figure 23.17 The simulation trace for safe distance scenario

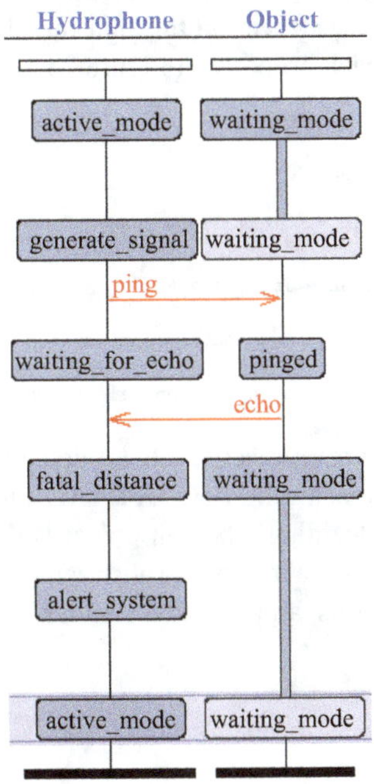

Figure 23.18 The simulation trace for safe distance scenario

the moment the *ping* is sent to the moment the *echo* is received. The distance can be estimated using the following formula:

$$\textit{One-way time} = \frac{\textit{time of echo reception} - \textit{time of ping emission}}{2}$$

Distance = Speed of sound in water × One-way time

We divide the total round-trip time by two to isolate the one-way travel time from the hydrophone to the object. This allows us to estimate the object's distance using the speed of sound in water. Several verification properties were defined and validated using the UPPAAL model checker to ensure the deadlock freedom, liveness, and internal consistency of the real-time object detection model. The property (23.8) ensures that the system avoids deadlocks and never reaches a state without further progress. The property (23.9) checks that triggering the alert system is always possible under some execution path, ensuring liveness.

To ensure correctness based on timing, the properties defined in properties (23.10) and (23.11) verify that the echo return time is accurately related to object classification. An echo received within 60 time units corresponds to a fatal distance (23.10), while an echo arriving after 60 time units is treated as a safe distance (23.11). The correctness of system responses is further validated through properties defined in properties (23.12) and (23.13), which confirm that alerts are raised only when appropriate and suppressed when no threat is present. In Figure 23.19, it shows that there is a green circle against all these properties, which means that the model satisfied all these properties. These verification properties confirm that the model behaves as expected in terms of safety, correctness, and distinguishing between safe and hazardous object proximity. The verification through the UPPAAL verifier is depicted by the green circle against each property in Figure 23.19

```
A[] not deadlock                                                  (23.8)

E<> Hydrophone.alert_system                                       (23.9)

A[] (Hydrophone.fatal_distance imply t <= 60)                     (23.10)

A[] (Hydrophone.safe_distance imply t > 60)                       (23.11)

Hydrophone.fatal_distance --> Hydrophone.alert_system             (23.12)

Hydrophone.safe_distance --> not Hydrophone.alert_system          (23.13)
```

Hence, this model captures a reactive object detection system where real-time constraints are critical, demonstrating how UPPAAL can be used to verify timing-sensitive behavior, such as deciding when an object is dangerously close, based on how fast the echo returns. This is a real-world abstraction of proximity detection used in AUVs, naval defense systems, and obstacle avoidance algorithms.

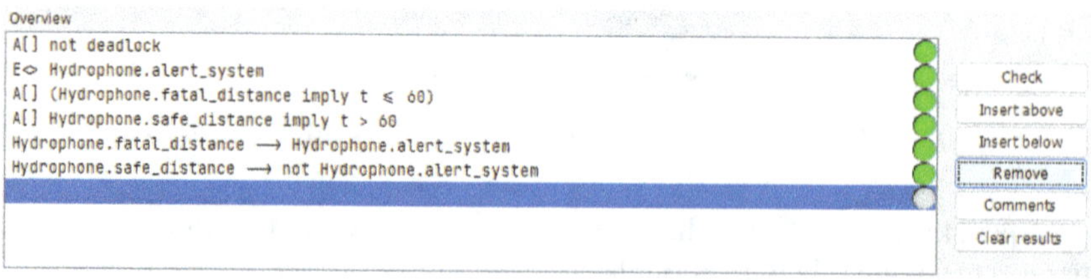

Figure 23.19 Verification properties of the third scenario

23.5 Conclusion

This work effectively illustrates how UPPAAL can be used to simulate and validate the behaviors of SONAR-based subsystems in real time. Each model captures key elements of the timing, synchronization, and safety of acoustic signals in progressively more complicated scenarios, from full-duplex communication and real-time object identification to passive listening. Critical criteria like deadlock-freedom and restricted echo reception periods are rigorously confirmed through the use of timed automata and formal verification. Finally, the models demonstrate how well UPPAAL supports the design of time-sensitive and safety-critical communication protocols by providing an organized basis for the examination of underwater acoustic systems. Incorporating these aspects into a real-time simulator will enable a more comprehensive evaluation of system performance in realistic scenarios. This will help refine timing constraints and improve the dependability of safety-critical underwater acoustic communication systems. Future work will focus on considering the mobility of the sensor nodes and the characteristics of the underwater environment in a real-time simulator to evaluate the modeled scenarios in a comprehensive way.

References

[1] Cerqueira R, Trocoli T, Neves G, *et al.* A novel GPU-based SONAR simulator for real-time applications. *Computers & Graphics*, 2017;68:66–76. https://www.sciencedirect.com/science/article/pii/S0097849317301371.

[2] Kumar NS, Kumar GS, Sivan S, *et al.* Formal verification of a MAC protocol for underwater sensor networks. *IEEE Access*, 2023;11:111846–111859.

[3] Larsen KG, Larsson F, Pettersson P, *et al.* Efficient verification of real-time systems: Compact data structure and state-space reduction. In: *Proceedings Real-Time Systems Symposium*; 1997. pp. 14–24.

[4] Quijano S, Varshosaz M, and Wasowski A. Modeling and safety analysis of autonomous underwater vehicles behaviors. In: *2024 IEEE International Conference on Software Testing, Verification and Validation Workshops (ICSTW)*; 2024. pp. 63–67.

[5] Wang J, Wang C, and Cheng T. Active SONAR reverberation suppression based on beam space data normalization. In: *2017 IEEE International Conference on Signal Processing, Communications and Computing (ICSPCC)*; 2017. pp. 1–4.

[6] Ferrère T, Maler O, Ničković D, *et al.* From real-time logic to timed automata. *Journal of the ACM*, 2019;66(3):1–31. https://doi.org/10.1145/3286976.

Index

www.ingramcontent.com/pod-product-compliance
Lightning Source LLC
LaVergne TN
LVHW011839200926
848484LV00002B/380
* 9 7 8 1 8 3 9 5 3 9 6 8 8 *